MW01221891

# THE MOLECULAR BIOLOGY AND PATHOLOGY OF ELASTIC TISSUES

The Ciba Foundation is an international scientific and educational charity (Registered Charity No. 313574). It was established in 1947 by the Swiss chemical and pharmaceutical company of CIBA Limited—now Ciba-Geigy Limited. The Foundation operates independently in London under English trust law.

The Ciba Foundation exists to promote international cooperation in biological, medical and chemical research. It organizes about eight international multidisciplinary symposia each year on topics that seem ready for discussion by a small group of research workers. The papers and discussions are published in the Ciba Foundation symposium series. The Foundation also holds many shorter meetings (not published), organized by the Foundation itself or by outside scientific organizations. The staff always welcome suggestions for future meetings.

The Foundation's house at 41 Portland Place, London W1N 4BN, provides facilities for meetings of all kinds. Its Media Resource Service supplies information to journalists on all scientific and technological topics. The library, open five days a week to any graduate in science or medicine, also provides information on scientific meetings throughout the world and answers general enquiries on biomedical and chemical subjects. Scientists from any part of the world may stay in the house during working visits to London.

Ciba Foundation Symposium 192

# THE MOLECULAR BIOLOGY AND PATHOLOGY OF ELASTIC TISSUES

1995

JOHN WILEY & SONS

Chichester · New York · Brisbane · Toronto · Singapore

Published in 1995 by John Wiley & Sons Ltd,
Baffins Lane, Chichester,
West Sussex PO19 1UD, England

Telephone: National 01243 779777
International (+44) 1243 779777



*Other Wiley Editorial Offices*

John Wiley & Sons, Inc., 605 Third Avenue,
New York, NY 10158-0012, USA

Jacaranda Wiley Ltd, 33 Park Road, Milton,
Queensland 4064, Australia

John Wiley & Sons (Canada) Ltd, 22 Worcester Road,
Rexdale, Ontario M9W 1L1, Canada

John Wiley & Sons (SEA) Pte Ltd, 37 Jalan Pemimpin #05-04,
Block B, Union Industrial Building, Singapore 2057

Suggested series entry for library catalogues:
Ciba Foundation Symposia

Ciba Foundation Symposium 192
xi + 361 pages, 87 figures, 15 tables

A catalogue record for this book is
available from the British Library

ISBN 0 471 95718 6
Typeset in 10/12pt Times by Dobbie Typesetting Limited, Tavistock, Devon.
Printed and bound in Great Britain by Biddles Ltd, Guildford.
This book is printed on acid-free paper responsibly manufactured from sustainable forestation,
for which at least two trees are planted for each one used for paper production.

# Contents

*Symposium on The molecular biology and pathology of elastic tissues, held at the Windsor Golf and Country Club, Nairobi, Kenya, 1–3 November 1994*

*This symposium was based on proposals made by Joseph Mungai, and Leslie Robert and Bob Mecham*

*Editors: Derek J. Chadwick (Organizer) and Jamie A. Goode*

# Participants

**A. J. Bailey** Muscle & Collagen Research Group, Division of Molecular and Cellular Biology, University of Bristol, Langford, Bristol BS18 7DY, UK

**C. D. Boyd** Department of Surgery, New Jersey's University of Health Sciences, Robert Wood Johnson Medical School, One Robert Wood Johnson Place, CN 19 New Brunswick, NJ 08903-0019, USA

**J. M. Davidson** Department of Pathology, Vanderbilt University School of Medicine, C3321 Medical Center North, Nashville, TN 37232-2561, USA

**M. E. Grant** Department of Biochemistry & Molecular Biology, School of Biological Sciences, University of Manchester, 2.205 Stopford Building, Oxford Road, Manchester M13 9PT, UK

**A. Hinek** Division of Cardiovascular Research, The Hospital for Sick Children, 555 University Avenue, Toronto, Ontario, Canada M5G 1X8

**H. M. Kagan** Department of Biochemistry, Boston University School of Medicine, 80 East Concord Street, Boston, MA 02118-2394, USA

**F. W. Keeley** Division of Cardiovascular Research, The Hospital for Sick Children, 555 University Avenue, Toronto, Ontario, Canada M5G 1X8

**J. K. Kimani** Department of Human Anatomy, University of Nairobi, PO Box 30197, Nairobi, Kenya

**K. Lapis** Institute of Pathology & Experimental Cancer Research, Semmelweis University Medical School, Ulloi ut 26, H-1085 Budapest, Hungary

**T. J. Mariani** (*Ciba Foundation Bursar*) Division of Dermatology, Department of Internal Medicine, Washington University Medical Center, The Jewish Hospital, 216 South Kingshighway Boulevard, St Louis, MO 63110, USA

**R. P. Mecham** Department of Cell Biology & Physiology, Washington University School of Medicine, Box 8228, 660 South Euclid Avenue, St Louis, MO 63110-1093, USA

**J. M. Mungai** Commission for Higher Education, PO Box 54999, Nairobi, Kenya

**T. Ooyama** Division of Immunology and Blood Transfusion, Tokyo Metropolitan Geriatric Hospital, 35-2 Sakae-machi, Itabashi-ku, Tokyo 173, Japan

**W. C. Parks** Division of Dermatology, The Jewish Hospital, 216 South Kingshighway Boulevard, St Louis, MO 63110, USA

**I. Pasquali-Ronchetti** Biomedical Sciences Department, Università degli Studi di Modena, Via Campi 287, I-41100 Modena, Italy

**R. A. Pierce** Division of Dermatology, Department of Internal Medicine, Washington University Medical Center, The Jewish Hospital, 216 South Kingshighway Boulevard, St Louis, MO 63110, USA

**L. Robert** (*Chairman*) 7 Rue Lully, 94440 Santeny, France

**J. Rosenbloom** Department of Anatomy and Histology, School of Dental Medicine, University of Pennsylvania, 4001 Spruce Street, Philadelphia, PA 19104-6003, USA

**L. Y. Sakai** Research Department, Shriners Hospital for Crippled Children, 3101 SW Sam Jackson Park Road, Portland, OR 97201, USA

**L. B. Sandberg** Department of Veterans Affairs, Jerry L. Pettis Memorial Veterans Hospital, 11201 Benton Street, Loma Linda, CA 92357, USA

**B. Starcher** Department of Biochemistry, University of Texas Health Center, PO Box 2003, Tyler, TX 75710-2003, USA

**A. M. Tamburro** Facoltà di Scienze, Università degli Studi della Basilicata, Prot 856/94, Via N. Sauro 85, I-85100 Potenza, Italy

**J. Uitto** Department of Dermatology, Jefferson Medical College, 233 South 10th Street, BLSB 450, Philadelphia, PA 19107, USA

**D. W. Urry** Laboratory of Molecular Biophysics, UAB School of Medicine, 525 Volker Hall VH 300, 1670 University Boulevard, Birmingham, AL 35294-0019, USA

**C. P. Winlove** Centre for Biological & Medical Systems, Physiological Flow Studies Group, Imperial College of Science Technology & Medicine, London SW7 2AZ, UK

# Preface

Although based in London, the Ciba Foundation is truly an international organization whose mission is to foster scientific cooperation globally. For this reason, we have a worldwide panel of scientific advisers drawn from most countries where there is substantive scientific research. We encourage our advisers to visit our headquarters in Portland Place whenever opportunities present and it was during such a visit that our scientific adviser for Kenya, Professor Joseph Mungai, suggested we should consider holding one of our symposia in Nairobi. Professor Mungai is an anatomist with particular interests in the elastic tissues of animals and it seemed a very happy coincidence, therefore, that we received almost simultaneously symposium proposals in this area from him and (entirely independently) from Dr Leslie Robert (France) and Professor Bob Mecham (USA). Given the crucial role of elastin in ageing and pathology, the stimulus of the cloning of the genes for the soluble elastin precursor (tropoelastin) and for several microfibrillar proteins, and the discovery of elastin receptors, the proposal was clearly very timely with research in the area progressing rapidly.

It gives me very great pleasure to acknowledge publicly the enthusiastic encouragement and help which Professor Mungai and his colleagues Professor Kimani and Miss Shah gave the Foundation in the organization of the symposium. I should also like to thank them and Dr Davy Koech, Director of KEMRI (the Kenya Medical Research Institute) for their assistance and advice with the subsequent open meeting, a collaborative venture with KEMRI and the Kenyan Commission for Higher Education (of which Professor Mungai is Secretary). We were delighted to have the opportunity for the Foundation to re-visit the African continent for an occasion which was memorable not only for the excellence and camaraderie of the scientific discourse but equally for the warmth and enthusiasm of our hosts.

Derek J. Chadwick
*Director, The Ciba Foundation*

# Chairman's introduction

Leslie Robert

*Laboratoire de Biologie Cellulaire, Université Paris VII, 3 Place Jussieu, F-75005 Paris, France*

About 500–600 million years ago, during the great Cambrian explosion, many new variants of life appeared. Among these were the oxygen-producing cyanobacteria. The progressive enrichment with oxygen of what was previously a reducing atmosphere continued during the Precambrian era. Because most of the previously existing species were adapted to an oxygen-free atmosphere, many (if not most) of these disappeared. New mutations allowed some species to adapt to an aerobic way of life, and these were therefore able to develop and diversify further.

Diploblasts, which developed in the aquatic milieu, gave rise to triploblasts. These had the further developmental potential offered by the possession of the mesenchyme, which had already made an appearance in a primitive form in the sponges. Many sponges contain collagen fibres and glycoproteins, some of which are very similar to fibronectin.

Soon after the Cambrian radiation, the chordates appeared, followed by the appearance of the neural tube and notochord. Vertebrates turned out to be very successful in colonizing all possible ecological niches, from the depths of the sea to the skies. Their extraordinary success was to a large extent a consequence of their perfect adaptation to aerobic life. This involved the creation of conditions maximizing accessibility of their cells to oxygen, which was so efficiently used by their mitochondria to generate energy in the form of ATP. Oxygen, gathered first through gills and later through lungs, was propelled by the developing cardiovascular system to the most remote parts of the body. Without these extraordinary inventions, no further increase in size, or the specialization of organ systems, would have been possible.

We can consider it to be an extraordinary coincidence that the elastin gene appeared at the same time. Lampreys (agnatha, notochord throughout life) possess a not-too-dissimilar protein, lamprin (as has been shown by Fred Keeley). As far as we can tell, subsequent development of the elastin gene occurred rapidly, from teleost fishes to terrestrial quadrupeds. This will be discussed by several participants at this meeting who have been pioneers in this field, especially Joel Rosenbloom and Charles Boyd.

Elastin is, in evolutionary terms, much younger than most collagens; the only similarity between these two types of protein is that both possess high proportions of glycine, a similarity that confused some of the early workers.

The physicochemical properties of elastin are very unusual, and its discovery was apparently the result of its extraordinary resistance to chemical, physical and enzymic attack, at least in aqueous media. Mörner, a 19th-century German chemist, noticed that when tissues were treated with calcium hydroxide (a very aggressive hydrolytic agent), a yellowish fibrous material (elastin) resisted this treatment. Eulenberger, in 1836, isolated elastic tissue from vascular walls using boiling water. Interestingly, the resistant properties of elastin collapse in the presence of organic solvents. As we showed some thirty years ago, in aqueous-organic solvents, mild alkali treatment will completely hydrolyse insoluble elastin fibres in peptides ($\kappa$-elastin) which were extensively used for biochemical and pharmacological studies, and which also found their way into industrial applications. This observation opened the way for a physicochemical interpretation of the structure and elasticity of elastin that is largely based on strong hydrophobic interactions between the predominantly aliphatic amino acid side chains. The elucidation of elastin's tertiary and quaternary structure, which will be recalled by Dan Urry, Tony Tamburro and Peter Winlove, also explained its strong affinity for $Ca^{2+}$ and lipids. The progressive deposition of $Ca^{2+}$ and lipids in elastic fibres with ageing is part of the atherosclerotic process and will also be discussed during this meeting.

The isolation of the first proteolytic enzyme that rapidly degrades elastin, pancreatic elastase, by J. Balo and I. Banga in Budapest in the 1950s, began a period of rapid increase in interest in this curious protein. Several other elastases have been identified and characterized. Some, such as polymorphonuclear leukocyte elastase, play important roles in pathological processes, for instance, lung emphysema. The elastin–elastase inhibitor balance hypothesis (increased elastolytic damage to tissues through lack of elastase inhibitor), proposed after the discovery of genetic deficiencies in elastase inhibitors ($\alpha_1$Pi) by Klas Bertil Laurell, further stimulated elastin research and resulted in an upsurge of synthetic elastase inhibitor patents.

The elucidation of the structure of elastin was initially a difficult task. The first breakthrough was the identification and characterization of the cross-linking amino acids, desmosine and isodesmosine, by Miles Partridge, whose recent death deprived this field of one of its founders. Allen Bailey will recall this crucial step in the elucidation of elastin's structure. The production of a soluble precursor, tropoelastin, from the aorta of pigs raised on a copper-free diet in Bill Carnes' laboratory, enabled Larry Sandberg to produce the first amino acid sequences of tropoelastin. The important observation was the presence of tetra-, penta- and hexapeptide repeating sequences, which further demonstrated the peculiar nature of elastin. Polymers made with such repeating units in Dan Urry's laboratory had many of the interesting properties of elastin.

Because of its extraordinary resistance, elastin was initially considered to be inert. For this reason, it was only recently that the study of elastin cell biology was initiated in the laboratories of Bob Senior and Bob Mecham, and also in ours. These studies have largely focused on the fibroblasts of the bovine ligamentum nuchae and the smooth muscle cells of large blood vessels, but chondrocytes and skin fibroblasts have also been used. These systems have enabled us to study the regulation of elastin biosynthesis and the mechanism of elastin fibre deposition, a field pioneered by Bob Mecham, who will talk about this important aspect of elastin cell biology. In our laboratory, we demonstrated an interaction between cells and elastic fibres that is mediated by a membrane receptor and an inducible adhesive component of cell membranes. This mechanism is important in physiological processes such as the regulation of blood vessel wall tension and also in tumour cell metastasis. The demonstration of the elastin receptor on several cell types enabled us to explore (with M. P. Jacob and T. Fülöp) its biological role. However, most of our recent knowledge of elastin has come from gene sequencing. The demonstration of alternative exon usage in several species means that cells can secrete hundreds of possible tropoelastin isoforms.

The degradation of elastic fibres is a hallmark of vascular ageing. Degradation products, such as elastin peptide fragments, are present in the circulation and may well play an important role in the pathogenesis of some age-dependent diseases. Loss of tissue elasticity, however, is not always accompanied by loss of elastic fibres. Increased elastin fibre deposition in senile or solar elastosis, as well as in some forms of breast cancer, still presents an enigma to elastin biochemistry.

It appears, therefore, that the phylogenetic appearance of elastin was vital in order for vertebrates to develop highly efficient respiratory and circulatory systems. This advantage is offset by the relatively rapid decay of elastic fibres, which lose their elasticity with time and which are also involved in several of the most common diseases of old age. This demonstrates once more the fact that the 'invention' of new genes serves mainly the success of the individual in terms of reproductive success, not for long survival. Thus, elastin biology is an area where pharmacology should come in with original, innovative research to alleviate the very widespread disorders involving elastic tissues and which commonly result in early mortality, as in the case of cardiovascular and pulmonary diseases. Let us hope that the papers presented at this symposium will encourage young researchers to enter this fascinating field of biological research and develop it further.

# Molecular biophysics of elastin structure, function and pathology

Dan W. Urry, Chi-Hao Luan and Shao Qing Peng

*Laboratory of Molecular Biophysics, School of Medicine, The University of Alabama at Birmingham, VH 300, Birmingham, AL 35294-0019, USA*

*Abstract*. Owing to the presence of the recurring sequence XPGX′ (where X and X′ are hydrophobic residues), the molecular structure of the sequences between cross-links in elastin is viewed primarily as a series of β-turns which become helically ordered by hydrophobic folding into β-spirals, which in turn assemble hydrophobically into twisted filaments. Both hydrophobic folding and assembly occur when the temperature is raised above $T_t$, the onset of an inverse temperature transition. Using poly[$f_v$(VPGVG),$f_x$(VPGXG)] (where $f_v$ and $f_x$ are mole fractions with $f_v + f_x = 1$ and X is now any of the naturally occurring amino acid residues), plots of $f_x$ versus $T_t$ result in a new hydrophobicity scale based directly on the hydrophobic folding and assembly processes of interest. With the reference values chosen at $f_x = 1$, the most hydrophobic residues of elastin, Tyr (Y) and Phe (F), have low values of $T_t$, −55 and −30 °C, respectively, and the most hydrophilic residues, Glu ($E^-$), Asp ($D^-$) and Lys ($K^+$), have high values of 250, 170 and 120 °C, respectively. Raising the average value of $T_t$ for a chain or chain segment from below to above physiological temperature drives hydrophobic unfolding and disassembly; lowering $T_t$ does the reverse. This $\Delta T_t$ mechanism has been used reversibly to interconvert many energy forms and is used here to explain initiating events of elastogenesis, pulmonary emphysema, solar elastosis and the paucity of elastic fibres in scar tissue. In general, oxidation and/or photolysis convert(s) hydrophobic residues into polar residues with the consequences of irreversibly raising $T_t$ to above 37 °C, hydrophobic unfolding and disassembly (fibre swelling), and greater susceptibility to proteolysis.

*1995 The molecular biology and pathology of elastic tissues. Wiley, Chichester (Ciba Foundation Symposium 192) p 4–30*

The study of the polypentapeptide of elastin, $(Val^1\text{-}Pro^2\text{-}Gly^3\text{-}Val^4\text{-}Gly^5)_n$ or poly(VPGVG), and many of its analogues and chemical modifications thereof, has led to significant insight into the factors that control hydrophobic folding/unfolding and assembly/disassembly. This has occurred just prior to the growing appreciation that hydrophobic folding can be the primary process in protein structure formation. It has been possible for us to substantiate the understanding developed on the poly(VPGVG) model by designing compositional variants, fashioning these protein-based polymers (high molecular weight polymers of repeating peptide sequences) into cross-linked elastic bands and driving contraction (hydrophobic folding and assembly) to produce the useful mechanical work of cyclically lifting and lowering a weight. The input energies shown to be capable of driving contraction and relaxation of these elastic bands are thermal, chemical, pressure, electrochemical and light energy. In other words, we have designed elastic protein-based polymers using the basic poly(VPGVG) structure to achieve thermo-, chemo-, baro-, electro- and photomechanical transduction (for a review, see Urry 1993). Also, the elastic protein-based polymers have been designed to convert light energy and electrochemical energy into the chemical energy of increased proton concentration (that is, to perform photo- and electrochemical transduction as occurs in photosynthesis and in the first energy conversion step of oxidative phosphorylation).

Because we have achieved this degree of functional protein engineering by starting with the most striking repeating sequence of elastin, it seems appropriate to ask what insights this knowledge may provide towards our understanding of elastin itself. Such an effort begins with knowledge of primary structure and is followed by the determination of molecular structure and the development of the principles that are fundamental to structure formation, function and dysfunction.

### *Primary structure and hydrophobicity plots*

The first substantial information on primary structure of elastin came from Sandberg, Gray and their co-workers (Gray et al 1973, Sandberg et al 1985), who identified in porcine elastin repeating peptide sequences occurring between potential cross-linking sequences. This occurred some years after Partridge (1969) had identified desmosine and isodesmosine and Franzblau & Lent (1969) had identified lysinonorleucine as the principal cross-links, all of which derived from the side-chains of lysine residues. The primary structures for human (Indik et al 1987), bovine (Yeh et al 1987) and porcine (Sandberg et al 1985) elastins are given in Fig. 1 in terms of single-residue hydrophobicity plots using the $T_t$-based hydrophobicity scale which we developed using poly(VPGVG) (Urry 1993). This repeating sequence is apparent as the longest and most regular repeating sequence seen in the hydrophobicity plots for bovine and porcine elastin in Fig. 1; following the Sandberg nomenclature, it is indicated as the W4 sequence. Other repeating sequences are apparent in Fig. 1, for example, the W6 repeat, $(Ala^1\text{-}Pro^2\text{-}Gly^3\text{-}Val^4\text{-}Gly^5\text{-}Val^6)_n$. Also apparent are the periodic Lys ($K^+$) residues seen as negative deflections. Four lysines, two from each of two chains, combine to form desmosine and isodesmosine cross-links, and two lysines combine to form lysinonorleucine cross-links. The development of the $T_t$-based hydrophobicity scale is a central

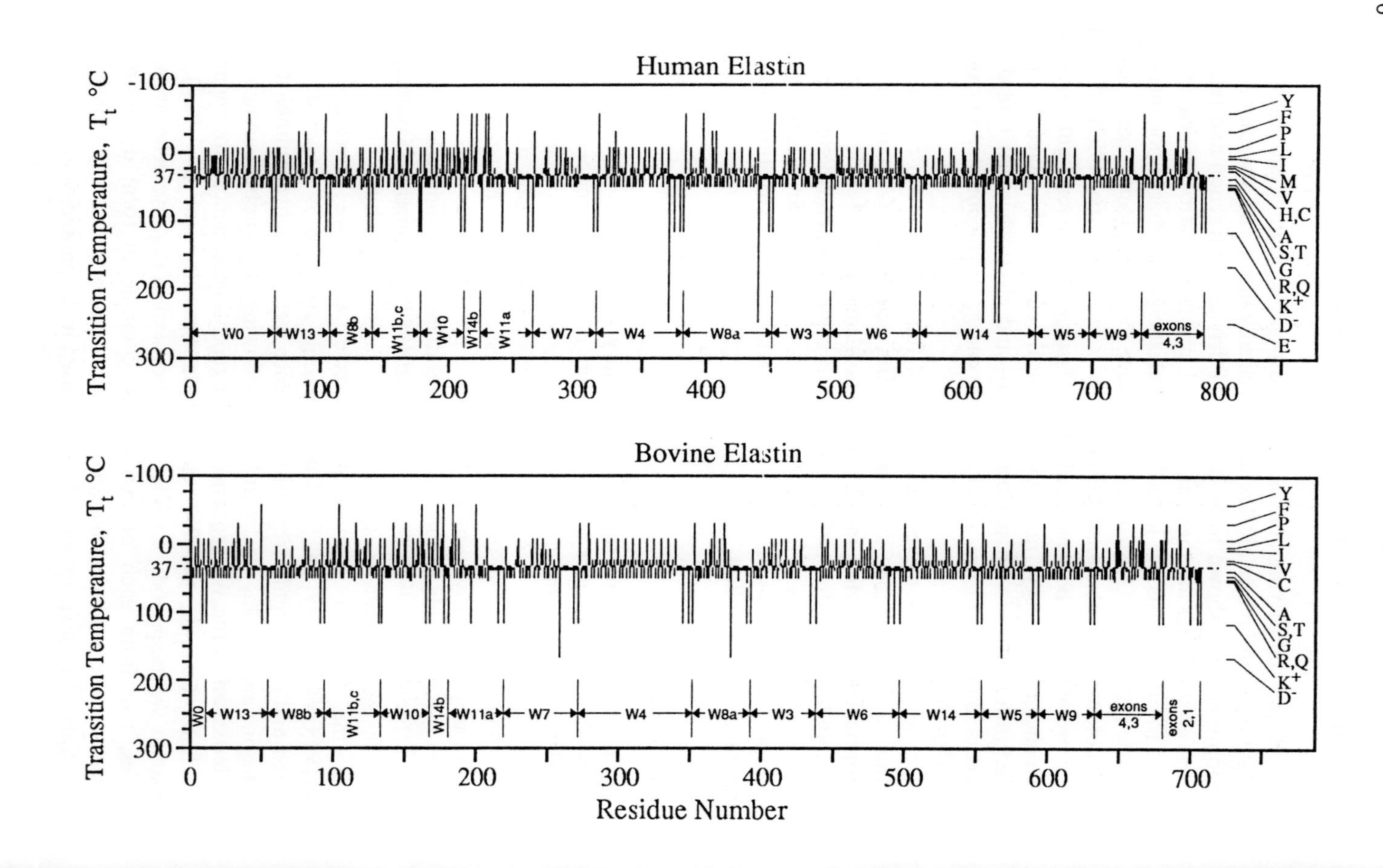
Human Elastin
Bovine Elastin
Transition Temperature, $T_t$ °C
Residue Number
W0
W13
W8b
W11b,c
W10
W14b
W11a
W7
W4
W8a
W3
W6
W14
W5
W9
exons 4,3
exons 2,1
Y
F
P
L
I
M
V
H,C
C
A
S,T
G
R,Q
$K^+$
$D^-$
$E^-$

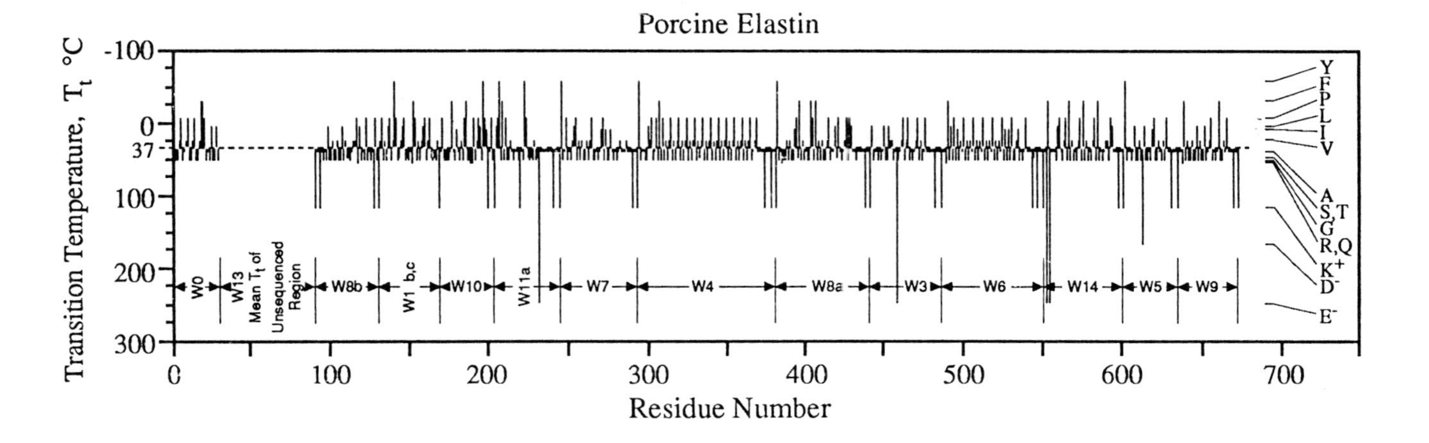


FIG. 1. Single-residue hydrophobicity plots for human, bovine and porcine tropoelastins using the $T_t$-based hydrophobicity scale listed in Table 1. The plots are given as deflections from the 37 °C line: the larger the upward deflection the more hydrophobic the residue, and the greater the negative deflection the more polar the residue. The single-letter codes for the amino acid residues are given on the right-hand side at the point of maximum deflection. Along the bottom of each plot, in addition to the amino acid residue number, is given the sequence according to the Sandberg nomenclature. Perhaps most striking is the repeating VPGVG sequence so apparent in the W4 sequence of the bovine and porcine proteins. There is also the repeating APGVGV hexapeptide of W6 most apparent in the human and porcine proteins, and throughout the non-cross-link sequences there is a prominent occurrence of PG, which is the most common means of inserting a $\beta$-turn. Finally, there are the regularly recurring negative deflections to 120 °C, the position of the charged lysine ($K^+$) residues which become involved in the cross-links.

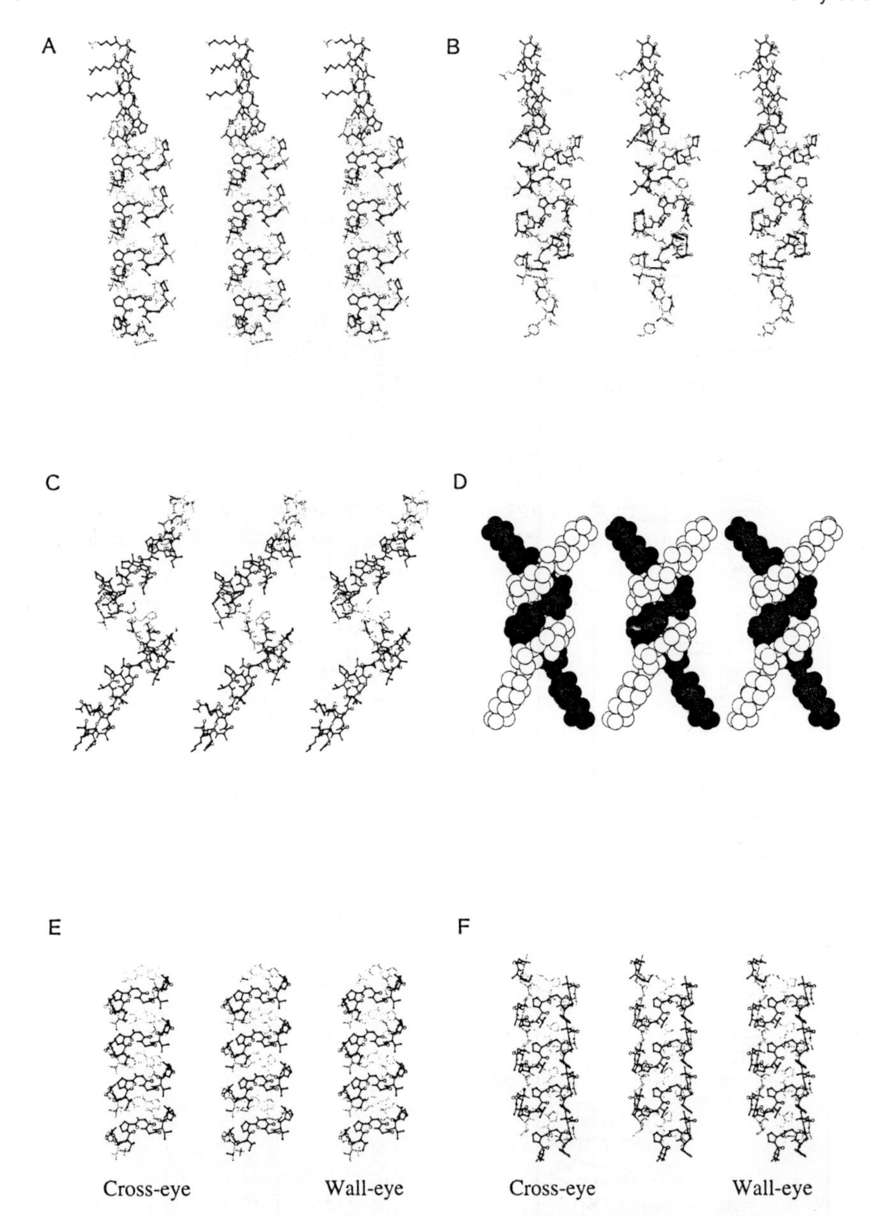


FIG. 2. Proposed molecular structures, shown in stereo perspective, for particular sequences of elastin. (A) Bovine W4; (B) human W4; (C) human W6 (single chain); (D) human W6 (association of a pair of chains shown in space-filling representation); (E) the polytetrapeptide, $(VPGG)_n$; and (F) the polynonapeptide $(VPGFGVGAG)_n$. In each case for the three structures, the left-hand pair are for cross-eye viewing and the right-hand pair are for wall-eye viewing. (C.-H. Luan & D. W. Urry, unpublished results.)

component in our understanding of the factors that control hydrophobic folding and unfolding.

### *β-spiral structures for repeating peptide sequences of elastin*

As a key step in moving towards an understanding of elastin structure, function and pathology, the proposed and working-model β-spiral molecular structures are presented in Fig. 2 for the bovine and human W4 sequences, the human W6 sequence and the repeating consensus sequences—poly(VPGG) and poly(VPGFGVGAG). These structures are based on extensive physical characterizations of the chemically synthesized protein-based polymers and on molecular mechanics and dynamics calculations.

### *Elastogenesis and inverse transitions*

As may be judged from its amino acid composition, elastin is a hydrophobic protein which has many hydrophobic residues (those with upward deflections from the 37 °C line of Fig. 1) and an extreme paucity of charged residues (those with the large negative deflections in Fig. 1, e.g. the Glu [$E^-$], Asp [$D^-$] and Lys [$K^+$] residues). A fibrous protein (e.g. myosin) has quite the inverse distribution. Accordingly, elastogenesis, the process of elastic fibre formation, is reasonably expected to be directed by hydrophobic folding and assembly. In fact, each of the high polymers of the repeat sequences, the precursor protein (tropoelastin) and the fibre fragmentation product (α-elastin) is soluble in water at a sufficiently low temperature but exhibits a phase separation of hydrophobic folding and assembly as the temperature is raised above a critical temperature (designated as $T_t$). As the polypeptide part of the system is characterized by an increase in order with increase in temperature, this hydrophobic folding and assembly transition is called an inverse temperature transition (Urry 1992).

### *The $\Delta T_t$ mechanism for controlling hydrophobic folding and assembly*

Since the polypeptide chains are unfolded and disassembled when the temperature is below $T_t$ (the onset temperature for the transition), but have completed their folding and assembly when the temperature is some 15–20 °C above $T_t$, the control of $T_t$ becomes the means with which to control whether the fibre forms or whether it dissociates to the extent allowed by the cross-links that may have formed. In short, instead of raising the temperature to drive the hydrophobic folding and assembly transition, in the $\Delta T_t$ mechanism the transition temperature, $T_t$, is lowered from above to below the working temperature to drive the hydrophobic folding and assembly transition isothermally (Urry 1992, 1993).

### *Consequences of the $\Delta T_t$ mechanism of hydrophobic folding and assembly for structure, function and pathology*

If the value of $T_t$ for a set of chains is above 37 °C, fibre formation will not occur. This can happen, for example, when there is too much hydroxylation of proline residues. If the fibre is already formed with cross-links in place and an external process such as oxidation occurs to raise the value of $T_t$ above 37 °C, the fibre will swell, with a marked loss of elastic strength (i.e. a decrease in elastic modulus), and extend with loss of function. In addition, it has been our experience that when the synthetic model elastic fibres are hydrophobically folded and assembled they are resistant to proteolytic degradation, but when the $T_t$ is raised above 37 °C, as occurs when a carboxamide breaks down to a carboxylate, proteolytic digestion occurs. Thus, we propose this to be a factor, for example, in the aetiology of pulmonary emphysema. Furthermore, should photo-oxidation occur, the value of $T_t$ would be raised and again there would be swelling followed by proteolytic degradation with the result being solar elastosis.

## Molecular structures for sequences between cross-links

The most extensively characterized elastic sequence of elastin is that of the prominent (VPGVG) repeat of the bovine W4 sequence. This molecular structure, given in stereo pairs in Fig. 2A, includes the lysine-containing, alanine-rich, α-helical cross-linking sequence at the C-terminus. The VPGVG repeat is characterized by a Type II $Pro^2$-$Gly^3$ β-turn that has a $Val^1$ C–O···H–N $Val^4$ 10-atom hydrogen-bonded ring. On raising the temperature in water through the transition temperature range, the series of β-turns wrap up into a helical structure, called a β-spiral, by optimizing intramolecular hydrophobic contacts in which the β-turns, with hydrophobic contacts, function as spacers between turns of the spiral. Two or more of these β-spiral structures, again by hydrophobic association, supercoil to give twisted filaments as observed in electron micrographs of negatively stained incipient aggregates (Urry 1992).

The calculated molecular structure of the human W4 sequence, which does not exhibit such a well-defined repeat, is given in Fig. 2B. We have synthesized, polymerized, cross-linked and physically characterized this sequence; we find that it also becomes a predominantly entropic elastomer at temperatures above the inverse temperature transition (Gowda et al 1995).

The human W6 sequence and the component APGVGV consensus sequence have also been synthesized, polymerized and physically characterized, and were found to be inelastic (Rapaka et al 1978a). The conformation of this quite-rigid structure, on the basis of extensive nuclear magnetic resonance and computational characterization, is shown in Fig. 2C as stereo pairs of a single chain. The single chain is seen to form an intertwined structure in space-filling stereo perspective (Fig. 2D) (C.-H. Luan & D. W. Urry, unpublished results). This rigid intertwining of chains is considered to be an important aligning element of elastogenesis prior to the formation of the natural cross-links.

Also included in Fig. 2 are the proposed and working-model structures of elastic poly(VPGG) (Fig. 2E) (Luan et al 1991) and elastic poly(VPGFGVGAG) (Fig. 2D) (C.-H. Luan & D. W. Urry, unpublished results), respectively. These repeats are less common but they have also been synthesized, polymerized, physically and computationally characterized, cross-linked and found to be elastomeric.

Because of the prominent recurrence of the Pro-Gly sequence apparent in Fig. 1, the repeat sequences of elastin are thought to form $\beta$-spiral structures. So far, the repeating sequences that contain VPG... have been found to be elastic, whereas those that contain APG... are inelastic.

**Hydrophobic folding and assembly (an inverse temperature transition)**

The elastin repeat peptides have the correct balance of hydrophobic (apolar) and polar (peptide) moieties to increase in order in water as the temperature is raised through a transition temperature range. This is unambiguously demonstrated by the cyclic analogue, cyclo$(APGVGV)_2$, of the repeating (APGVGV) sequence of W6. Just as with poly(APGVGV), cyclo$(APGVGV)_2$ is soluble in water at a sufficiently low temperature, and it aggregates as the temperature is raised through the transition temperature range. Whereas poly(APGVGV) shows twisted filaments in the transmission electron micrographs of negatively stained aggregates (Urry 1982), cyclo$(APGVGV)_2$ reversibly forms crystals (Urry et al 1978). These molecules go from being randomly distributed in solution at low temperature to being precisely positioned in a crystal when the temperature is raised. The polypeptides unambiguously increase in order as the temperature is increased. Quite similarly, poly(VPGVG) is seen in the electron micrographs to form twisted filaments (Urry 1982), and the cyclic structure, cyclo$(VPGVG)_3$, which has a nearly identical conformation, is found in the crystal state to associate by hydrophobic contacts between molecules within the stacks of molecules (in analogy to intramolecular folding to form the $\beta$-spiral) and between stacks of molecules (in analogy to intermolecular assembly to form twisted filaments) (Cook et al 1980). Similarly, tropoelastin and $\alpha$-elastin go from a random distribution in solution to parallel aligned twisted filaments as the temperature is raised through the transition temperature range (Urry 1982).

The general finding of folding and assembly into regular structures as the temperature is raised, which is confirmed by the crystallization of cyclo$(APGVGV)_2$ and cyclo$(VPGVG)_3$, indicates that the dominant inter- and intramolecular interactions of elastin are hydrophobic, a result that can be

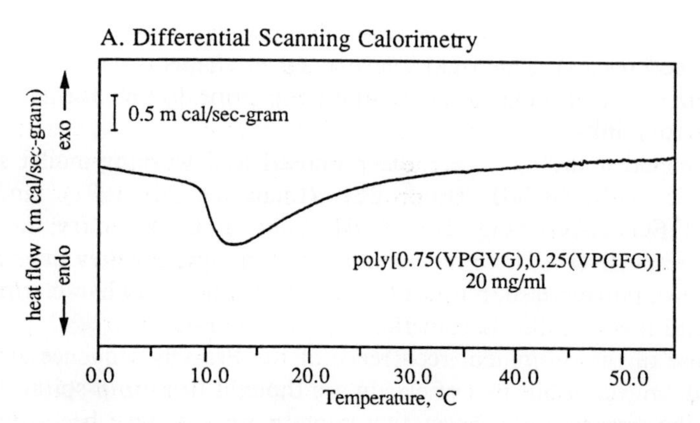
A. Differential Scanning Calorimetry
0.5 m cal/sec-gram
heat flow (m cal/sec-gram)
exo
endo
poly[0.75(VPGVG),0.25(VPGFG)]
20 mg/ml
0.0
10.0
20.0
30.0
40.0
50.0
Temperature, °C

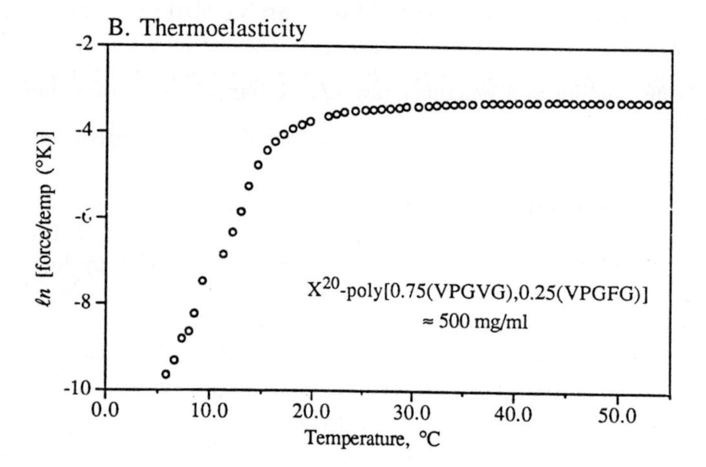
B. Thermoelasticity
ℓn [force/temp (°K)]
-2
-4
-6
-8
-10
X20-poly[0.75(VPGVG),0.25(VPGFG)]
≈ 500 mg/ml
0.0
10.0
20.0
30.0
40.0
50.0
Temperature, °C

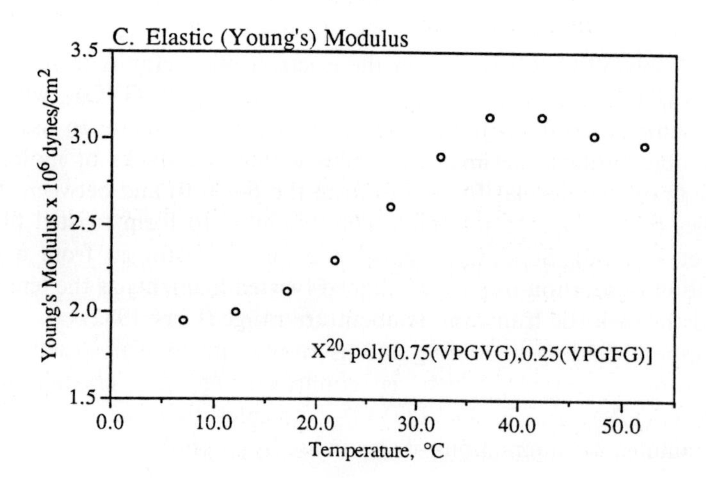
C. Elastic (Young's) Modulus
Young's Modulus x 10^6 dynes/cm^2
3.5
3.0
2.5
2.0
1.5
X20-poly[0.75(VPGVG),0.25(VPGFG)]
0.0
10.0
20.0
30.0
40.0
50.0
Temperature, °C

predicted from the amino acid composition so apparent in Fig. 1. This increase in order with temperature arises because there is structured water surrounding the hydrophobic side chains when they are dissolved at low temperature that becomes less-ordered bulk water during hydrophobic association as the temperature is raised through the transition. The entropy (the disorder) of the entire system (polypeptide or protein plus water) increases with increasing temperature in keeping with the second law of thermodynamics; but there is a larger increase in disorder of the water part of the system and a smaller, though very important, increase in order for the polypeptide part of the system. That the folding and assembly occurs as a transition is apparent from Fig. 3A, where the transition is seen to be endothermic. Heat is required to destructure the water of hydrophobic hydration. Because our focus is on the polypeptide folding and function and because there is an increase in order instead of the usual decrease in order with temperature, this is called an inverse temperature transition. Accordingly, the folding and assembly of the structures of Fig. 2 are the result of an inverse temperature transition to increased order with increased temperature (Urry 1992).

## The $T_t$-based hydrophobicity scale

The temperature for the onset of the inverse temperature transition is designated as $T_t$. An easy measure of $T_t$ is shown in Fig. 4A. The onset of aggregation when the temperature is raised results in an onset of light scattering, and the temperature at which half-amplitude is achieved in a plot of turbidity versus temperature is defined as $T_t$. As is apparent in Fig. 4A, the value of $T_t$ changes with amino acid composition in a systematic way.

Using the general formula poly[$f_v$(VPGVG),$f_x$(VPGXG)], where $f_v$ and $f_x$ are mole fractions, $f_v + f_x = 1$, and X is any amino acid, we see that plots of $f_x$ versus $T_t$ result in straight lines (Fig. 4B) such that it becomes possible to

FIG. 3. (A) Differential scanning calorimetry using a 20 mg/ml sample of poly[0.75(VPGVG),0.25(VPGFG)] showing an endothermic transition beginning just below 10 °C. This shows the inverse temperature transition. (B) Thermoelasticity curve for 20 Mrad $\gamma$-irradiation cross-linked poly[0.75(VPGVG),0.25(VPGFG)] where the resulting elastic band, held at fixed length, is seen to contract with the dramatic development of elastic force as the temperature is raised through the range of the inverse temperature transition for hydrophobic folding and assembly. Above the transition, the plot of *ln*(force/temperature) versus temperature is seen to approach a zero slope (particularly above 37 °C), which is classically interpreted to indicate a dominantly entropic elastic force. (C) Elastic (Young's) modulus for the cross-linked elastic strip, $X^{20}$-[0.75(VPGVG),0.25(VPGFG)], where the elastic modulus of the same strip is seen to increase as the temperature is raised through the range of the inverse temperature transition. The hydrophobic folding and assembly that occur on passing through the inverse temperature transition contribute to the elastic force.

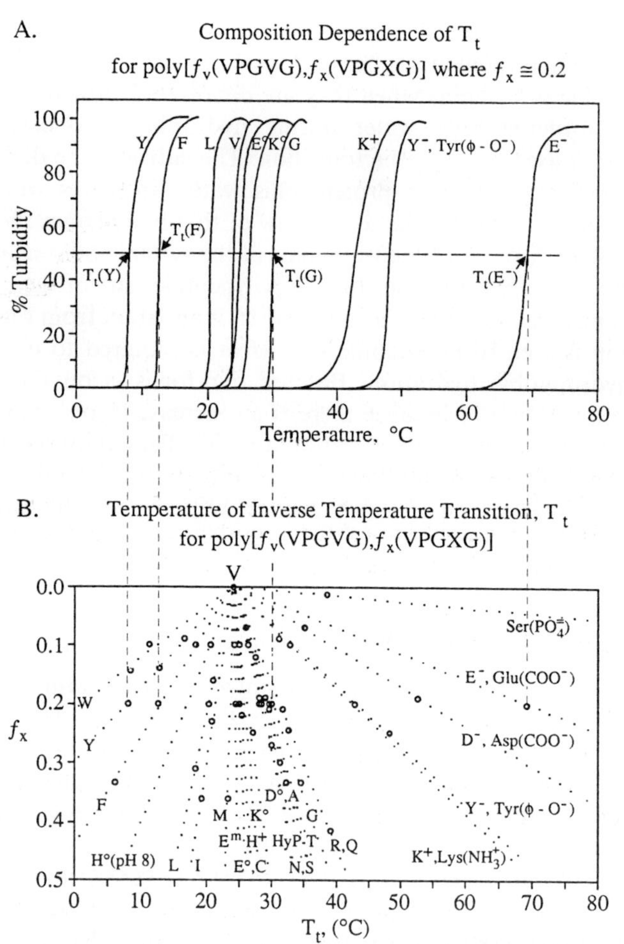


FIG. 4. (A) Measurement of $T_t$ and composition dependence of $T_t$ for poly[$f_v$(VPGVG),$f_x$(VPGXG)] where $f_x = 0.2$. $T_t$ is taken as the temperature at which 50% turbidity is obtained. Clearly, as the substituted residue becomes more hydrophobic, the value of $T_t$ lowers and, as the residues become more polar, the value of $T_t$ rises. Note the very large effect of side chain ionization. These values of $T_t$, along with those for different values of $f_x$ for all of the naturally occurring amino acid residues, are plotted as $f_x$ versus $T_t$ in (B), where the plots for each amino acid residue define a straight line which when extrapolated to $f_x = 1$ gives a measure of the relative hydrophobicities as given in Table 1. (Adapted with permission from Urry 1993.)

extrapolate to $f_x = 1$ and obtain a set of relative $T_t$ values. It is obvious that an increase in hydrophobicity causes a decrease in $T_t$ and a decrease in hydrophobicity causes an increase in $T_t$ such that a hydrophobicity scale is produced which, for the first time, is dependent on the hydrophobic folding and assembly process of interest (Urry et al 1992).

The $T_t$-based hydrophobicity scale, which we have developed using the most striking repeating sequence of elastin as the host sequential polypeptide, is given in Table 1. This scale was used for the hydrophobicity plots for human, bovine and porcine elastin in Fig. 1.

Hydrophobic folding and assembly in proteins is now taking on greater significance. Dill et al (1993) have developed a theoretical foundation for cooperative hydrophobic folding transitions which, of course, is what we have been describing for the elastin sequences. Dobson et al (1994), in their studies of hen lysozyme, show hydrophobic collapse to be the initial folding step, followed by the sorting out of the best secondary structure that can occur given the hydrophobic domain that forms. Furthermore, it is becoming more apparent that the periodicity at which hydrophobic residues appear in a sequence can dictate the secondary structure that will result. Accordingly, the presence of a hydrophobic residue every third and/or fourth residue dictates an amphiphilic $\alpha$-helix; the occurrence of hydrophobic side chains at alternating residues dictates $\beta$-chains, etc. The hydrophobic side chain periodicity of poly(VPGVG) is as required for a $\beta$-turn, residues $i$ and $i+3$, coupled with an extended $\beta$-like chain which gives the $\beta$-spiral structure.

## The $\Delta T_t$ hydrophobic paradigm for protein folding and function

As seen in Fig. 4, raising the temperature above $T_t$ (the temperature for the onset of the inverse temperature transition) drives aggregation, which occurs by hydrophobic folding and assembly. In the $\Delta T_t$ hydrophobic paradigm the temperature is kept constant and the value of $T_t$ is changed. Lowering $T_t$ from above to below physiological temperature drives hydrophobic unfolding and disassembly.

Importantly, there are many ways by which the value of $T_t$ can be changed: (1) the value of $T_t$ decreases as the concentration and chain length of the protein-based polymer are increased; (2) changing the composition of the protein-based polymer changes the value of $T_t$; (3) increasing the concentration of salts decreases the value of $T_t$; (4) introduction of organic solutes such as sodium dodecylsulfate, guanidine hydrochloride, urea, etc. increases the value of $T_t$; (5) increasing the side chain charge markedly increases $T_t$; (6) phosphorylation, e.g. of a serine side chain, most dramatically increases the value of $T_t$; (7) application of pressure, particularly with aromatic residues present, increases the value of $T_t$; (8) reduction of a bound prosthetic group, such as a nicotinamide, markedly lowers $T_t$; (9) absorption of light by a bound chromophore, which thereby converts from a *trans* to a *cis* isomer, or which in other ways makes the polymer more polar, raises $T_t$; and (10) salt neutralization of a charged side chain markedly decreases the value of $T_t$ (Urry 1993). In general, any change that causes chains to become more polar and less hydrophobic will raise the

**TABLE 1 $T_t$-based hydrophobicity scale for proteins**

| *Residue X* | | *$T_t$(°C), linearly extrapolated to $f_x = 1$* | *Correlation coefficient* |
|---|---|---|---|
| Lys(NMeN, reduced)[a] | | −130 | 1.000 |
| Trp | (W) | −90 | 0.993 |
| Tyr | (Y) | −55 | 0.999 |
| Phe | (F) | −30 | 0.999 |
| His (pH 8) | ($H^o$) | −10 | 1.000 |
| Pro | (P)[b] | (−8) | calculated |
| Leu | (L) | 5 | 0.999 |
| Ile | (I) | 10 | 0.999 |
| Met | (M) | 20 | 0.996 |
| Val | (V) | 24 | reference |
| Glu($COOCH_3$) | ($E^m$) | 25 | 1.000 |
| Glu(COOH) | ($E^o$) | 30 | 1.000 |
| Cys | (C) | 30 | 1.000 |
| His (pH 4) | ($H^+$) | 30 | 1.000 |
| Lys($NH_2$) | ($K^o$) | 35 | 0.936 |
| Pro | (P)[c] | 40 | 0.950 |
| Asp(COOH) | ($D^o$) | 45 | 0.994 |
| Ala | (A) | 45 | 0.997 |
| HyP | | 50 | 0.998 |
| Asn | (N) | 50 | 0.997 |
| Ser | (S) | 50 | 0.997 |
| Thr | (T) | 50 | 0.999 |
| Gly | (G) | 55 | 0.999 |
| Arg | (R) | 60 | 1.000 |
| Gln | (Q) | 60 | 0.999 |
| Lys($NH_3^+$) | ($K^+$) | 120 | 0.999 |
| Tyr($\phi$-$O^-$) | ($Y^-$) | 120 | 0.996 |
| Lys(NMeN, oxidized)[a] | | 120 | 1.000 |
| Asp($COO^-$) | ($D^-$) | 170 | 0.999 |
| Glu($COO^-$) | ($E^-$) | 250 | 1.000 |
| Ser($PO_4^=$) | | 1000 | 1.000 |

[a]NMeN is for *N*-methyl nicotinamide pendant on a lysyl side chain, i.e. *N*-methyl nicotinate attached by amide linkage to the ε-$NH_2$ of Lys; the reduced state is *N*-methyl-1,6-dihydronicotinamide.
[b]The calculated $T_t$ value for Pro comes from poly(VPGVG) when the experimental values of Val and Gly are used. This hydrophobicity value of −8 °C is unique to the β-spiral structure where there is hydrophobic contact between the $Val^1 \gamma CH_3$ and $Pro^2 \beta CH_2$ moieties.
[c]The experimental value determined from poly [$f_v$(VPGVG),$f_P$(PPGVG)].
$T_t$ = temperature of inverse temperature transition for poly[$f_v$(VPGVG),$f_x$(VPGXG)].
Adapted from Urry et al (1992).

value of $T_t$ and favour unfolding and disassembly (i.e. will cause swelling when cross-linked).

## Consequences of the $\Delta T_t$ hydrophobic paradigm for elastin structure formation, function and pathology

All of the ways by which the value of $T_t$ can be changed can either enhance fibre formation (by lowering $T_t$) or interfere with fibre formation (by raising $T_t$). As the fibre is formed, lowering $T_t$ can stabilize the fibre with increased elastic modulus and improved function, or raising the value of $T_t$ can lead to fibre swelling with decrease in elastic modulus, with increased incidence of tearing and, in general, with loss of function and enhanced susceptibility to proteolytic degradation.

### *Elastogenesis and wound repair*

Those factors which lower $T_t$ will favour fibre formation, e.g. increasing precursor protein concentration, conversion of lysine to aldehydes and the addition of salts. But the well-known post-translational modification of prolyl hydroxylation, as suggested early on (Urry et al 1979), would interfere with elastin fibre formation (see the relative values of $Pro^c$ of $-8\,°C$ and HyP of $50\,°C$ in Table 1). That prolyl hydroxylation interferes with fibre formation in culture has been shown by Franzblau and colleagues (Barone et al 1985). In wound repair there is a high level of prolyl hydroxylase, which ensures abundant high-quality collagen due to hydroxylation of many of the prolyl residues of collagen (Rapaka et al 1978b). The same enzyme hydroxylates tropoelastin (Bhatnagar et al 1978), raising $T_t$ above physiological temperature and preventing fibre formation. This provides an explanation for the paucity of elastic fibres in scar tissue. Furthermore, in the assembly process there are specific charged residues that need to be neutralized, for instance by side-chain ion pairing, during assembly. In particular, the human W4 sequence has a Glu residue that must be neutralized to bring the chain segment $T_t$ sufficiently below $37\,°C$.

## Function: the development of entropic elasticity

The remarkable durability of mammalian elastic fibres, each of which is capable of some 2 000 000 000 demanding stretch/relaxation cycles in the aortic arch during its lifetime, requires that the elastomeric force be dominantly entropic. In Figs 3B and C, force development is seen to occur as the temperature is raised above $T_t$ and through the range of the inverse temperature transition; that is, elastomeric force develops over the temperature range in which fibre entropy must decrease and for which all of the foregoing discussion shows structure formation to occur by hydrophobic folding and assembly. By the classical argument of a near zero slope in the plot of $ln(f/T)$ versus temperature curve in Fig. 3B, once above the transition temperature range (at 37 °C or higher) the elastic force is dominantly entropic. Since the preceding arguments have not suggested the formation of a random chain network as the temperature is raised through the range of the inverse temperature transition, but rather argue that there is an increase in order, the random chain network theory of rubber elasticity does not seem to provide a relevant mechanism.

Clearly, there are changes in solvent entropy on extension, as water of hydrophobic hydration would form around the hydrophobic groups exposed by stretching. An experimental basis for this contribution to the elastomeric force, however, is hard to demonstrate. For example, addition of ethylene glycol causes the heat of the inverse temperature transition to approach zero (implying that the entropy change for the inverse temperature transition approaches zero), but instead of a loss of entropic elastic force, entropic elastic force increases on addition of ethylene glycol (Luan et al 1989).

The mechanism of elasticity that arose from the study of poly(VPGVG) is one of a decrease in entropy on extension arising from a damping of internal chain dynamics on extension. This has been called the librational entropy mechanism of elasticity (Urry 1991, Wasserman & Salemme 1990).

## Elastic fibre pathology

Elastin is perhaps most noteworthy for its capacity to provide tissues with the elasticity they require and yet to do so with relatively limited pathology and associated disease. This is particularly true in view of the finding that it has such a low turnover in the body. The half-life for elastin in the body, as determined by three independent methods, is of the order of 70 years or more (Powell 1990, Rucker & Tinker 1977).

### *Pulmonary emphysema*

The usual description of pulmonary emphysema is a biochemical process without a significant biophysical component. In the genetic anomaly of deficiencies in $\alpha_1$-antitrypsin, which occurs in 2% of those with pulmonary emphysema (Crystal 1991), the lung is the site in the body at which the deficiency is expressed. It is generally thought that the primary process results from unchecked proteolytic digestion of elastin. But the primary disability occurs in the lungs and not in the elastic arteries where the demand for elastic function is so extreme. In smokers, the occurrence of pulmonary emphysema is greatly increased, yet there does not seem to be any difference between smokers and non-smokers in the levels of $\alpha_1$-proteinase inhibitor (Luisetti et al 1992, Janoff 1983, 1985), neither do the levels of inactivated $\alpha_1$-proteinase inhibitor appear to differ between the groups (Afford et al 1988).

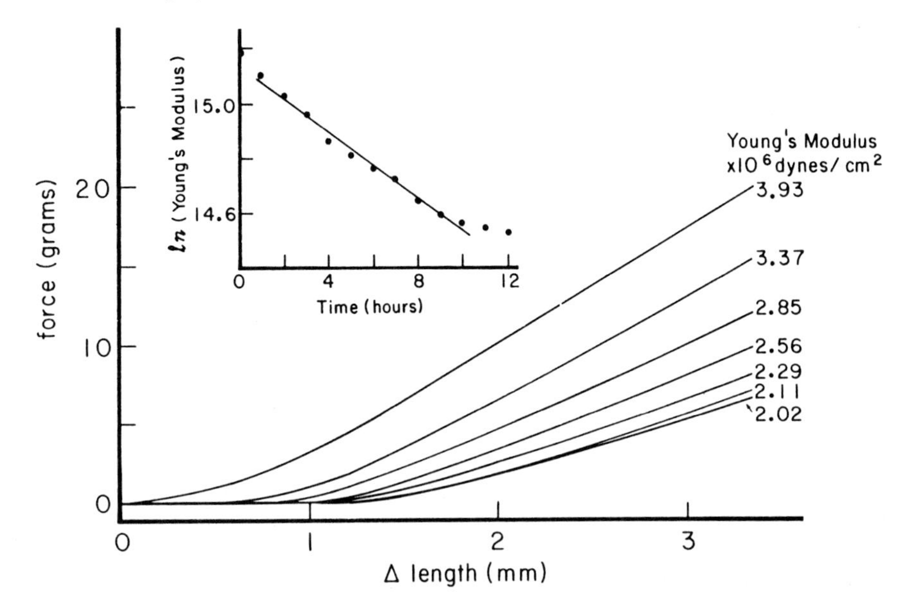


FIG. 5. Stress/strain curves (development of elastic restoring force resulting from an increase in length) for a strip of purified ligamentum nuchae elastin followed during the time course for exposure to a xanthine oxidase superoxide-generating system. The resulting superoxide and hydrogen peroxide cause the fibre to increase in length (seen by the greater extension required before resistance to extension is observed as an increase in force) and cause the slope to decrease (a measure of the capacity to recover from deformation). Inset: a plot of *ln*(Young's modulus) versus time exposed to the superoxide-generating system, which shows a near-linear loss over a 10 h period. The conclusion is that oxidation causes elastic fibre extension and loss of function, i.e. loss of the capacity to provide a restoring force after extension. Oxidative bursts from macrophages attacking foreign particles in the lung would be expected to cause similar damage to lung elastic fibres, as would any oxidative atmospheres drawn into the lung. (Reproduced with permission from Urry 1988.)

Furthermore, premature infants exposed to positive pressure ventilation with increased oxygen levels have a high incidence of lung disease with associated elastic fibre destruction. On the basis of the $\Delta T_t$ mechanism, as proposed previously (Urry et al 1980), it is here again argued that the primary insult is oxidation that results in attached polar moieties; these lead to an increase in $T_t$ and unfolding of the protein. This has been demonstrated experimentally (Fig. 5) (Urry 1988). In our view, it is then the partially unfolded protein that is susceptible to proteolytic digestion—otherwise, why would the lung be the tissue most affected by the genetic defect and why would this happen to protein otherwise astounding for its resistance to proteolytic degradation?

*Actinic (solar) elastosis*

Premature degeneration of elastic tissue of the skin due to solar damage caused by prolonged exposure to sunlight (actinic or solar elastosis) can also be explained by the $\Delta T_t$ mechanism. There are only a few residues in the elastin of the elastic fibre that can absorb near ultraviolet light: tyrosine (Y), phenylalanine (F) and the cross-links desmosine and isodesmosine. The former two are the most hydrophobic residues of elastin (see Fig. 1 and Table 1) with $T_t$ values of $-55$ and $-30\,°C$, respectively; the latter two act as cross-links to constrain swelling and dissolution and, when present, add to effective chain length and hydrophobicity. The loss of these residues, and particularly their conversion to more polar residues, raises the value of $T_t$ of the chain segments involved.

In particular, Guay et al (1985) and previously Baurain et al (1976) found that the exposure of elastolysates to 254 nm-wavelength light for one hour degraded all of the tyrosines, one-half of the phenylalanines, all of the desmosines and three-quarters of the isodesmosines, and the photoproducts include aspartic acid (D) from tyrosine and lysine ($K^+$) from the cross-links. The change in $T_t$ for tyrosine breakdown to aspartic acid is $+225\,°C$. This could be expected to cause the chain segment in which it occurred to disassemble hydrophobically and unfold and, according to our proposed mechanism of solar elastosis, to become susceptible to proteolytic degradation. In general, any photoproducts with higher values of $T_t$ than the parent residue would shift the involved chain segment towards disassembly, unfolding and proteolytic degradation, with the result being premature degeneration of elastic tissue.

*Acknowledgements*

This work was supported in part by the Office of Naval Research, contract N00014-89-J-1970 and by the National Institutes of Health, grant HL29578. The authors are pleased to acknowledge Richard Knight of the Auburn University Nuclear Science Center for carrying out the $\gamma$-irradiation cross-linking.

## References

Afford SC, Burnett D, Campbell EJ, Cury JD, Stockley RA 1988 The assessment of alpha 1 proteinase inhibitor form and function in lung lavage fluid from healthy subjects. Biol Chem Hoppe-Seyler 369:1065–1074

Barone LM, Faris B, Chipman SD, Toselli P, Oakes BW, Franzblau C 1985 Alteration of the extracellular matrix of smooth muscle cells by ascorbate treatment. Biochim Biophys Acta 840:245–254

Baurain R, Larochelle J-F, Lamy F 1976 Photolysis of desmosine and isodesmosine by ultraviolet light. Eur J Biochem 67:155–164

Bhatnagar RS, Rapaka RS, Urry DW 1978 Interaction of polypeptide models of elastin with prolyl hydroxylase. FEBS Lett 95:61–64

Cook WJ, Einspahr HM, Trapane TL, Urry DW, Bugg CE 1980 Crystal structure and conformation of the cyclic trimer of a repeat pentapeptide of elastin, cyclo-(L-valyl-L-prolylglycyl-L-valylglycyl)$_3$. J Am Chem Soc 102:5502–5505

Crystal RG 1991 Alpha 1-antitrypin deficiency: pathogenesis and treatment. Hosp Pract 26:81–94

Dill KA, Fiebig FM, Chan HS 1993 Cooperativity in protein-folding kinetics. Proc Natl Acad Sci USA 90:1942–1946

Dobson CM, Evans PA, Radford SE 1994 Understanding how proteins fold: the lysozyme story so far. Trends Biochem Sci 19:31–37

Franzblau C, Lent RW 1969 Studies on the chemistry of elastin. In: Shaw EN, Hirs CHW, Koenig DF, Popenoc EA, Siegelman HW (eds) Brookhaven symposium in biology number 21: structure, function and evolution in proteins, vol 2, p 358–377

Gowda DC, Luan C-H, Furner RC et al 1995 Synthesis and characterization of human elastin $W_4$ sequence. Int J Pept Protein Res, in press

Guay M, Lagace M, Lamy F 1985 Photolysis and ozonolysis of desmosine and elastolytic peptides. Conn Tiss Res 14:89–107

Gray WR, Sandberg LB, Foster JA 1973 Molecular model for elastin structure and function. Nature 246:461–466

Indik Z, Yeh N, Ornstein-Goldstein N et al 1987 Alternative splicing of human elastin mRNA indicated by sequence analysis of cloned genomic and complementary DNA. Proc Natl Acad Sci USA 84:5680–5684

Janoff A 1983 Biochemical links between cigarette smoking and pulmonary emphysema. J Appl Physiol 55:285–293

Janoff A 1985 Elastases and emphysema. Current assessment of the protease–antiprotease hypothesis. Am Rev Respir Dis 132:417–433

Luan CH, Jaggard J, Harris RD, Urry DW 1989 On the source of entropic elastomeric force in polypeptides and proteins: backbone configurational vs. side chain solvational entropy. Int J Quantum Chem Symp 16:235–244

Luan CH, Chang DK, Parker TM, Krishna NR, Urry DW 1991 $\beta$-spiral conformations of the elastomeric polytetrapeptides, $(VPGG)_n$ and $(IPGG)_n$, by 2D NMR, and molecular mechanics studies. Int J Quantum Chem Symp 18:183–198

Luisetti M, Piccioni PD, Donnetta AM, Bulgheroni A, Peone V 1992 Protease–antiprotease imbalance: local evaluation with bronchoalveolar lavage. Respiration (suppl 1) 59:24–27

Powell JT 1990 Dilatation through loss of elastin. In: Greenhalgh RM, Marrick JA (eds) The cause and management of aneurysms. W. B. Saunders, London, p 89–96

Partridge SM 1969 Elastin, biosynthesis and structure. Gerontologia 15:85–100

Rapaka RS, Okamoto K, Urry DW 1978a Non-elastomeric polypeptide models of elastin: synthesis of polyhexapeptides and a cross-linked polyhexapeptide. Int J Pept Protein Res 11:109–127

Rapaka RS, Renugopalakrishnan V, Urry DW, Bhatnagar RS 1978b Hydroxylation of proline polytripeptide models of collagen: stereochemistry of polytripeptide–prolyl hydroxylase interaction. Biochemistry 17:2892–2898

Rucker RB, Tinker D 1977 Structure and metabolism of arterial elastin. Int Rev Exp Pathol 17:1–47

Sandberg LB, Leslie JG, Leach CT, Torres VL, Smith AR, Smith DW 1985 Elastin covalent structure as determined by solid phase amino acid sequencing. Pathol Biol 33:266–274

Urry DW 1982 Characterization of soluble peptides of elastin by physical techniques. Methods Enzymol 82:673–716

Urry DW 1988 Entropic elastic processes in protein mechanisms. II. Simple (passive) and coupled (active) development of elastic forces. J Protein Chem 7:81–114

Urry DW 1991 Thermally driven self-assembly, molecular structuring and entropic mechanisms in elastomeric polypeptides. In: Balaram P, Ramaseshan S (eds)

Molecular conformation and biological interactions. Indian Academy of Sciences, Bangalore, p 555–583
Urry DW 1992 Free energy transduction in polypeptides and proteins based on inverse temperature transitions. Prog Biophys Mol Biol 57:23–57
Urry DW 1993 Molecular machines: how motion and other functions of living organisms can result from reversible chemical changes. Angew Chem Int Ed Engl 32:819–841
Urry DW, Long MM, Sugano H 1978 Cyclic analog of elastin polyhexapeptide exhibits an inverse temperature transition leading to crystallization. J Biol Chem 253:6301–6302
Urry DW, Sugano H, Prasad KU, Long MM, Bhatnagar RS 1979 Prolyl hydroxylation of the polypentapeptide model of elastin impairs fiber formation. Biochem Biophys Res Commun 90:194–198
Urry DW, Bhatnagar RS, Sugano H, Prasad KU, Rapaka RS 1980 A molecular basis for defective elastic tissue in environmentally induced lung disease. In: Bhatnagar RS (ed) Molecular basis of environmental toxicity. Ann Arbor Science Publishers, Ann Arbor, MI, p 515–530
Urry DW, Gowda DC, Parker TM et al 1992 Hydrophobicity scale for proteins based on inverse temperature transitions. Biopolymers 32:1243–1250
Wasserman ZR, Salemme FR 1990 A molecular dynamics investigation of the elastomeric restoring force in elastin. Biopolymers 29:1613–1631
Yeh H, Ornstein-Goldstein N, Indik Z et al 1987 Sequence variation of bovine elastin mRNA due to alternative splicing. Collagen Relat Res 7:235–247

## DISCUSSION

*Keeley:* What would be the differences in the properties of an elastic polypeptide if instead of a PGVGV or PGLGV repeat it had a GGLGV or GGVGV repeat?

*Urry:* We have synthesized GGLGV, actually as the GLGVG permutation, and polymerized it to form poly(GLGVG) of greater than 50 kDa molecular mass. In our hands it has not been possible to form this material into a viscoelastic or elastic state, but rather it forms a more rigid structure.

The essential feature for ideal (entropic) elasticity and viscoelasticity, in our view, is to have a hydrophobically folded structure in which the particular chain segments are not constrained by hydrogen bond formation. Instead, short chain segments, while maintaining structurally non-constraining hydrophobic contacts, are free to undergo librational (rocking) motions, that is, tonal oscillations, that become damped, or decreased in amplitude, as the chain is extended.

This is consistent with the broader perspective that is emerging for protein folding in which hydrophobic folding is the primary event. This perspective has recently received major support from both theoretical and experimental lines of reasoning. The theoretical basis comes from a study by Dill et al (1993), in which they used a string of hydrophobic and polar beads. They showed that if a pair of hydrophobic beads associate and by doing so can bring a second pair of hydrophobic beads into proximity, these will quickly fall into place, as would a similar third pair and so on. For example, if you take a chain and make a turn, then, as the hydrophobic groups match up, cooperative hydrophobic folding occurs which would give transitions of the sort that we've been seeing for over 15 years with poly(VPGVG).

The experimental basis is also moving forward. Perhaps the best results to quote here are those from Chris Dobson and colleagues (1994) on hen lysozyme. They studied the renaturation of hen lysozyme using three different methodologies. First they used circular dichroism (CD) in the far UV range, with stopped-flow methods in order to achieve very short times. CD is usually interpreted to give the secondary structure: the amount of α-helix and β-sheet and so on. Within 3–4 ms they found that the native CD pattern was recovered, suggesting that within this time the folding is completed. In order to look at which of the peptide NH hydrogens had become shielded by hydrogen bonding, they did a pulsed hydrogen-labelling study, using $^2H_2O$ labelling and electrospray mass spectrometry. They found that all of the NH hydrogens exchanged; none was shielded by hydrogen bonding. This raised the question as to what was holding these segments of peptides together to produce something like a natural CD pattern. Then they did the critical experiment. Lysozyme has six tryptophans, so they looked at tryptophan fluorescence intensity, because tryptophan in a hydrophobic milieu has a very different fluorescence intensity. They found out that the first thing that happens is a hydrophobic collapse of the tryptophans and the other hydrophobic residues, and then there is stabilizing of the best secondary structure that fits with this hydrophobic folding. It appears that it is more the periodicity of hydrophobic residues that dictates the protein-folded structure than the propensity to form the hydrogen bonds of an α-helix or β-sheet. For example, the β-chain would be driven by alternating hydrophobic residues. So you could simply deduce what sort of secondary structure will arise from the hydrophobic domain that would form.

*Mecham:* Another interesting observation from the lysozyme work was the discovery that there is a large number of hydrophobic domains on the surface of lysozyme (McKenzie & White 1981). This is consistent with your finding that these domains don't necessarily bury themselves inside proteins; they can be in positions where they can interact with solvent.

*Urry:* If I recall Chris Dobson's presentation a short time ago in Cambridge (personal communication), lysozyme has a very strong hydrophobic domain and also a secondary hydrophobic domain. Some of the hydrophobic residues may be exposed at the surface because their expression of hydrophobicity is suppressed by proximal charged residues that destructure the water of hydrophobic hydration. But it's an issue of hydrophobic residues dehydrating and collapsing, and the protein then assumes a secondary structure that fits well with the hydrophobic interactions. As I understand it, the dominant driving force for folding, in the case of lysozyme, is the formation of a hydrophobic domain, followed by the sorting out of the best secondary structure that fits with that domain. For example, the presence of hydrophobic residues every third or fourth residue for a long sequence results in an $\alpha$-helix. You can deduce secondary structures by looking at hydrophobic periodicities perhaps more successfully than by conformational energy calculations that emphasize secondary structure formation in the absence of water.

*Tamburro:* Do you think that the $\beta$-spiral structure is still stable in sequences where PG sequences are substituted by GG sequences? I ask this because, according to our results (Tamburro et al 1990), the $\beta$-turns are much less stable in the latter case.

*Urry:* It is the PG insert in $\beta$-turns that really helps to set up the $\beta$-spirals. We see this very well in the repeating sequences that have the PG in them. When we look at a sequence of an elastin peptide, it is very important first to find out whether it is an elastic sequence or not, and then to decide what sort of folding pattern it has. One of the points I would like to make here is that poly(VPGVG) was a wonderful system which gave us a new insight into the nature of entropic elasticity. It did this because it was so regular. Within each pentamer is a 1,4 pairing of hydrophobic side chains, which could suggest an $\alpha$-helix, but then there is an alternating 4,6 pairing, which suggests it could be a $\beta$-sheet. It can't form both. Instead, it forms what we believe to be the $\beta$-spiral, which allows for both features as well as inter-turn and inter-spiral hydrophobic contacts. The essential issue is that entropic elasticity occurs in proteins where there is dominantly a hydrophobic side chain assembly which is such that it leaves the backbone rather free to move.

*Tamburro:* According to your model of entropic elasticity for poly(VPGVG), the librations occur in the suspended dipeptide segments VG—that is, outside the $\beta$-turns. However, when PG is substituted with GG, librations occur inside the $\beta$-turn (Villani & Tamburro 1993), and I suggest that these librations could be at the origin of the instability of GG-containing $\beta$-turns. Would you comment on this?

*Urry:* Poly(VPGG), for example, is entropically elastic (Urry et al 1986). We believe it forms an entropic elastomer because one of its peptide moieties, the GV peptide moiety, has a greater librational entropy than the others (Luan et al 1991). But I would like to make elasticity more general than that. One of the structures I showed in my presentation was of the type III domain of fibronectin. It's analogous or homologous to the domain for which the crystal structure has been determined in human tenascin (Leahy et al 1992) and these domains are homologous to the repeating domains in titin (Trinick 1992). Titin is the third filament in muscle: it is a filament over a micron long with a molecular mass in excess of a million, it's elastic and yet each of these repeating sequences is folded into a sort of hydrophobically collapsed $\beta$-barrel structure. In our view, elasticity is conferred on the structure by the ability of the backbone to librate and have entropy due to non-restricting hydrophobic side-chain contacts, and then on extension the amplitudes of the backbone motions are damped.

*Rosenbloom:* I was impressed by the fact that if you introduce just two charged residues per 100 into tropoelastin you drastically change its properties. The molecule has 35 charged lysines; have you looked at the effect of putting the same lysines periodically in these hydrophobic structures? What's going to happen during biosynthesis? You've got a tremendously charged protein.

*Urry:* The charged residues do destroy the water of hydrophobic hydration, but a charged lysine is not nearly as polar as a charged glutamic acid. Also, as you get more water of hydrophobic hydration you shift the pK of the charged residues. The first question is therefore: how many of those 35 lysines are really charged, or have their pK values been lowered due to interactions with the water of hydrophobic hydration? Our best example in this regard concerns aspartic acid, not lysine, but we do have substantial data on lysine (Urry et al 1994a). The pK of aspartic acid is 3.9. If you take five phenylalanines per six pentamers and place them as far from one aspartic acid residue per 30mer as possible, the pK shifts to 6.7. If you then move the phenylalanines as close as possible, the pK shifts to 10.1: changing hydrophobicity can cause enormous pK shifts (Urry et al 1994b). In this way it is possible to put in certain kinds of energy and get out chemical work, for example driving a proton gradient. We talk about driving hydrophobic folding, but we can also shift pKs by changing hydrophobicity or hitting the molecule with light (C. J. Heimbach & D. W. Urry, unpublished results), or by oxidation or reduction of a side chain (L. C. Hayes & D. W. Urry, unpublished results). Absorption of a photon by a chromophore, for example, can shift the pK of an aspartic acid residue, removed by 50 bonds from the chromophore, by five pH units (D. W. Urry, C. J. Heimbach & S. Q. Peng, unpublished results).

The other factors that will probably be very important in this regard are pairing by positively and negatively charged side chains and the conversion of the lysine positive charge to an aldehyde, catalysed by lysyl oxidase. The enzyme is called an 'oxidase', but this conversion actually causes an increase in the hydrophobicity of the system, which is a significant way to facilitate the hydrophobic folding.

*Robert:* So the crucial question really concerns the environment of tropoelastin when it comes out of the cell.

*Urry:* Just as we showed that salts markedly change the temperature of the transition when the charged species are present, ion pairing is a very good way to neutralize the effect of the charged species of raising the temperature of the hydrophobic folding transition.

*Kagan:* During the process of lysine cross-linking, 30 out of approximately 48 lysines per 1000 residues in tropoelastin are oxidized. Are there temporally dependent, oxidation-dependent changes in local conformation? That is, you might be driving fibril formation by the catalysis of lysine oxidation, since positively charged amino groups are lost in the process. This would be expected to favour hydrophobic interactions between tropoelastin monomers.

*Urry:* Yes, we are very interested in the sequence of events question. I doubt that there's enough hydrophobicity to shift the pKs of lysine very far. The important issues, we believe, are ion pairing and how soon lysyl oxidase begins to convert the lysines to aldehydes. Then you can begin to have better folding and assembly for fibre formation. There is also the issue of the glutamic and aspartic acid residues, and whether these can pair, for example, with the residual lysines to reduce the effect of these charges.

*Robert:* You have said that the hydrophobic effect on the charge is much more important than the charge–charge interaction. When the microfibril captures tropoelastin, the first thing that will happen is a charge–charge interaction, so the question is really not completely answered. When does lysyl oxidase come in, and what can be the conformation of the tropoelastin when it reacts with this enzyme?

*Urry:* The energy gained by a charge–charge interaction in normal water is minimal due to the high dielectric constant of the water; that is why the charges formed in the first place. When the charges are under the influence of a hydrophobic domain, however, the effective dielectric constant of the water of hydrophobic hydration is much lower and the energy of the interaction between charges is enhanced.

*Mungai:* What is the biophysical explanation for the phenomenon that when you boil chicken, the wing remains folded? There is an elastic ligament in the anterior border of the wing. Similarly, when you boil the neck of a chicken, the neck remains curled. There are short elastic ligaments on the posterior aspects of cervical vertebrae. Why aren't these ligaments affected by boiling?

*Urry:* That's an interesting question. As the temperature is raised, hydrophobic assembly is enhanced, and the elastic nature of the material is improved. However, if the temperature is raised for too long, some denaturation will occur, so the elastic properties of the tissue depend on how long the temperature is raised. If you go back and re-do the Hoeve & Flory (1974) studies, whether or not you obtain data indicating an ideal entropic elastomer depends on how fast you go up to the high temperature and how long you stay there before you go back. If you go up at just the right speed the slope of the plot of *ln*(force/temperature) versus temperature becomes zero and the protein appears to be an ideal entropic elastomer. If you go up very slowly, you find that it turns back over, becoming a negative slope because of a loss of elastic force at higher temperature. Then when you go back down you've irreversibly lowered the force, and so you have effectively denatured your protein; part of that denaturation as observed for cross-linked poly(VPGVG) is a racemization that has occurred to the L-residues at elevated temperature.

*Rosenbloom:* What's the timescale?

*Urry:* We do our experiments overnight, for example, under computer control.

*Sandberg:* When Bill Gray and I first put together a model for tropoelastin, we thought there was α-helicity in the molecule (Gray et al 1973). What is your feeling about that?

*Urry:* I think that those cross-linking sequences are α-helical, and that you and Bill hit it right on the head when you saw that relationship between the lysines. It's not going to be too handy to have those lysines charged, but I don't think even that is a big problem. Once these lysines are converted to aldehydes, I'm convinced that an α-helix results. In fact, when we did the CD patterns early on α-elastin in the coacervated state (Urry et al 1969, Starcher et al 1973), we got a reasonable amount of α-helix (about 20%), which is not far from what those cross-linking sequences might do.

*Sandberg:* So conversion of lysines to aldehydes would enhance the α-helix formation?

*Urry:* Yes.

*Pasquali-Ronchetti:* I will show later on that the elastic fibre is a rather complex structure made of different proteins. Do you think that proteins or proteoglycans inside the elastic fibre can modulate its elasticity? Some of these proteins have a high affinity for ions. Do you think that local release of ions can influence the physiological behaviour of the protein?

*Urry:* Yes; especially if there is a charged residue present. For example, it worried us for some time that the polymer of W4 (Sandberg & Davidson 1984) is not folded at 37 °C unless you neutralize the Glu residue, so charge can be important in this regard.

It is interesting, however, that in elastogenesis the fibre grows with age. This tells us that, in spite of all of the other proteins that are present, tropoelastin is still finding tropoelastin. It still appears to go in, sort its way through the milieu of extracellular matrix and find elastin with which to cross-link, even though the elastin may be coated with microfibrillar proteins.

*Robert:* The secondary heart effect could be very strongly influenced by the occurrence of splice or sequence variants of elastin. This effect is very important in the age-dependent decline of the circulation, where elastin gets less and less elastic. This process could take place either very rapidly or much more slowly, depending on the sequence differences in elastin. Are there any data on the sequences which would substantiate such a claim? The effects you found with $Na^+$ and $Ca^{2+}$ were not as dramatic as those we have found (Jacob et al 1983). We tried to saturate elastin peptides with cholesterol, coacervate down and determine radioactive cholesterol. With $Na^+$ we got a very low saturation curve, but with $Ca^{2+}$ we got a constant increase in cholesterol uptake. According to your data, can any detailed picture be drawn of how lipids interact with elastin?

*Urry:* Hydrophobicity does affect the elastic modulus; it effects the rate at which you can snap back on extension. This is shown when the 20 Mrad γ-irradiation cross-linked matrix of poly(VPGVG) is compared with that of poly(IPGVG): the former is a transparent material with an elastic modulus of about $1 \times 10^6$ dynes/cm$^2$, and the latter forms an opaque tough matrix with an elastic modulus in the range $4\text{–}8 \times 10^6$ dynes/cm$^2$. It has a stress–strain curve that looks like that of the femoral artery.

*Robert:* Concerning the $Ca^{2+}$ and lipid effect: is there any detailed picture you can draw of how lipids interact with the $\beta$-spirals for instance?

*Urry:* I can do little more than make the point that the associated $\beta$-spirals form a syncitium of spiralling hydrophobic ribbons, not a big hydrophobic core. You have the hydrophobic folding of a $\beta$-spiral, and then you have hydrophobic ridges on those $\beta$-spirals, which then associate to form twisted filaments. So the hydrophobic regions are filamentous in their own right. I assume that the hydrophobic portions of the lipid will interdigitate into those hydrophobic filamentous regions.

*Starcher:* If I understood you correctly, you said that oxidative changes in the elastin could alter the folding of the elastin and make it more susceptible to proteolysis. Over the years there have been two different approaches to the production of emphysema; one of those has been proteolytic damage and the other has been oxidative damage. Are you saying that these are not mutually exclusive events?

*Urry:* No, they're sequential. I believe oxidative damage occurs first and then the proteolytic damage is second.

*Starcher:* It's not necessarily damage; you're saying that there can actually be changes in some of the residues.

*Urry:* Anything that makes a residue more polar—an oxidation process. I wish the data on degradation were better, but they are not bad. If we take our cross-linked polypentapeptide and implant it in the peritoneal cavity, it can remain there for weeks or months, and no fibrous capsule forms around it, and no degradation occurs—it stays beautifully clear, as though the body didn't know it existed. Grace Picciolo has shown that when you add this material to macrophages it actually subdues their oxidative bursts that would ultimately lead to unfolding and degradation. For drug delivery we want to load these things with drug and control the rate of release of the drug. Asparagine, when included in the polypentapeptide, will act as a chemical clock, where it breaks down to aspartic acid with a half-life that is determined by its nearest neighbours. If you put it in as VPGNG, its half-life is approximately 18 days. So now if we put in just two asparigines in a 100 residues and have that nice clean transparent sheet, and implant that in the peritoneal cavity, after about a month you find about 50% of the material remaining. Asparagine breaks down to aspartic acid and this positive charge then causes it to swell; presumably it can degrade in the peritoneal cavity once it's swollen but not while it's in the contracted, folded state.

Analogously, oxidative damage, whether due to oxidative atmospheres directly or due to the superoxide and hydrogen peroxide bursts from macrophages attacking a foreign particle, raises the temperature of the hydrophobic folding transition and causes the elastic fibre to swell and become susceptible to proteolytic degradation.

*Hinek:* There's no doubt that this hydrophobic interaction plays a very important role in the initiation of elastic fibre formation. This is an extracellular event: why don't elastin molecules interact with each other intracellularly? Why don't they coacervate?

*Urry:* The simplest answer is that while the elastin peptides are in the cell the lysines are still all charged. Lysyl oxidase functions extracellularly to convert those to aldehydes, thus increasing the hydrophobicity.

*Robert:* Maybe part of the story is that it's not alone: as Bob Mecham has shown, inside the cell elastin is accompanied by the receptor protein.

*Kimani:* It has been suggested that elastin can bind catecholamines (Powis 1973). Have you tested the effect of catecholamines on the hydrophobic contractions/relaxations in elastin?

*Urry:* No. There are a set of molecules—choline, *N*-acetylputrescine, GABA, betaine, and a number of transmitter-related molecules—that do change the temperature of the hydrophobic transition. They are related to fibre assembly for the orbweb of spiders (Tillinghast & Townley 1986), for example. Thus, in addition to chaperones, there could be molecular species in the cell which raise the temperature of the hydrophobic folding transition to prevent association happening intracellularly.

## References

Dill KA, Fiebig FM, Chan HS 1993 Cooperativity in protein-folding kinetics. Proc Natl Acad Sci USA 90:1942–1946

Dobson CM, Evans PA, Radford SE 1994 Understanding how proteins fold: the lysozyme story so far. Trends Biochem Sci 19:31–37

Gray WR, Sandberg LB, Foster JA 1973 Molecular model for elastin structure and function. Nature 246:461–466

Hoeve CAJ, Flory PJ 1974 Elastic properties of elastin. Biopolymers 13:677–688

Jacob MP, Hornebeck W, Robert L 1983 Studies on the interaction of cholesterol with soluble and insoluble elastin. Int J Biol Macromol 5:275–278

Leahy D, Hendrickson W, Aukhil I, Joshi P, Erickson HP 1992 X-ray crystal structure and cell adhesion studies of the RGD-containing FN III domain from tenascin. Mol Biol Cell 3(suppl):A1

Luan CH, Chang DK, Parker TM, Krishna NR, Urry DW 1991 $\beta$-spiral conformations of the elastomeric polytetrapeptides, $(VPGG)_n$ and $(IPGG)_n$, by 2D NMR, and molecular mechanics studies. Int J Quantum Chem Symp 18:183–198

McKenzie HA, White FH 1981 Lysozome and $\alpha$-lactalbumin: structure, function and relationship. Adv Protein Chem 41:173–315

Powis G 1973 Binding of catecholamines to connective tissue and the effect upon the responses of blood vessels to noradrenaline and to nerve stimulation. J Physiol 234:145–162

Sandberg LB, Davidson JM 1984 Elastin and its gene. Pept Protein Rev 3:169–226

Starcher BC, Saccomani G, Urry DW 1973 Coacervation and ion binding studies on aortic elastin. Biochim Biophys Acta 310:481–486

Tamburro AM, Guantieri V, Pandolfo L, Scopa A 1990 Synthetic fragments and analogues of elastin. II. Conformational studies. Biopolymers 29:855–870

Tillinghast EK, Townley MA 1986 The independent regulation of protein synthesis in the major ampullate glands of *Araneus caraticus* (keyserling). J Insect Physiol 32:117–123
Trinick J 1992 Understanding the functions of titin and nebulin. FEBS Lett 307:44–48
Urry DW, Starcher B, Partridge SM 1969 Coacervation of solubilized elastin effects a notable conformational change. Nature 222:795–796
Urry DW, Harris RD, Long MM, Prasad KU 1986 Polytetrapeptide of elastin: temperature-correlated elastomeric force and structure development. Int J Pept 28:649–660
Urry DW, Peng SQ, Gowda DC, Parker TM, Harris RD 1994a Comparison of electrostatic-induced and hydrophobic-induced pKa shifts in polypentapeptides: the lysine residue. Chem Phys Lett 225:97–103
Urry DW, Gowda DC, Peng SQ, Parker TM, Jing NJ, Harris RD 1994b Nanometric design of extraordinary hydrophobic-induced pKa shifts for aspartic acid: relevance to protein mechanisms. Biopolymers 34:889–896
Villani V, Tamburro AM 1993 Conformational analysis by molecular mechanics energy minimizations of the tetrapeptide Boc-Gly-Leu-Gly-Gly-NMe, a recurring sequence of elastin. J Chem Soc Perkin Trans 2:1951–1961

# Ultrastructure of elastin

Ivonne Pasquali-Ronchetti, Claudio Fornieri, Miranda Baccarani-Contri and Daniela Quaglino

*Department of Biomedical Sciences, General Pathology, via Campi 287, 41100 Modena, Italy*

*Abstract.* Almost all structural studies on elastin have been done in higher vertebrates, in which it is organized as an extracellular network of branched fibres which vary from fractions f microns to several microns in diameter. By conventional electron microscopy, elastin appears amorphous. By both freeze-fracture and negative staining on cryosections, it can be resolved as beaded filaments 5 nm in diameter forming a 3D meshwork that, upon stretching, becomes oriented in the direction of the force applied. This filamentous aggregation of elastin molecules is confirmed *in vitro* by the observation that its soluble precursor, tropoelastin, shows a strong tendency to associate into short 5 nm-thick filaments that, with time, become longer and aggregate into bundles of various dimensions. If chemically fixed and embedded, these aggregates appear amorphous and identical to natural elastin fibres. The tendency of tropoelastin to aggregate into 4–5 nm-thick beaded filaments, which then associate into 12 nm-thick filaments forming a 3D network, has been observed by atomic force microscopy for recombinant human tropoelastin. Therefore, the amorphous structure of elastin seems to be a technical artefact. Apart from elastin-associated microfibrils, which are always present at the periphery of growing elastic fibres and probably have a role more complex than being a scaffold for tropoelastin aggregation *in vivo*, the elastic fibres seem to be composed of several matrix constituents, which are different in different organs and change with age and in pathological conditions. This is demonstrated by immunocytochemical studies on ultrathin sections.

*1995 The molecular biology and pathology of elastic tissues. Wiley, Chichester (Ciba Foundation Symposium 192) p31–50*

## The elastic network

Almost all structural studies on elastin have been performed in higher vertebrates, where it is organized into an extracellular network of interwoven fibres. It is these fibres that are responsible for the peculiar elasticity of connective tissue. The elastin network is most abundant in tissues subject to stretching, such as blood vessels, lung, skin and elastic cartilage. The distribution and 3D organization of elastic fibres vary in different organs, mainly in accordance with the orientation of stress forces applied to each tissue.

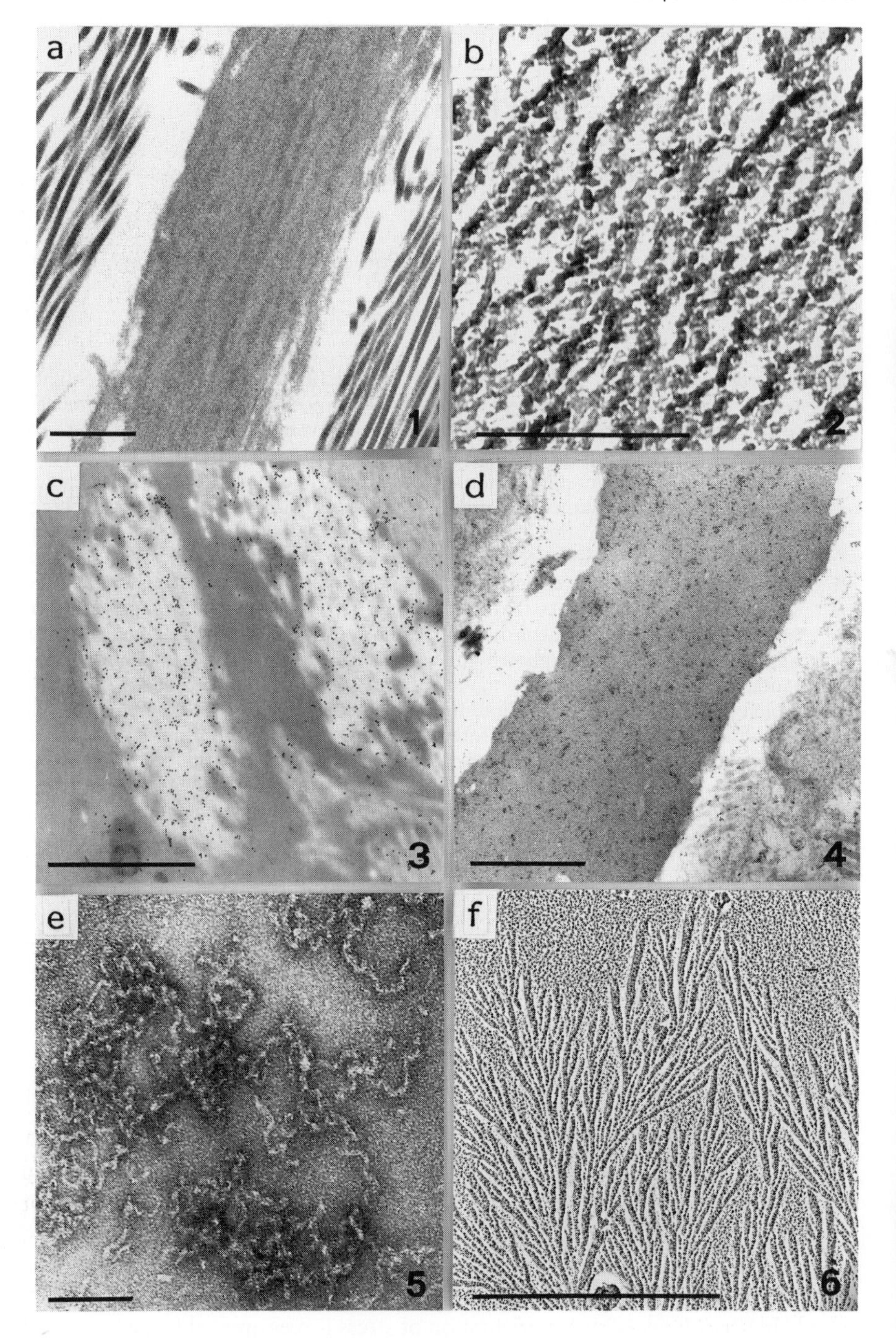
a
1
b
2
c
3
d
4
e
5
f
6

In blood vessels, the elastin network has been extensively studied by both optical and electron microscopy. The 3D organization of the elastin polymer in the vessel wall has been investigated by scanning electron microscopy, after exhaustive digestion of proteoglycans, cells and other soluble compounds by formic acid (Ushiki 1992) or mild NaOH treatment (Shimada et al 1993). The network consists of fenestrated laminae forming layers concentric to the vessel and separated by spindle-shaped smooth muscle cells, small and loose collagen bundles, and proteoglycans. The medial elastic laminae vary in thickness from 0.5 to 4 μm, have an irregular contour, are branched and are generally oriented along the long axis of adjacent smooth muscle cells. They are frequently anastomized by fine lateral fibres connecting the various concentric layers (Smith 1976). When the longitudinal direction is taken into account, the concentric fenestrated sheets form a cylindrical network able to absorb the intraluminal pressure and, at least in certain vessels, the orientation of fibres in the adjacent elastic layers forms angles extending over a complete 180° range (Smith 1976). Davies (1993) has described in mouse aortic media 'contractile-elastic units' formed by lateral branching of the elastic fibres leaving the laminae. These were strictly connected to the surface of smooth muscle cells; from these attachment points, cell filaments appear to span the cell and connect the opposite portions of the cell membrane. Therefore, smooth muscle cells and elastin seem to form a contractile unit spanning from one elastic lamina to the next, with different orientation, and giving to the whole tunica media a continuous functional organization (Davies 1993). These results are in agreement with those we obtained in the ageing rat aorta, where we showed that contacts between elastic fibres and cells are lost in very old animals (Fornieri et al 1992, Quaglino et al 1993). The volume of elastin in the aortic wall is around 20–50%, depending on the animal and on the vessel region, and can be influenced by chemicals that interfere with elastin assembly and cross-linking. For instance, vitamin C, which induces a dramatic dose-dependent reduction in elastin mRNA and secretion *in vitro* (Quaglino et al 1989), has been shown to cause a rearrangement of the elastin

---

FIG. 1. (a) The elastic fibre in the human dermis consists of amorphous elastin and electron-dense material organized into longitudinal strands (bar = 1 μm). (b) Freeze-fractured elastin appears to consist of beaded filaments forming a 3D network (bar = 0.1 μm). (c) Gold particles reveal that osteopontin is localized inside dermal human elastic fibres. Ultrathin sections were incubated with specific antibodies and the immunoreaction was revealed by gold-conjugated secondary antibody (bar = 1 μm). (d) Gold particles show that heparan sulfate is present within the elastic fibre of human aorta. The immunoreaction was performed on ultrathin sections and revealed by gold-conjugated secondary antibody (bar = 1 μm). (e) Recombinant human tropoelastin molecules aggregate into 12 nm-thick flexible filaments when incubated at 30 °C (bar = 0.1 μm). (f) Aggregates of recombinant human tropoelastin, incubated at 50 °C and observed by rotary shadowing (bar = 1 μm).

network when administered in excess *in vivo*; the process is probably mediated by an increase in the degree of prolyl hydroxylation of tropoelastin, which interferes with fibrillogenesis or stabilization of the polymer (Quaglino et al 1991). When both chickens and rats were fed a diet containing lathyrogens, which are specific inhibitors of lysyl oxidase activity, elastic fibres appeared wider and much more polymorphous than in controls, as assessed by incorporation of large amounts of proteoglycans into the polymer (Fornieri et al 1987). Alterations in the formation of the elastic fibres were also observed upon addition to the diet of penicillamine, which inhibits the formation of desmosine cross-links by binding to the aldehyde precursors: the elastic fibres of aortas from chicks treated for 15 and 30 days were much more numerous and smaller than in controls (Pasquali-Ronchetti et al 1986). These results were later confirmed in the dermis of humans suffering from systemic sclerosis or Wilson's disease who had been treated with penicillamine (Pasquali-Ronchetti et al 1989a, 1989b).

In skin, both light and electron microscopic studies showed that elastic fibres are particularly abundant in the reticular dermis, where they form a loose network with prevalent orientation parallel to the skin surface. The distribution, number and size of fibres vary among different species and with age. In the whole human dermis, the density of elastin was calculated at $2.14 \pm 1.09\%$ (Uitto et al 1983), in agreement with data obtained by chemical analysis of elastin through desmosine and isodesmosine content. Recently, by light microscopy on semithin sections, the fractional volume of elastin was calculated to be about 2.5 and 1% in the human adult reticular and papillar dermis, respectively; the mean diameter of the elastic fibres ranged from 1 to 3 $\mu$m in the reticular dermis and was about 0.5 $\mu$m in the papillary dermis (Vitellaro-Zuccarello et al 1994). In a recent electron microscopic study performed in our laboratory, fibres smaller than 0.1 $\mu$m could be seen and the elastin density in the human dermis varied between about 4 and 7% depending on the age of the subject (Quaglino et al 1995).

In lung, elastic fibres are present in the blood vessel walls and in the interstitium of the whole respiratory tree, where they help to restore the lung structure to its original shape and volume during expiration. Lung elastic fibres, which account for 1.0–2.5% of the dry weight of the lung, are highly branched and the majority of them are less than 0.1 $\mu$m thick and can be seen only under the electron microscope by the use of stains, such as tannic acid, which renders them electron opaque.

Elastin is also present in several other tissues and organs, where its density may be very high (such as in sheep or bovine ligamentum nuchae) or very low (such as in tendons). In ligamentum nuchae, elastin forms extremely long 10–20 $\mu$m-thick fibres which run in parallel and are separated by collagen bundles and, in the adult, by occasional cells. In rat tail tendon, elastic fibres are scarce and exhibit different organization in different areas: mature fibres are present in the para-, epi- and endotendineum. In the endotendineum, they are small and are formed by scarce amorphous elastin associated with bundles of microfibrils (Caldini et al 1990). In the palmar aponeurosis of normal adult subjects, elastic fibres either consist of 0.5–1.0 $\mu$m-thick amorphous fibres, with rare branching and a few associated microfibrils near cells, or they are spread among collagen fibres in the form of small bundles of microfibrils with a narrow central portion occupied by amorphous elastin (Pasquali-Ronchetti et al 1993a). Thin elastic fibres, 0.1–1.0 $\mu$m thick, form a thin lace-like sheet in the serosa covering the peritoneum in the mouse urinary bladder, and a loose network in the human ductus deferens (Ushiki & Murakumo 1991). In the elastic cartilage of monkey auricle, elastin is interspersed among collagen bundles in the interterritorial zone, where it forms polymorphous aggregates that roughly surround each chondron (Shimada et al 1993).

## The elastic fibre

### *The amorphous component*

Elastic fibres can be visualized by light microscopy after they are stained with specific stains, such as orcein, Weigert's resorcin–fuchsin, Gomori's aldehyde fuchsin (whether preceded or not by potassium permanganate oxidation), or Verhoeff's iron haematoxylin stain. On semithin sections of epoxy resin-embedded specimens, elastic fibres can be distinguished from collagen by simple staining with toluidine blue (Vitellaro-Zuccarello et al 1994). By all these stains, elastic fibres appear as long branching strips without any apparent substructure.

By conventional transmission electron microscopy, at low magnification, elastic fibres appear to be made of homogeneous material which appears electron transparent when stained with uranyl acetate and lead citrate (Fig. 1a), or electron dense when stained with tannic acid or silicotungstic acid. At high magnification, 10–12 nm-wide microfibrils are seen at the periphery of the fibres; moreover, poorly defined electron-dense strips can be recognized, running within the amorphous elastin and oriented roughly parallel to the fibre itself. Therefore, the elastic fibre appears to have a composite structure: the amorphous material is true elastin and the microfibrils seem to be made of various glycoproteins, but the chemical nature of the electron-dense longitudinal strips within the fibres is still unknown.

By freeze-fracture electron microscopy, elastin does not appear homogeneous. The elastic fibres of bovine ligamentum nuchae, frozen in liquid nitrogen and fractured, appear to consist of a 3D network of 5 nm-thick beaded filaments, which can be oriented in the direction of the fibre by stretching (Pasquali-Ronchetti et al 1979) (Fig. 1b). This methodological approach is highly conservative as it allows the observation of unfixed and naturally hydrated specimens. This may be very important in the study of the elastin structure. In fact, given the highly hydrophobic nature of the polymer, together with its high degree of hydration and extreme deformability, the structure of elastic fibres may be strongly affected by procedures tending to dehydrate the specimen. The macromolecular organization of elastin filaments is also suggested by three further lines of evidence. First, prolonged fixation of the elastin polymer by osmic acid in the presence of tannic acid, followed by conventional dehydration and embedding, revealed that under these conditions the elastic fibres are much wider than after conventional fixation and consist of a network of interwoven filaments (Pasquali-Ronchetti et al 1993b). Second, intact fragments of elastic fibres, small enough to be directly observed under the electron microscope, together with negatively stained cryosections of frozen elastic fibres, revealed that elastin consists of 5 nm-thick filaments crossing each other and packed together within the fibre (Gotte et al 1974, Fornieri et al 1982). Third, one of the properties of isolated elastin fragments and of the soluble precursor tropoelastin is their great tendency to aggregate *in vitro* at physiological temperatures (Volpin & Pasquali-Ronchetti 1977, Bressan et al 1986). This self-assembly, which perhaps mimics the aggregation of tropoelastin molecules on the growing fibres *in vivo*, has been shown to occur *in vitro* by lateral apposition of discrete 5 nm-thick filaments, giving rise to compact bundles identical to those obtained by mechanical fragmentation of natural elastic fibres; moreover, these bundles, when conventionally fixed, dehydrated, embedded in epoxy resin and sectioned, exhibit homogeneous structure identical to that of natural elastic fibres treated in the same way (Bressan et al 1986, Pasquali-Ronchetti et al 1993b). Therefore, the homogeneous appearance of elastin in conventional transmission electron microscopy is very likely an artefact induced by chemical treatment of this very peculiar protein.

*Elastin-associated microfibrils*

Microfibrils approximately 12 nm wide are associated with the elastic fibres; they are more numerous and particularly evident around immature fibres, where it has been suggested that they function as a scaffold for the deposition of tropoelastin molecules (Ross et al 1977, Daga-Gordini et al 1987). Microfibrils are made of a series of proteins and glycoproteins, among which are the fibrillins (Sakai et al 1986). A number of proteins and glycoproteins were suggested to be microfibrillar by the observation that microfibrils reacted with antibodies directed to those molecules (Gibson et al 1989, 1991, Kobayashi et al 1989, Bressan et al 1983a, Horrigan et al 1992). Moreover, the acidic nature of some of these proteins was thought to play a role in the assembly of the basic tropoelastin molecules (Horrigan et al 1992). However, apart from in tissues with a high elastin content, microfibrils are widely distributed in a series of tissues where elastin is almost absent (Gibson et al 1989). The oxytalan fibrils in the skin and the eye (Cotta-Pereira et al 1976, Daga-Gordini et al 1990) or similar structures in tendon (Caldini et al 1990) consist of microfibril bundles with very little, if any, associated elastin. This indicates that the composition of microfibrils may vary and that their role is probably much more complex than that of simply being a scaffold for the deposition of elastin. The complex role of microfibrils also appears in Marfan's syndrome, where mutations of a gene coding for a protein from the fibrillin family induce alterations in several tissues and organs, such as bone and eye, in which elastin is practically absent. In the light of the present data, it seems reasonable to think that what appear to be microfibrils at the electron microscope level are actually aggregates of proteins and glycoproteins which can differ among organs and in different regions of the same organ. Even if incomplete, their amino acid sequence seems to indicate that at least some of the microfibrillar proteins contain repeated motifs present also in unrelated molecules (Rosenbloom et al 1993). This could have at least two important consequences: first, immunolocalization with antibodies against these portions may give misleading results; second, the presence or absence of one or more of these motifs might change the biological role of the resulting structures.

*Mature elastic fibres*

Apart from microfibrils, conventional electron microscopic studies have shown that all mature elastic fibres contain strips of electron-dense materials oriented in the direction of the fibre. The simplest explanation is that these are residues of microfibrils enclosed within the elastin polymer. However, the finding that these strips are more abundant in human skin of old individuals compared with young ones, coupled with the observation that elastin synthesis and fibrogenesis stop at maturity, makes this statement seem rather too simplistic.

From a series of studies performed in our laboratory on tissue samples taken from various organs and species, it appears that elastic fibres are rather more complex. From immunocytochemical studies, the elastic fibres of different tissues appear to contain a number of matrix components which, apart from in a few cases, have never been identified by chemical isolation procedures. This is probably because of the exhaustive treatments used to purify elastin, which destroy or extract associated molecules. Unfortunately, immunocytochemistry usually involves antibodies able to recognize specific epitopes and not the whole macromolecule. Therefore, if the same motifs are present on different molecules, as is common for the extracellular matrix constituents, misleading results may ensue. In a series of studies conducted in our laboratory, the same antibodies were shown to localize to the structures they were specific for, and some of them, in addition, to the elastic fibres. The immunoreactions were performed on ultrathin sections and therefore were able to reveal the internal composition of the elastic fibre. Apart from reacting with antibodies recognizing elastin and desmosine, dermal elastic fibres were positive for decorin and biglycan (Baccarani-Contri et al 1990) and were highly positive for osteopontin-recognizing antibodies (Fig. 1c). There is also indirect evidence that elastic fibres contain hyaluronan, as anti-tropoelastin antibodies are much more reactive towards the elastic fibres after digestion of serial sections with *Streptomyces* hyaluronidase (Baccarani-Contri et al 1990). Skin elastic fibres are also slightly positive for chondroitin-4-sulfate, whereas elastic fibres from human aortas are also positive for chondroitin-6-sulfate and highly positive for heparan sulfate (Fig. 1d). It should be mentioned that heparan sulfate was extracted from elastic fibres of the aorta purified by chemical treatments and collagenase (Dalferes et al 1987). The small elastic fibres in human palmar aponeuroses stain positive for chondroitin-4-sulfate and chondroitin-6-sulfate and are also slightly positive for fibronectin. From the data we obtained on different tissues by the use of antibodies specific to various components of the extracellular matrix, we postulate that elastic fibres have a complex composition that varies among different tissues; their composition seems to depend on several factors, among which the most significant is the composition of the extracellular matrix at the moment of their assembly. This hypothesis is further substantiated by the observation that the composition of elastic fibres changes in some pathological conditions. For instance, an accurate study on the elastic fibres of the dermis of pseudoxanthoma elasticum patients revealed that in addition to the above-mentioned glycosaminoglycans, the apparently normal elastic fibres in patients also stained positive for vitronectin and fibronectin (Baccarani-Contri et al 1994a). Moreover, the altered fibres were also positive for osteonectin, bone sialoprotein and alkaline phosphatase, which, on the contrary, were never found within normal elastic fibres (Baccarani-Contri et al 1994b). The same proteins and glycoproteins could be recognized in the extracellular matrix of the patients' dermis in form of huge aggregates, never present in the control dermis, which can be interpreted as abnormal products of dermal fibroblasts in this peculiar genetic disease (Baccarani-Contri et al 1994a).

We recently performed a careful study on the nature of the glycosaminoglycan chains present within the elastic fibres in the aorta of a subject with Menkes disorder, a genetic disease characterized by a copper transport defect affecting copper-dependent enzymes, one of which is lysyl oxidase, whose activity is reduced to about 10% of normal (Pasquali-Ronchetti et al 1994). The elastic fibres in the aorta were wide and polymorphous by incorporation of huge amounts of proteoglycans whose glycosaminoglycan chains were predominantly chondroitin-4- and -6-sulfates. Heparan sulfate, on the contrary, was more abundant in normal than in Menkes elastic fibres. Although it is not known whether proteoglycans change in Menkes aorta, this finding indicates that inhibition of lysyl oxidase leads to strict association between under-cross-linked elastin and matrix proteoglycans and that chondroitin sulfate is preferred to heparan sulfate, which, in contrast, seems to be a normal component of the elastic fibres (Dalferes et al 1987).

## Tropoelastin assembly *in vitro*

As mentioned earlier, tropoelastin molecules show an increasing tendency to aggregate *in vitro* as the temperature is increased towards physiological levels

(Bressan et al 1983b). This phenomenon, called coacervation, was also observed with elastin peptides (Volpin et al 1976a) and with synthetic polypeptides made of repeats of the tropoelastin hydrophobic sequences (Volpin et al 1976b, Long et al 1980). At present, two different types of macromolecular assembly have been recognized. Up to physiological temperatures, tropoelastin and its derivatives form filaments, about 5 nm thick, which, by lateral aggregation, give rise to bundles of various dimensions resembling mechanical fragments of native elastic fibres. At temperatures higher than 50 °C, the filament aggregates are still present, but a new form of aggregation is predominant: banded fibres, with a major period of about 50 nm, and numerous subperiods of reproducible length, indicate the great tendency of the molecules to self-aggregate in an ordered, staggered manner similar to collagen (Bressan et al 1983b). Recently, we had the opportunity of studying the aggregation properties of recombinant human tropoelastin by electron microscopy and atomic force microscopy. Tropoelastin (kindly provided by Dr J. Rosenbloom) was dispersed in Tyrode's solution and examined by transmission electron microscopy after negative staining with uranyl acetate or rotary shadowing with platinum–carbon. In the first case, a drop of tropoelastin suspension was placed on a carbon-coated copper grid and immediately stained with 1% uranyl acetate in water; after air drying the specimen was observed in the electron microscope. In the second case, the drop of tropoelastin suspension was spread on the surface of freshly cleaved mica and left to dry; the specimen was covered with a 1.5 nm-thick platinum–carbon film from an angle of 11° under rotation and then by a 10 nm-thick carbon film from above. The replica was detached onto water, collected on a copper grid and observed. For atomic force microscopy, a drop of the tropoelastin suspension was placed on carbon-coated copper grids or on freshly cleaved mica and air dried. The microscope used was an NSIII atomic force microscope (Digital Instruments). In all cases, before processing, tropoelastin solutions were incubated for 60 min up to 24 h at 20, 30 or 60 °C. The results confirmed that tropoelastin molecules associate into discrete filaments of variable length that tend to aggregate forming rather ordered filament bundles. However, recombinant human tropoelastin behaved differently from chick tropoelastin previously examined by negative staining (Bressan et al 1986). First of all, at 20 °C, with both negative staining and rotary shadowing, tropoelastin appears as roundish 15 nm-wide particles with a tendency to form arborescent clusters. At higher temperatures, various tropoelastin aggregates could be seen by negative staining: short, extremely flexible filaments, 12 nm wide, spread on the carbon film (Fig. 1e) and long, rather rigid 12 nm-thick filaments, forming an interlacing loose network on the support. At high magnification, each filament appeared longitudinally split into two subunits, of 3–5 nm thickness. This is the best resolution so far obtained for tropoelastin coacervates. Atomic force microscopy showed that tropoelastin aggregates consist of a 3D network of filaments, 12 nm thick, each of which appears to be split into two filament subunits, 4–5 nm thick. This suggests a basic aggregation of two molecules of tropoelastin along their major axis, into short filaments, which join together forming long filaments and then aggregate into bundles that appear to be organized into a loose 3D array. Another interesting feature was observed by rotary shadowing of high-temperature coacervates of recombinant human tropoelastin. When spread on freshly cleaved mica, coacervates of tropoelastin at 60 °C formed an arborescent crystal-like structure made of needles about 12 nm thick (Fig. 1f). The arborescent organization of tropoelastin at both low and high temperatures may explain its tendency to give rise to fractal structures. Whether or not this behaviour is maintained *in vivo* is a matter for speculation and for further studies. Macromolecules in the extracellular space, mostly microfibril constituents, seem to play a pivotal role at least in the macromolecular assembly of the elastic fibre as a whole.

*Acknowledgements*

The authors are grateful to Dr G. Mori and Dr D. Guerra for their valuable technical help. Grants from Italian MURST (40% and 60%) and Italian CNR (Progetto Finalizzato Ingegneria Genetica).

## References

Baccarani-Contri M, Vincenzi D, Cicchetti F, Mori G, Pasquali-Ronchetti I 1990 Immunochemical localization of proteoglycans within normal elastin fibre. Eur J Cell Biol 53:305–312

Baccarani-Contri M, Vincenzi D, Cicchetti F, Pasquali-Ronchetti I 1994a Immunochemical identification of abnormal constituents in the dermis of pseudoxanthoma elasticum patients. Eur J Histochem 38:111–123

Baccarani-Contri M, Taparelli F, Boraldi F, Pasquali-Ronchetti I 1994b Mineralization in PXE is induced by proteins with high affinity for calcium. XIVth FECTS meeting, Lyon, Abstr. N3

Bressan GM, Castellani I, Colombatti A, Volpin D 1983a Isolation and characterization of a 115000 Da matrix-associated glycoprotein from chick aorta. J Biol Chem 258:13262–13267

Bressan GM, Castellani I, Giro GM, Volpin D, Fornieri C, Pasquali-Ronchetti I 1983b Banded fibers in tropoelastin coacervates at physiological temperatures. J Ultrastruct Res 82:335–340

Bressan GM, Pasquali-Ronchetti I, Fornieri C, Mattioli F, Castellani I, Volpin D 1986 Relevance of aggregation properties of tropoelastin to the assembly and structure of the elastic fibers. J Ultrastruct Mol Struct Res 94:209–216

Caldini EG, Caldini N, DePasquale V et al 1990 Distribution of elastic system fibres in the rat tendon and its associated sheaths. Acta Anat 139:341–348

Cotta-Pereira G, Rodrigo FG, Bittencourt-Sampaio S 1976 Oxytalan, elaunin, and elastic fibers in the human skin. J Invest Dermatol 66:143–148

Daga-Gordini D, Bressan GM, Castellani I, Volpin D 1987 Fine mapping of tropoelastin-derived components in aorta of developing chick embryo. Histochem J 19:623–632

Daga-Gordini D, Castellani I, Volpin D, Bressan GM 1990 Ultrastructural immunolocalization of tropoelastin in the chick eye. Cell Tissue Res 260:137–146
Dalferes ER Jr, Radhakrishnamurthy B, Ruiz HA, Berenson GS 1987 Composition of proteoglycans from human atherosclerotic lesions. Exp Mol Pathol 47:363–376
Davies EC 1993 Smooth muscle cell to elastic lamina connections in developing mouse aorta. Role in aortic medial organization. Lab Invest 68:89–99
Fornieri C, Baccarani-Contri M, Quaglino D Jr, Pasquali-Ronchetti I 1987 Lysyl oxidase activity and elastin–glycosaminoglycan interactions in growing chick and rat aortas. J Cell Biol 105:1463–1469
Fornieri C, Pasquali-Ronchetti I, Edman AC, Sjostrom M 1982 Contribution of cryotechniques to the study of elastin ultrastructure. J Microsc 125:87–93
Fornieri C, Quaglino D Jr, Mori G 1992 The role of the extracellular matrix in age-related rat aorta modifications—ultrastructural, morphometric and enzymatic evaluations. Arterioscler Thromb 12:1008–1016
Gibson MA, Kumaratilake JS, Cleary EG 1989 The protein components of the 12-nanometer microfibrils of elastic and nonelastic tissues. J Biol Chem 264:4590–4598
Gibson MA, Sandberg LB, Grosso LE, Cleary EG 1991 Complementary DNA cloning establishes microfibril-associated glycoprotein (MAGP) to be a discrete component of the elastin-associated microfibrils. J Biol Chem 266:7596–7601
Gotte L, Giro GM, Volpin D, Horne RW 1974 The ultrastructural organization of elastin. J Ultrastruct Res 46:23–33
Horrigan SK, Rich CB, Streeten BW, Li ZY, Foster JA 1992 Characterization of an associated microfibril protein through recombinant DNA techniques. J Biol Chem 267:10087–10095
Kobayashi R, Tashima Y, Masuda H et al 1989 Isolation and characterization of a new 36-kDa microfibril-associated glycoprotein from porcine aorta. J Biol Chem 264:17437–17444
Long MM, Rapaka RS, Volpin D, Pasquali-Ronchetti I, Urry DW 1980 Spectroscopic and electron micrographic studies on the repeat tetrapeptide of tropoelastin: Val-Pro-Gly-Gly. Arch Biochem Biophys 201:445–452
Pasquali-Ronchetti I, Fornieri C, Baccarani-Contri M, Volpin D 1979 The ultrastructure of elastin revealed by freeze-fracture electron microscopy. Micron 10:89–99
Pasquali-Ronchetti I, Fornieri C, Baccarani-Contri M, Quaglino D Jr, Caselgrandi E 1986 Effect of DL-penicillamine on the aorta of growing chickens: ultrastructural and biochemical studies. Am J Pathol 124:436–447
Pasquali-Ronchetti I, Guerra D, Quaglino D Jr et al 1989a Dermal elastin in systemic sclerosis. Effect of D-penicillamine treatment. Exp Clin Rheumatol 7:373–383
Pasquali-Ronchetti I, Quaglino D Jr, Baccarani-Contri M, Hayek J, Galassi G 1989b Dermal alterations in patients with Wilson's disease treated with D-penicillamine. J Submicrosc Cytol 21:131–139
Pasquali-Ronchetti I, Guerra D, Baccarani-Contri M et al 1993a A clinical, ultrastructural and immunochemical study on Dupuytren's disease. J Hand Surg 18B:262–269
Pasquali-Ronchetti I, Baccarani-Contri M, Fornieri C, Mori G, Quaglino D Jr 1993b Structure and composition of the elastin fibre in normal and pathological conditions. Micron 24:75–89
Pasquali-Ronchetti I, Baccarani-Contri M, Young RD, Vogel A, Steinmann B, Royce PM 1994 Ultrastructural analysis of skin and aorta from a patient with Menkes disease. Exp Mol Pathol 61:36–57
Quaglino D Jr, Zoia O, Kennedy R, Davidson JM 1989 Ascorbate affects collagen and elastin mRNA in pig skin fibroblast cultures. Eur J Cell Biol 49 (suppl 28):45

Quaglino D Jr, Fornieri C, Botti B, Davidson JM, Pasquali-Ronchetti I 1991 Opposing effect of ascorbate on collagen and elastin deposition in the neonatal rat aorta. Eur J Cell Biol 54:18–26

Quaglino D Jr, Fornieri C, Nanney LB, Davidson JM 1993 Extracellular matrix modifications in rat tissues of different ages. Correlations between elastin and collagen type I mRNA expression and lysyl oxidase activity. Matrix 13:481–490

Quaglino D Jr, Bergamini G, Boraldi F, Pasquali-Ronchetti I 1995 Ultrastructural and morphometrical evaluations on normal human dermal connective tissue. Br J Dermatol, in press

Rosenbloom J, Abrams WR, Mecham R 1993 Extracellular matrix. 4. The elastic fiber. FASEB J 7:1208–1218

Ross R, Fialkow PJ, Altman LK 1977 The morphogenesis of elastic fibers. Adv Exp Biol Med 79:7–17

Sakai LY, Keene DR, Engvall E 1986 Fibrillin, a new 350-kD glycoprotein, is a component of extracellular microfibrils. J Cell Biol 103:2499–2509

Shimada T, Sato F, Zhang L, Ina K, Kitamura H 1993 Three-dimensional visualization of the aorta and elastic cartilage after removal of extracellular ground substance with a modified NaOH maceration method. J Electron Microsc 42:328–333

Smith P 1976 A comparison of the orientation of elastin fibers in the elastic laminae of the pulmonary trunk and aorta of rabbits using the scanning electron microscope. Lab Invest 35:525–529

Uitto J, Paul JL, Brockley K, Pearce RH, Clark JG 1983 Elastin fibers in human skin: quantitation of elastic fibers by computerized digital image analyses and determination of elastin by radioimmunoassay of desmosine. Lab Invest 49: 499–505

Ushiki T 1992 Preserving the original architecture of elastin components in the formic acid-digested aorta by an alternative procedure for scanning electron microscopy. J Electron Microsc 41:60–63

Ushiki T, Murakumo M 1991 Scanning electron microscopic studies of tissue elastin components exposed by a KOH-collagenase or simple KOH digestion method. Arch Histol Cytol 54:427–436

Vitellaro-Zuccarello L, Cappelletti S, Dal Pozzo-Rossi V, Sari-Gorla M 1994 Stereological analysis of collagen and elastic fibers in the normal human dermis: variability with age, sex and body region. Anat Rec 238:153–162

Volpin D, Pasquali-Ronchetti I 1977 The ultrastructure of high temperature coacervates of elastin. J Ultrastruct Res 61:295–302

Volpin D, Pasquali-Ronchetti I, Urry DW, Gotte L 1976a Banded fibers in high temperature coacervates of elastin peptides. J Biol Chem 251:6871–6873

Volpin D, Urry DW, Pasquali-Ronchetti I, Gotte L 1976b Studies by electron microscopy of the structure of coacervates of synthetic polypeptides of tropoelastin. Micron 7:193–198

## DISCUSSION

*Winlove:* The dimensions for the fibrils and the intrafibrillar spacings you have determined microscopically are very consistent with the conclusions we reach from molecular probe measurements of the spaces within the fibrils (Winlove & Parker 1990). Do the proteoglycans associated with elastin have the same sort of functions in controlling fibril dimensions that they are known to have in the formation of collagen fibrils?

*Pasquali-Ronchetti:* I do not have direct evidence for this. When we worked with experimental lathyrism *in vivo*, we found large amounts of proteoglycans associated with the elastic fibre and we suggested that proteoglycans could have a role keeping the tropoelastin molecules in suspension (Fornieri et al 1987). Actually, lysyl oxidase inhibitors work by preserving the lysine groups (or the majority of them) on the tropoelastin molecules: in these conditions, a conspicuous number of proteoglycan molecules are found associated with elastin. We do not think that the elastin–proteoglycan association is an effect of the chemicals *per se*, as we obtained the same results with chemically different compounds and found the same association in Menkes disease, an inherited disorder of copper metabolism characterized by very low lysyl oxidase activity (Pasquali-Ronchetti et al 1994). Therefore, we think that there is a functional association between tropoelastin and proteoglycans, before the action of lysyl oxidase. Furthermore, it could be that some proteoglycans remain associated with the elastic fibre after elastin polymerization and cross-linking, favouring, for instance, its hydration.

*Robert:* Many of the protein components of proteoglycans are very hydrophobic—their interaction with elastin and inhibition of lysyl oxidase may involve protein–protein interactions rather than glycosaminoglycan–protein interactions.

*Pasquali-Ronchetti:* Yes, it could be a hydrophobic interaction among proteins. However, if this were the case, the glycosaminoglycan portion of the proteoglycans would remain trapped within the elastic fibre and could play some physical role.

*Robert:* But the highly charged glycosaminoglycan chains probably wouldn't inhibit such hydrophobic interactions.

*Urry:* Hydrophobic surfaces of proteins, as well as those capable of ion pairing, could certainly interact with the elastin and could provide a chaperoning type of interaction. Again, I'd come back to the point that the elastin–elastin interaction has got to be dominant, because you can displace the other interactions during the fibril growth.

*Kagan:* Some studies that we've done *in vitro* support the notion that the association of tropoelastin with proteoglycans may have something to do with the regulation of lysyl oxidase activity. We have shown that lysyl oxidase activity towards peptidyl lysine is suppressed by negatively charged ions vicinal to the susceptible lysine. These negatively charged ions may be intrinsic to the sequence of the oligopeptide regions being oxidized (Nagan & Kagan 1994), or they could be extrinsic and come from associated ligands that are binding to elastin (Kagan et al 1981). In particular, we showed that free fatty acids are very potent inhibitors of oxidation of both tropoelastin and insoluble elastin. In contrast, positively charged ligands stimulate the oxidation by lysyl oxidase. The association with proteoglycans, because of their high ionic density, might be playing that sort of role, keeping the lysines available in the appropriate orientation or even preventing fibril growth. The suggestion that proteoglycans may influence the timing and sequence of lysyl oxidation seems reasonable.

*Hinek:* The interaction between elastin and proteoglycans is extremely complicated. There are many other players, including the elastin binding proteins. We have shown that treatment of smooth muscle cells, fibroblasts or chondrocytes with galactosugars, or galactosugar-bearing proteoglycans such as chondroitin sulfate or dermatan sulfate, completely disrupts the organization of the elastic fibres (Hinek et al 1988, 1991). When we tried this with non-galactosugar proteoglycans such as heparan sulfate, as a control, nothing happened. Therefore, not only is there a charge effect of the proteoglycan, but also a certain specificity is necessary. We proposed that a complex is formed between tropoelastin and the 67 kDa elastin binding protein which is secreted and presented at the cell surface. Then galactosugars, protruding from the glycosylated microfibril proteins, bind to the lectin site of the elastin binding protein. Under the influence of galactosugars, the whole structure of elastin binding protein is changed and its affinity to elastin dramatically drops, so elastin is released. If such a thing happens prematurely, or if we have a high concentration of free galactosugars around the cells, it can completely disrupt the organization of the elastic fibres. Simply, elastin never reaches the microfibrils.

*Robert:* There is a problem with this, because elastin in young babies might well be in contact with lactose present in milk at high concentrations, and yet it seems to cross-link very nicely.

*Hinek:* We simply don't know whether the intact lactose present in the milk can reach the tissues that produce elastic fibres.

*Mecham:* Another issue to think about when we consider elastic fibre formation is that by the time the cell gets around to making an elastic fibre, the non-elastin extracellular matrix around it is well organized, including the proteoglycan matrix. Therefore, a large elastic fibre has to be built within this existing matrix. So the question is, does the elastic fibre just push this stuff apart as it forms, or does it coalesce around existing matrix components? Perhaps some of the proteoglycan that we can detect in elastic fibres is where the elastin has polymerized around the proteoglycan as the fibre is being made. The diameter of the elastic fibre is very large compared with other components of the extracellular matrix. The physical assembly of a structure this large makes for some interesting problems when other restricting proteins are present.

*Pasquali-Ronchetti:* I think that the elastic fibre is much larger than that we see in conventional electron micrographs. When we compare, for instance in ligamentum nuchae, the mean diameters of elastic fibres obtained by conventional fixation and embedding, and by freeze-fracture, we see that with the latter technique (which is more conservative) the fibres are significantly larger. The natural elastic fibre is highly hydrated and is probably a network with large meshes filled by water and other molecules. Its organization is completely altered by fixation, dehydration and embedding procedures. I also agree that the formation of elastic fibres must be more complicated than just tropoelastin molecules coming out of the cell and going freely through the matrix onto a preformed glycoprotein scaffold.

*Mecham:* There is interesting energetics involved if we are talking about pushing an existing matrix apart. It becomes another level of complexity in the assembly question.

*Davidson:* Does the ultrastructure of elastic fibres change when they are under tension? This is the way the fibre exists *in vivo*.

*Pasquali-Ronchetti:* Before I froze the fibres, I stretched them to 150% of their original length—this is the maximum I could obtain. I was hoping to be able to see modifications in the globular structure of each filament on freeze-fractured replicas, but the technique (the vacuum) was probably not good enough. Perhaps atomic force microscopy will help with this. By atomic force microscopy it is theoretically possible to visualize the organization of each filament in both the hydrated and dried states, in resting conditions and after stretching, without any chemical or physical treatment.

*Sandberg:* It's interesting that you found that chick tropoelastin looked different. We have previously done sequence work on the tryptic fragments and we saw a repeating tripeptide, VPG, present in the chick but not in other species (Smith et al 1981). Would this influence the structure?

*Urry:* It certainly could; I think it would produce a more rigid kind of structure, but we need to remember that the tripeptide sequence in collagen is not VPG, it's a GPX kind of dominant repeat.

*Pasquali-Ronchetti:* One answer to this question would be to look at the structures obtained *in vitro* by aggregation of tropoelastin molecules lacking specific exons.

*Uitto:* What expression system was used to produce the recombinant human elastin, and specifically, did the elastin undergo any post-translational modification, such as prolyl hydroxylation?

*Rosenbloom:* We produced it in *Escherichia coli.* To my knowledge there's no prolyl hydroxylation and there is a minimal amount of oxidation to just a few aldehydes. There's no other modification.

*Uitto:* This then raises the more general issue of the role of hydroxyproline in elastin, if it has any. Those of us who have been working on collagen have considered the presence of hydroxyproline in elastin almost coincidentally, since elastin goes through the same rough endoplasmic reticulum where collagen becomes hydroxylated by prolyl hydroxylase.

*Urry:* We have wondered about this issue. Early on, when we first saw the effects of hydroxyproline and we saw the shift in the temperature of the transition, we had the perspective that this must interfere with the hydrophobic assembly (Urry et al 1979). We saw that prolyl hydroxylase expression is increased in repair conditions. Subsequently, Karl Franzblau's group demonstrated in culture that if you enhance the prolyl hydroxylation by adding ascorbic acid, there is very little fibrous elastin formed and the elastin remains in solution, consistent with the view that it has raised the temperature of the transition and there is no longer hydrophobic folding and assembly to form some fibre (Barone et al 1985). There is some controversy about how much elastin there is in scar tissue, although it is generally stated that there is little or none. This could be explained as follows: when you have a break in a tissue, if your primary purpose is to sew that up and close the breach, it is best simply to optimize collagen formation; elastin is not going to help (Urry 1988). Perhaps that's what is happening here—you get the elastin out of the way and quickly sew up the break in the tissue with collagen.

*Keeley:* Some years ago, together with Dr Dorothy Johnson, I did some experiments investigating the role of hydroxyproline in the assembly of elastin, which we never published. We used chick aorta organ cultures in which the transition from soluble to insoluble elastin is normally very rapid. The aortic tissue was double-labelled with [$^3$H]valine and [$^{14}$C]proline in the presence or absence of 1,10-phenanthroline, which chelates iron and inhibits prolyl hydroxylase. The elastin produced in the presence of 1,10-phenanthroline contained no hydroxyproline, but there was no difference in the time of conversion of soluble to insoluble elastin in this tissue. We concluded that lack of hydroxylation of elastin did not affect assembly.

*Rosenbloom:* A couple of quick comments on hydroxyproline and the effects of mutations in collagen. The problem with the secretion of mutant collagen where you have a swelling of the endoplasmic reticulum is that it may affect the transit times for elastin as well. We showed a long time ago that many of the prolines can be hydroxylated; in fact fully half of them can be hydroxylated if you simply incubate them with prolyl hydroxylase *in vitro*. It's possible that as elastin is synthesized under these conditions it is actually being over-hydroxylated, and that this is what is going out into the media, because of the decreased transit time. This is an alternative explanation of what is happening, rather than a direct interaction between collagen and elastin. It is also rather curious that under normal circumstances where a smooth muscle cell, for example, is making collagen and elastin at the same time, that even though there is a high rate of hydroxylation in the collagen, the elastin doesn't have a greater percentage of prolines hydroxylated.

*Grant:* I feel we are dodging around the hydroxyproline question a little bit, and it is potentially quite important. Initially it was considered that hydroxylation was a random post-translational modification. Now I suspect that this probably isn't the case, but I don't have any evidence for it. Dan Urry has commented very clearly on how he would envisage that hydroxylation would have significant effects on the physical properties of elastic molecules. This begs the question; to what extent does hydroxylation influence morphology and function?

*Urry:* If I can remember the enzymic hydroxylation data correctly, 10% hydroxylation of poly(VPGVG) will raise the temperature of transition sufficiently to prevent folding at 37 °C (Urry et al 1979). The chemically synthesized hydroxylation was not as efficient in raising the temperature of transition as enzymic hydroxylation, presumably because the chemically synthesized molecule has an enantiomer problem.

*Sandberg:* We've had a detailed look at the hydroxylation of the elastin pentapeptide in species where it exists. In sheep, cattle and some diving mammals, we have found that hydroxylation appears to be organ-dependent; it's different in skin compared with aorta. We've also found that it changes with age. It's different between species: for instance, there's virtually no hydroxylation of the elastin pentapeptide in the aorta of the dolphin, whereas in bovine aorta, a significant amount of the elastin is hydroxylated. We can identify these differences because the thermolysin pattern is discernible when we separate the peptides on HPLC.

*Robert:* In which cells and in which species do the hydroxylation patterns of elastin change with age?

*Sandberg:* In all species studied, hydroxylation decreases with age.

*Lapis:* In certain tumour types there are large deposits of elastin, which sometimes occur in a multilayered form. Are these structurally different from normal elastin?

*Pasquali-Ronchetti:* I don't have much experience with tumours. I have looked at some breast carcinomas and carcinomas of lip (I. Pasquali-Ronchetti, unpublished results). In these cases the elastic fibres are not normal. They are partially degraded or destroyed; they sometimes appear very large and swollen and seem to be made of aggregates of interwoven filaments. It has to be noted that these altered elastic fibres remain positive for antibodies specific for elastin only in areas where apparently normal amorphous elastin is still recognizable.

*Boyd:* There are quite large differences between the electron micrographs that you have shown and those that Bob Mecham has published. Bob's pictures show a globular structure, not a filamentous one.

*Mecham:* I think our structures are fairly consistent with what Ivonne has seen. We see globules or globular fibres 5–7 nm in diameter. In that sense, we're in pretty close agreement. Where our results differ is mostly in the structure of tropoelastin on mica.

*Pasquali-Ronchetti:* We have tried deep etching; when we increased the etching time these structures became larger and larger. I don't think our vacuum structure is good enough and water may re-precipitate onto the frozen structures.

*Boyd:* Bob Mecham has made the point that the overall structure of elastin in the elastic fibre resembles ping pong balls in a jar, where these globules associate with each other in the same way as ping pong balls might do in a glass tube. Is that the sort of model you're thinking about?

*Pasquali-Ronchetti:* No, I think that these filaments are real structures. Molecules stick together forming beaded filaments, which can be observed by two completely different techniques, freeze fracture and negative staining.

*Starcher:* Tropoelastin has to get through this milieu of matrix to find other elastin molecules. Does the antibody you used for immunostaining differentiate between tropoelastin and insoluble elastin?

*Pasquali-Ronchetti:* No.

*Starcher:* Several years ago we made an antibody to the cross-linked area of elastin before it cross-links the lysine–alanine sequence. It was quite specific *in vitro*, in that it reacted only with tropoelastin. It would not react with any species of insoluble elastin or α-elastin. Using this antibody, Alek Hinek did some beautiful immunogold localization studies. We used copper-deficient chickens and we were expecting to show a large excess of tropoelastin. As a control we used older hens. It turned out that this antibody indicated more tropoelastin in the older animals. Is this something that can actually happen?

*Pasquali-Ronchetti:* There is probably tropoelastin in the extracellular space which doesn't aggregate into fibres. We have recently shown that in Menkes disease, a genetic form of lathyrism, there is a lot of tropoelastin spread throughout the matrix (Pasquali-Ronchetti et al 1994). However, in these conditions, matrix tropoelastin was associated with beautifully designed proteoglycans which we visualized with toluidine blue O or alcian blue. I have already mentioned this. It cannot be excluded that the tropoelastin–proteoglycan association is an artefact of fixation and dehydration; certainly, the matrix contains very nice spider-like proteoglycans with attached gold particles indicating the presence of tropoelastin associated with them.

*Davidson:* The extracellular matrix is a biochemically integrated unit. Altering one matrix component is going to influence the organization of others. Osteogenesis imperfecta and Marfan's syndrome are examples of matrix pathologies where we have been led down the garden path for several years by assuming that the matrix molecules act independently. Wound sites are another area where the biomechanics of the matrix are changing dramatically and are therefore suboptimal for the efficient organization of elastic fibres. We know that the protein is being synthesized there—the gene is being expressed—yet we don't see efficient accumulation of elastin under these circumstances. Thus, there are a number of areas of injury response where there's very inefficient assembly of elastic fibres.

*Tamburro:* I would like to emphasize that the initial formation of small globular aggregates, then of beaded filaments, then bundles of filaments, then twisted rope filaments and so on, is a general property shared by tropoelastin, α-elastin, sequential polypeptides corresponding to the pentapeptide repeating sequences of elastin, and also by short fragments of elastin such as a VGGVG pentapeptide.

It is a self-assembly property of elastin and of some small regions of elastin that is independent of the presence of other extracellular matrix macromolecules.

*Robert:* May I ask a question about osteopontin? We have known for many years that $Ca^{2+}$ increases steadily in elastin with age. Dan Urry showed that $Ca^{2+}$ fits beautifully in some of these $\beta$-turns. Does osteopontin play a role in calcification of elastin?

*Pasquali-Ronchetti:* Osteopontin has been observed to be present in tissue during bone differentiation and bone remodelling, although its physiological role is still under investigation. Osteopontin is very similar to uropontin, which is a molecule made by the epithelial cells of the kidney tubule. It has been suggested that it has a role in the prevention of crystal formation and growth. Given the great tendency of elastin to calcify *in vitro*, as shown by Urry and co-workers more than 20 years ago, we suggest that osteopontin is a molecule that could inhibit the precipitation of ions within the elastic fibre or, at least, work against the crystal growth.

*Robert:* So it might bring $Ca^{2+}$ to the fibre without really being an active ingredient. If that's the situation, how does it explain the fact that if elastin fibres interact with some fatty acid chains or with cholesterol esters, it dramatically increases the sites of $Ca^{2+}$ binding on the fibre (Jacob et al 1983)?

*Urry:* The thing that gives you the impression that $Ca^{2+}$ could bind immediately is simply the sequence. If there are alternating glycines, this allows the backbone carbonyls to wrap around and chelate the $Ca^{2+}$ reasonably well. Barry Starcher worked out the calcifying medium for the $\alpha$-elastin coacervate, and even for the synthetic peptide coacervates (Starcher & Urry 1973). It's very clear that the synthetic peptides and $\alpha$-elastin will calcify in a completely serum-free medium.

*Robert:* The lipids could work both ways. They could decrease the access of $Ca^{2+}$ to these sites by interacting with the hydrophobic sites and yet they increase $Ca^{2+}$ fixation.

*Urry:* In order for the elastin molecule to wrap round and bind the $Ca^{2+}$, it is necessary to disrupt the elastin hydrophobic interactions. If lipid is present, it will disrupt the elastin hydrophobic folding and self-assembly and the elastin backbone carbonyls will be freer to wrap around $Ca^{2+}$.

*Kimani:* Classical electron microscopy and biochemical analysis have revealed further that elastic fibres in the elephant aorta are surrounded by a band of microfibril-like material (McCullagh 1973). What is the functional basis of this finding, and how might it be related to the high incidence of atherosclerosis in this animal?

*Robert:* Years ago I went to Cambridge and I found lots of elephant aorta in Keith McCullagh's freezer; he had brought them over from Kenya. I was surprised to find that elephants sometimes die of atherosclerosis just as humans do, with very similar lesions, fragmentation and calcification of elastic laminae (McCullagh 1973, McCullagh et al 1973). What is the blood pressure of the elephant?

*Kimani:* The systolic blood pressure in the giraffe is in the region of 300 mmHg. I'm afraid I have not come across any publications on the blood pressure profile of the elephant.

*Robert:* My feeling at that time was that if the elephant were not such a cumbersome beast, it would be much better to study atherosclerosis in elephants than in humans.

**References**


Barone LM, Faris B, Chipman SD, Toselli P, Oakes BW, Franzblau C 1985 Alteration of the extracellular matrix of smooth muscle cells by ascorbate treatment. Biochim Biophys Acta 840:245–254

Fornieri C, Baccarani-Contri M, Quaglino D, Pasquali-Ronchetti I 1987 Lysyl oxidase activity and elastin–glycosaminoglycan interactions in growing chick and rat aortas. J Cell Biol 105:1463–1469

Hinek A, Wrenn DS, Barondes SH, Mecham RP 1988 The elastin receptor–galactoside binding protein. Science 239:1539–1541

Hinek A, Mecham RP, Keeley F, Rabinovitch M 1991 Impaired elastin fiber assembly is related to reduced 67 kDa elastin binding protein in fetal lamb ductus arteriosus and in cultured aortic smooth muscle cells treated with chondroitin sulfate. J Clin Invest 88:2083–2094

Jacob MP, Hornbeck W, Robert L 1983 Studies on the interaction of cholesterol with soluble and insoluble elastin. Int J Biol Macromol 5:275–278

Kagan HM, Tseng L, Simpson DE 1981 Control of elastin metabolism by elastin ligands: reciprocal effects on lysyl oxidase activity. J Biol Chem 256:5417–5421

McCullagh KG 1973 Studies on elephant aortic elastic tissue. I. Histochemistry and fine structure of the fiber. Exp Mol Pathol 18:190–201

McCullagh KG, Derouette S, Robert L 1973 Studies on elephant aortic elastic tissue. II. Amino acid analysis, structural glycoproteins and anigenicity. Exp Mol Pathol 18:202–213

Nagan N, Kagan HM 1994 Modulation of lysyl oxidase activity toward peptidyl lysine by vicinal dicarboxylic amino acid residues. Implications for collagen cross-linking. J Biol Chem 269:22366–22371

Pasquali-Ronchetti I, Baccarani-Contri M, Young RD, Vogel A, Steinmann B, Royce PM 1994 Ultrastructural analysis of skin and aorta from a patient with Menkes disease. Exp Mol Pathol 61:36–57

Smith DW, Sandberg LB, Leslie BH et al 1981 Primary structure of a chick tropoelastin peptide: evidence for a collagen-like amino acid sequence. Biochem Biophys Res Commun 103:880–885

Starcher B, Urry DW 1973 Specificity and site of calcium ion interactions with elastin. Bioinorgan Chem 3:107–114

Urry DW 1988 Entropic elastin processes in protein mechanisms. II. Simple (passive) and coupled (active) development of elastic forces. J Protein Chem 7:81–114

Urry DW, Sugano H, Prasad KU, Long MM, Bhatnagar RS 1979 Prolyl hydroxylation of the polypentapeptide model of elastin impairs fibre formation. Biochem Biophys Res Commun 90:194–198

Winlove CP, Parker KH 1990 Physicochemical properties of vascular elastin. In: Hukins DWL (ed) Connective tissue matrix, part 2. Macmillan, London p 167–198

# General discussion I

## Hydroxylation of the pentapeptide VGVPG in ovine elastin

*Sandberg:* I am going to describe work carried out in collaboration with Philip Roos and Tony Tamburro.

### *Primary protein structure of tropoelastin*

Within the primary structure of the elastin precursor, tropoelastin, are alternating hydrophobic and hydrophilic domains (Parks & Deak 1990, Parks et al 1993). The hydrophilic domains are characterized by large stretches of polyalanine interspersed with lysine residues destined to become cross-links, whereas the hydrophobic domains characteristically have numerous repeating structures, some of which are common to several species. I want to focus this talk on the repeating pentapeptide, valyl-glycyl-valyl-prolyl-glycine (VGVPG), and its hydroxylated derivative, valyl-glycyl-valyl-hydroxyprolyl-glycine (VGVHyPG). Our studies are on elastin, because in most species tropoelastin is unavailable. Because of its tightly cross-linked structure, insoluble elastin does not lend itself well to structural chemical analysis. We principally have two procedures at our disposal: classical amino acid analysis and fingerprinting techniques employing proteolytic enzymic cleavage.

### *Thermolysin digestion of elastin*

Of late, we have extensively explored digestion of elastin with thermolysin, which gives relatively simple fingerprint patterns by HPLC, by which the peptides can be well separated and where the repeating structures of the hydrophobic domains such as VGVPG give very tall peaks (Sandberg et al 1986). This happens because thermolysin cleaves at amino acids with bulky side-chains such as leucine and valine, in which elastin is rich. These amplified peaks allow for a very sensitive quantitative assay for elastin and, with this in mind, we have developed these assays around repeats of YG for the rat, VGVPG for sheep, pig and dolphin, and VAPG for the human (Sandberg et al 1986, 1990, Price et al 1993).

### *VGVPG and VGVHyPG are prominent peaks on the thermolysin-produced map*

We're fortunate that both hydroxylated and non-hydroxylated VGVPG are released by thermolysin digestion and that these two peptide species are well

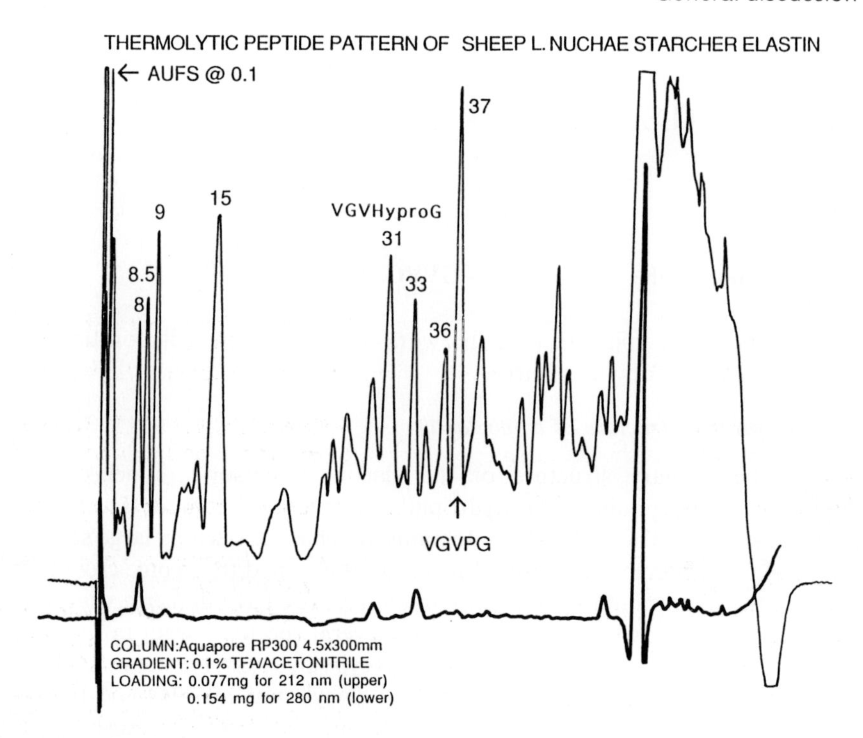


FIG. 1. Elution profile of a thermolysin digest of purified ovine aortic elastin showing the elution times of VGVHyPG (VGVHyproG, 31 min) and VGVPG (37 min). Monitoring is at 214 nm (thin line) and 280 nm (heavy line). The latter wavelength reflects locations of tyrosine and desmosines in the elution profile, presumably in peptide linkage.

resolved on our peptide maps. These are shown in the chromatogram from sheep elastin (Fig. 1). Because they elute as pure compounds, they can be quantitated by amino acid analysis to obtain an extinction coefficient for each when monitoring the peaks at 214 nm. This information can then be used to calculate a percentage hydroxylation of VGVPG for that particular elastin sample. This percentage is given in Table 1 for different organs and ages. It is interesting to note the differences in hydroxylation in various tissues, lung parenchyma being the highest observed thus far. Also, with age, the percentage hydroxylation appears to drop in all tissues. In the ligamentum nuchae, for example, the adult level is less than half of that in the newborn, suggesting that there is virtually no hydroxylation in the elastin produced in this organ after birth. When we compared pig aortic tropoelastin and insoluble elastin, we found the degrees of hydroxylation to be almost identical, 22.8 and 21.7%, respectively, confirming the reliability of the method. Interestingly, in the dolphin, aortic elastin VGVPG is hydroxylated less than 5% suggesting that

**TABLE 1 Percentage hydroxylation in insoluble sheep elastin**

| *Source of elastin* | *Newborn lamb* | *Adult sheep* |
|---|---|---|
| Aorta | 38.7 | 28.5 |
| Pulmonary artery | 46.4 | 36.1 |
| Ductus arteriosus | 47.4 | – |
| Lung parenchyma | 60.3 | 56.1 |
| Ligamentum nuchae | 16.9 | 8.0 |

these mammals may place different kinds of stresses on their aortas than do land mammals. Presumably, hydroxylation would produce a stiffer elastin because of increased hydrogen bonding.

### *Synthetic polyVGVHyPG: scanning electron microscopy*

A further look at molecular structure has been obtained by scanning electron microscopy of synthetic polyVGVHyPG. These results are shown in Fig. 2. Whereas some areas of fibrillar organization are seen as shown in Fig. 2A and reported previously by Urry et al (1974) for VGVPG, we also see areas of globular organization (grape-like structures) indicating a different type of association at the supramolecular level (Fig. 2B). The exact meaning here is unclear and we are in the process of looking at polymeric mixtures of the two pentapeptides to help us interpret these unique and interesting structures.

### *Conclusions*

We can state that specific residues of proline are hydroxylated in tropoelastin and elastin. The hydroxylation appears to be the same in the precursor protein tropoelastin as in mature elastin, in keeping with hydroxylation being a post-translational intracellular event. These similar numbers suggest that hydroxylation is not a random process, as regards the repeating pentapeptide VGVPG found in exon 18, but we do not have data on specific proline residues. However, the degree of hydroxylation appears to vary with each species studied, e.g. 28% for sheep aorta, 22% for pig and less than 5% for dolphin. In the sheep there was a wide variation in proline hydroxylation between elastins from different organs ranging from 17% in ligamentum nuchae to 60% in lung. In all instances, there was a decrease in hydroxylation with age. Finally, the appearance of globular structures in polyVGVHyPG on scanning electron microscopy suggests that there is a different supramolecular association within exon 18 when the proline of VGVPG becomes hydroxylated.

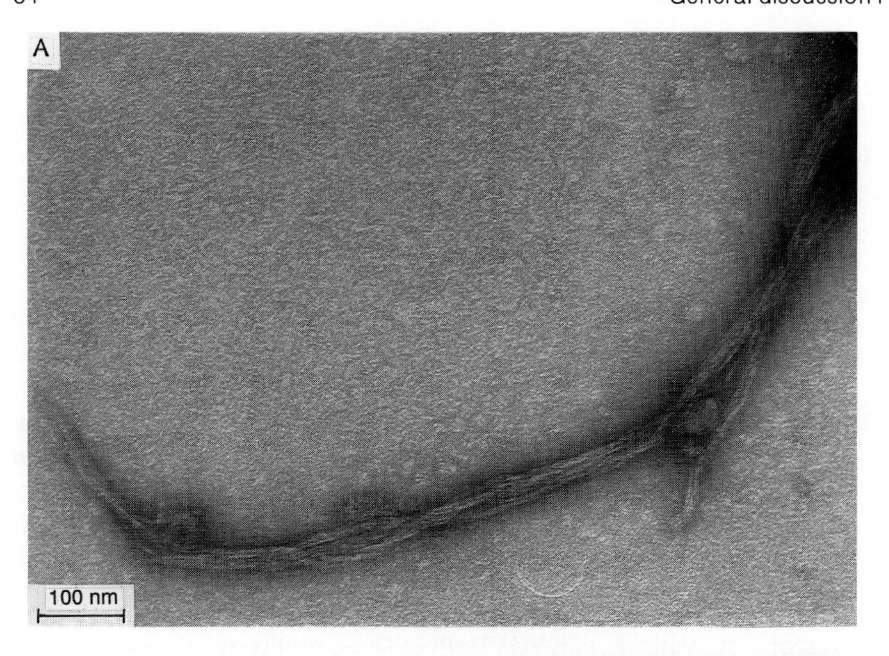
A
100 nm

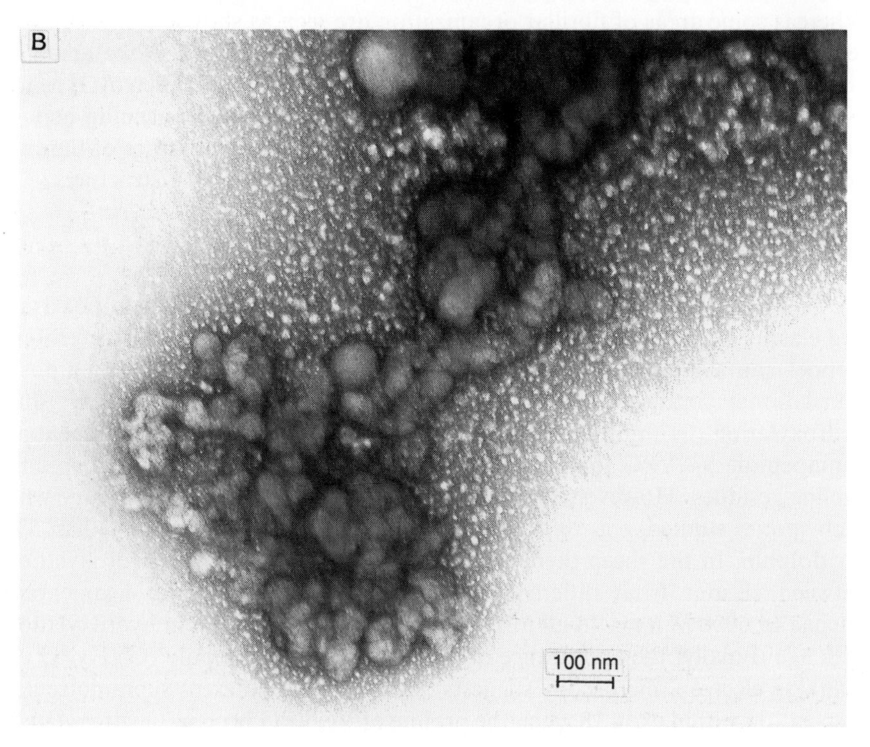
B
100 nm

*Speculations*

What does it all mean? I think we can speculate that hydroxylation of exon 18 prolines may modify the mechanical properties of the elastic fibre. As suggested above, more hydroxylation may produce a stiffer elastin due to increased hydrogen bonding between polypeptide chains. The different hydroxylation patterns may reflect the differing mechanical demands placed on different organs, e.g. aorta versus lung. Finally, with the recent emphasis on receptors and chaperone proteins (Hinek et al 1988), hydroxylation of proline in exon 18 repeating peptides may induce modification in the fibre assembly of that specific organ.

*Uitto:* Did you have a chance to analyse whether all hydroxyproline residues in elastin are in the form of *trans*-4-hydroxy-L-proline, the predominant isoform in collagens, or whether there are *trans*-3-hydroxy-L-prolines present as well? The reason I ask is that a 60% level of hydroxylation, as you noted in the tissues of newborn lambs, is pretty high. In fact, I wonder whether the sheep elastin cDNA sequence allows you to calculate whether there are enough prolyl residues in the appropriate positions to achieve a 60% level of hydroxylation. In the case of collagen, prolyl residues in the Y-position of the repeating Gly-X-Y sequence are hydroxylated by prolyl-4-hydroxylase, whereas *trans*-3-hydroxyproline is synthesized by a distinct enzyme, prolyl-3-hydroxylase, which recognises prolyl residues only in the X position if the Y position is already occupied by a 4-hydroxyproline.

*Sandberg:* Yes, there are different sequences. However, we don't have a way to differentiate these two isomers, so we can't really be certain as to which it is. We've assumed that it is all 4-hydroxyproline, because it occurs in the Y position.

*Urry:* One of the reasons we didn't proceed further with the characterization of all of the other sequences between cross-links (we have sequenced them all) is that we were always concerned about having an adequate basis for determining the required level of purity. There is no enzymic activity to test for and it is extremely difficult to remove a polymer of 100 repeats, but with one or two repeats with an error, from the remainder of the polymers.

The poly(VPGVG) model for the W4 sequence provides an example of the problem. You can synthesize it, dialyse it against 50 000 molecular weight membranes, verify amino acid composition, characterize it with a range of spectroscopic methods and it looks perfect. Yet one synthesis will have a transition temperature of 25 °C and another will have the transition at 40 °C.

---

FIG. 2. (A) Fibrillar structure of poly(VGVHyPG) as visualized by scanning electron microscopy at a concentration of 1 mg/ml in water at room temperature. Negative staining is by uranyl acetate. 100 nm bar is shown. (B) Globular structure of poly(VGVHyPG) visualized by scanning electron microscopy at a concentration of 2 mg/ml in water at 40 °C. Negative staining is by uranyl acetate. 100 nm bar is shown.

The sample with a 25 °C transition temperature forms a coacervate that is 400 mg/ml, and $\gamma$-irradiation cross-links it to form beautiful elastic sheets, but the sample with a 40 °C transition temperature forms a more dense coacervate of about 800 mg/ml. This does not cross-link, but instead remains as a viscoelastic soup. In further efforts to determine the source of this difference, we purchased amino acids from five different sources, carried out in parallel exactly the same steps of synthesis, and found that the transition temperatures varied widely.

Finally, we found that there is a crucial process of removing the D-residues during the synthesis. So with poly(GVGVP), the Val-Pro is synthesized and crystallized to form beautiful crystals (however, even here the first crop of crystals results in a polypentapeptide with a lower $T_t$ (onset temperature for the inverse temperature transition) than does the second crop from the same mother liquor). Then the Val-Gly is synthesized, crystallized and reacted to form Gly-Val-Gly, which is also crystallized. The dimer and trimer are then coupled to give the desired pentamer. The pentamer is then polymerized, most commonly with proline at the C-terminus as the proline will not racemize, and with *O*-nitrophenol as the activating group to get the highest molecular weights.

But only when we prepared poly(GVGVP) by microbial synthesis were we finally satisfied that we had the right stuff. When $(GVGVP)_{251}$ was expressed from *Esherichia coli* with its own ATG start codon and its own stop codon (so that the only contamination could be an initial methionine residue, which appeared to be processed off) and it had a 26 °C transition temperature, it was possible to conclude that the syntheses with the low $T_t$ were correct.

Also, with the polyhexapeptide, poly(APGVGV), the criterion of $T_t$ dependence on concentration could be developed to establish purity.

*Tamburro:* We also took the usual precautions to avoid racemization during the synthesis of the pentapeptide GVGVHyP. The poly-condensation step was carried out using hydroxyproline as the C-terminal residue, because pirrolidine imino acids are known to forbid racemization. Furthermore, the circular dichroism spectra of our poly(GVGVHyP) in both water and trifluorethanol are very similar to those of your polypentapeptide in the same solvents (Urry et al 1974). Additionally, the temperature effect on the circular dichroism spectrum in water is very similar. As you know, circular dichroism is strongly dependent on the racemization.

*Urry:* The difficulty with using circular dichroism for assessing racemization is that these measurements must be taken in the absorption region, and that the signal : noise ratio is poor due to the low ellipticity exhibited by poly(GVGVP). It is possible, however, to use optical rotatory dispersion with a high concentration of polymer outside the absorption region, that is, at wavelengths longer than 240 nm. Interestingly, the racemized samples exhibit the more negative optical rotary dispersion curves.

What was the temperature of the transition?

*Tamburro:* At the concentration I used (that is 0.1 mg/ml) there was no discernible aggregation in the hydroxylated polymer.

*Urry:* In a good poly(VPGVG) 0.02 mg/ml will actually coacervate with a transition temperature of about 40 °C (Urry et al 1985). The usual spectroscopic methods were not adequate to tell us the difference between a good and a poor sequence, it was the temperature of the transition that was really crucial.

*Davidson:* If we compare nuchal ligament and lung elastic fibres, it seems that the elastic fibre can tolerate an order of magnitude difference in the degree of hydroxylation, yet can still form what appear to be equivalent fibres at the ultrastructural level. Is this a reasonable conclusion to draw from these studies?

*Sandberg:* If we compare the two tissue extremes, newborn lung parenchymal elastin and adult ligamentum nuchal elastin—delicate lace-like fibres in the former and a very strong rubber-band-like structure in the latter—even though the fibres appear similar at the ultrastructural level, there is a world of difference in functionality. Thus, hydroxylation has something to do with the supramolecular organization of the fibre and probably the way the total aggregate behaves as an elastomer.

*Urry:* I should add that we did the hyroxyproline studies (in collaboration with Raj Bhatnagar) before we had a full understanding of what one had to do to take care of the racemization issues. You need to reach a high concentration limit to do the transition temperature comparisons adequately. More recently, in determining the $T_t$-based hydrophobicity scale, we did determine the $T_t$ in the hydrophobicity plot for hydroxyproline, and it's about 50 °C, if I recall correctly.

*Rosenbloom:* What you have told us, Dan, could have a fairly dramatic effect on the transition temperature, particularly in this pentapeptide which may be playing a critical role in nucleation.

*Urry:* I would think elastin that has 50% hydroxyprolines is going to have difficulty aggregating to form fibres at 37 °C.

*Rosenbloom:* There is a way of looking at this. In the sheep nuchal ligament the hydroxyproline content was extremely low. Therefore it is possible to prepare tropoelastin from ligament cells and from smooth muscle cells in the aorta that's going to vary tremendously in hydroxylation. Have you done anything like this by accident or design?

*Mecham:* No I haven't, but if I remember the sequencing data from the pig, there were hydroxyprolines scattered throughout the molecule, it's not as if position 18 is the only place you find hydroxylation. When we sequenced through those hydroxylation sites, we found that the prolines weren't completely hydroxylated—perhaps 30% remained as prolines. Is there a reason for this?

*Sandberg:* I think we're going to have to look at a number of other hydroxylation sites to really say what's happening. Perhaps we should look at the folding structures of those, build models and see whether or not there are any differences.

*Urry:* I think that's a very good point. We've synthesized the monomers of each of the hydrophobic regions, and then made the high polymers. Some of those polymers of sequences between cross-links have very low $T_t$s, causing real problems with solubility. One may in fact like to have some hydroxylation to raise the $T_t$s of these more-hydrophobic sequences and get some solubility. Hydroxylation in the more hydrophobic regions would assist fibre assembly, especially in the phenylalanine-containing monopeptide repeats which result in a very hydrophobic sequence. With that sequence you might want to raise the $T_t$ somewhat to allow fibre assembly to proceed more effectively.

## References

Hinek A, Wrenn DS, Mecham RP, Barondes SH 1988 The elastin receptor is a galactoside binding protein. Science 239:1539–1541

Parks WC, Deak SB 1990 Tropoelastin heterogeneity: implications for protein function and disease. Am J Resp Cell Mol Biol 2:399–406

Parks WC, Pierce RA, Lee KA, Mecham RP 1993 Elastin. Adv Mol Cell Biol 6:133–181

Price LSC, Roos PJ, Shively VP, Sandberg LB 1993 Valyl-alanyl-prolyl-glycine (VAPG) serves as a quantitative marker for human elastins. Matrix 13:307–311

Sandberg LB, Wolt TB, Leslie JG 1986 Quantitation of elastin through measurement of its pentapeptide content. Biochem Biophys Res Commun 136:672–678

Sandberg LB, Roos PJ, Pollman MJ, Hodgkin DD, Blankenship JW 1990 Quantitation of elastin in tissues and culture. Problems related to the accurate measurement of small amounts of elastin with special emphasis on the rat. Connective Tissue Res 25:1–10

Urry DW, Long MM, Cox BA, Ohnishi T, Mitchell LW, Jacobs M 1974 The synthetic polypentapeptide of elastin coacervates and forms filamentous aggregates. Biochim Biophys Acta 371:597–602

Urry DW, Trapane TL, Iqbal M, Venkatachalam CM, Prasad KU 1985 C13 NMR relaxation studies demonstrate an inverse temperature transition in the elastin polypentapeptide. Biochem 24:5182–5189

# Structure of the elastin gene

Joel Rosenbloom, William R. Abrams, Zena Indik, Helena Yeh, Norma Ornstein-Goldstein and Muhammad M. Bashir

*Department of Anatomy and Histology, School of Dental Medicine, University of Pennsylvania, Philadelphia, PA 19104, USA*

*Abstract.* The isolation and characterization of cDNAs encompassing the full length of chicken, cow, rat and human elastin mRNA have led to the elucidation of the primary structure of the respective tropoelastins. Large segments of the sequence are conserved but there are also considerable variations which range in extent from relatively small alterations, such as conservative amino acid substitutions, to variation in the length of hydrophobic segments and large-scale deletions and insertions. In general, smaller differences are found among mammalian tropoelastins and greater ones between chicken and mammalian tropoelastins. Although only a single elastin gene is found per haploid genome, the primary transcript is subject to considerable alternative splicing, resulting in multiple tropoelastin isoforms. Functionally distinct hydrophobic and cross-link domains of the protein are encoded in separate exons which alternate in the gene. The introns of the human gene are rich in Alu repetitive sequences, which may be the site of recombinational events, and there are also several dinucleotide repeats, which may exhibit polymorphism and, therefore, be effective genetic markers. The 5′ flanking region is G+C rich and contains potential binding sites for numerous modulating factors, but no TATA box or functional CAAT box. The basic promoter is contained within a 136 bp segment and transcription is initiated at multiple sites. These findings suggest that the regulation of elastin gene expression is complex and takes place at several levels.

*1995 The molecular biology and pathology of elastic tissues. Wiley, Chichester (Ciba Foundation Symposium 192) p 59–80*

The physiological function of many tissues in multicellular organisms requires that they possess elastic properties. Thus, during systole the work of the heart is absorbed by expansion of the great vessels, which then recoil elastically during diastole, maintaining the blood pressure and assuring continuous perfusion of the tissues. Similarly, under normal circumstances inspiration is an active, energy-requiring process, whereas expiration is a passive one because of the elastic recoil of the respiratory tree. Several unrelated proteins have evolved elasticity, including resilin in arthropods (Andersen 1971), abductin in molluscs (Kelly & Rice 1967), an elastomer in octopus (Shadwick & Gosline 1981) and elastin in vertebrates (Sage & Gray 1979). Phylogenetic studies have shown

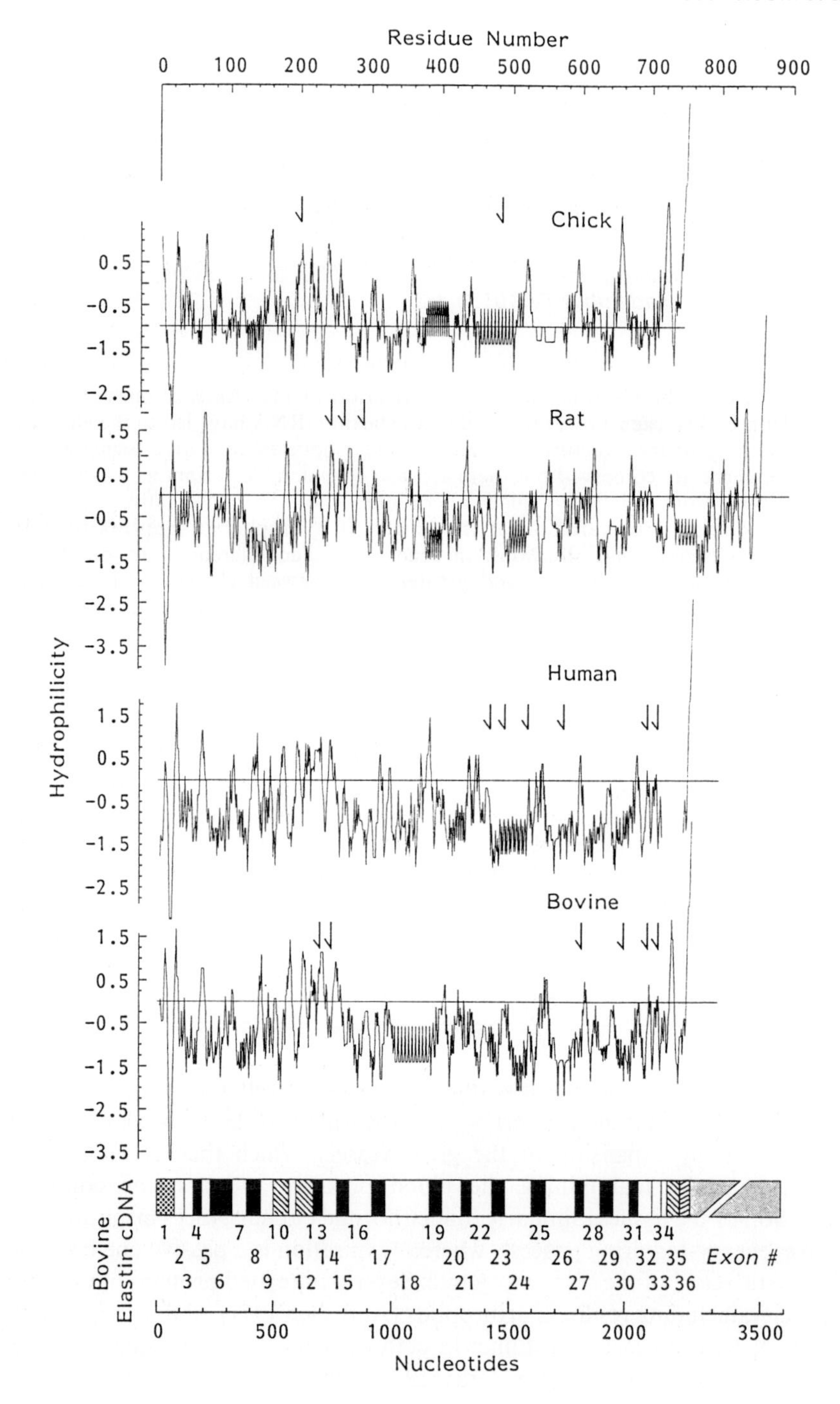

Residue Number
0 100 200 300 400 500 600 700 800 900
Chick
Rat
Human
Bovine
Hydrophilicity
0.5
-0.5
-1.5
-2.5
1.5
-3.5
Bovine Elastin cDNA
1 4 7 10 13 16 19 22 25 28 31 34
2 5 8 11 14 17 20 23 26 29 32 35
3 6 9 12 15 18 21 24 27 30 33 36
Exon #
0 500 1000 1500 2000 3500
Nucleotides

that elastin appeared at some point after the divergence of the cyclostome and gnathostome lines and is found in all vertebrate species, including the cartilaginous fish, but not in invertebrates. Within vertebrate tissues, elastin is found in the extracellular matrix in elastic fibres, which may constitute a small (2–4%) but important percentage of the dry weight (as in the skin) or greater than 50% (as in large arteries). By electron microscopy (Fahrenbach et al 1966, Greenlee et al 1966, Karrer & Cox 1961), elastic fibres are seen to be composed of two morphologically distinguishable components: (1) an amorphous fraction lacking any apparent regular or repeating structure, which constitutes 90% of the mature fibre and is composed exclusively of elastin; and (2) a microfibrillar component consisting of 10–12 nm-diameter fibrils, which are located primarily around the periphery of the amorphous component, but also to some extent interspersed within it. Although the exact composition of the microfibrils remains to be defined, they contain several glycoproteins (Cleary 1987), including fibrillin (Sakai et al 1986).

This paper describes the structure of the elastin gene and its relationship to the structure and function of the protein.

## Characterization of the elastin gene

### *cDNA analysis*

Although early work yielded only partial cDNA clones encoding chick and sheep elastin (Burnett et al 1982, Yoon et al 1984), improved techniques permitted the construction of cDNA clones encompassing the entire length of human, bovine, chick and rat elastin mRNA (Bressan et al 1987, Raju & Anwar 1987, Indik et al 1987, Yeh et al 1989, Pierce et al 1990). Homologous sequences to the tryptic peptides of porcine tropoelastin were also identified. Additionally, all the cDNAs encoded an unusual, highly conserved C-terminal segment, GGACLGLACGRKRK, which had not previously been identified by protein sequencing. The highly basic character of the C-terminus, as well as

---

FIG. 1. Diagrammatic representation of bovine elastin cDNA and hydropathy analysis of tropoelastins by the method of Kyte & Doolittle. The cDNA is divided into exons, which are numbered. In order to preserve a numbering system consistent with homologous sequences in the bovine gene, the most 3′ exon has been designated 36 in the human gene. Exons 34 and 35, found in all other species, are absent in the human gene. An attempt has been made to align the tropoelastin hydropathy diagrams, but because of insertions and deletions the alignment with the cDNA diagram is exact only for the bovine tropoelastin. Exons encoding hydrophobic sequences (□); exons encoding potential cross-linking sequences (■); exons having both a hydrophobic and a cross-linking character (▧); exon 1 encoding most of the signal sequence (⬚); exon encoding the C-terminus (▨); 3′ untranslated region (▩). A √ symbol marks the exons known to be subject to alternative splicing.

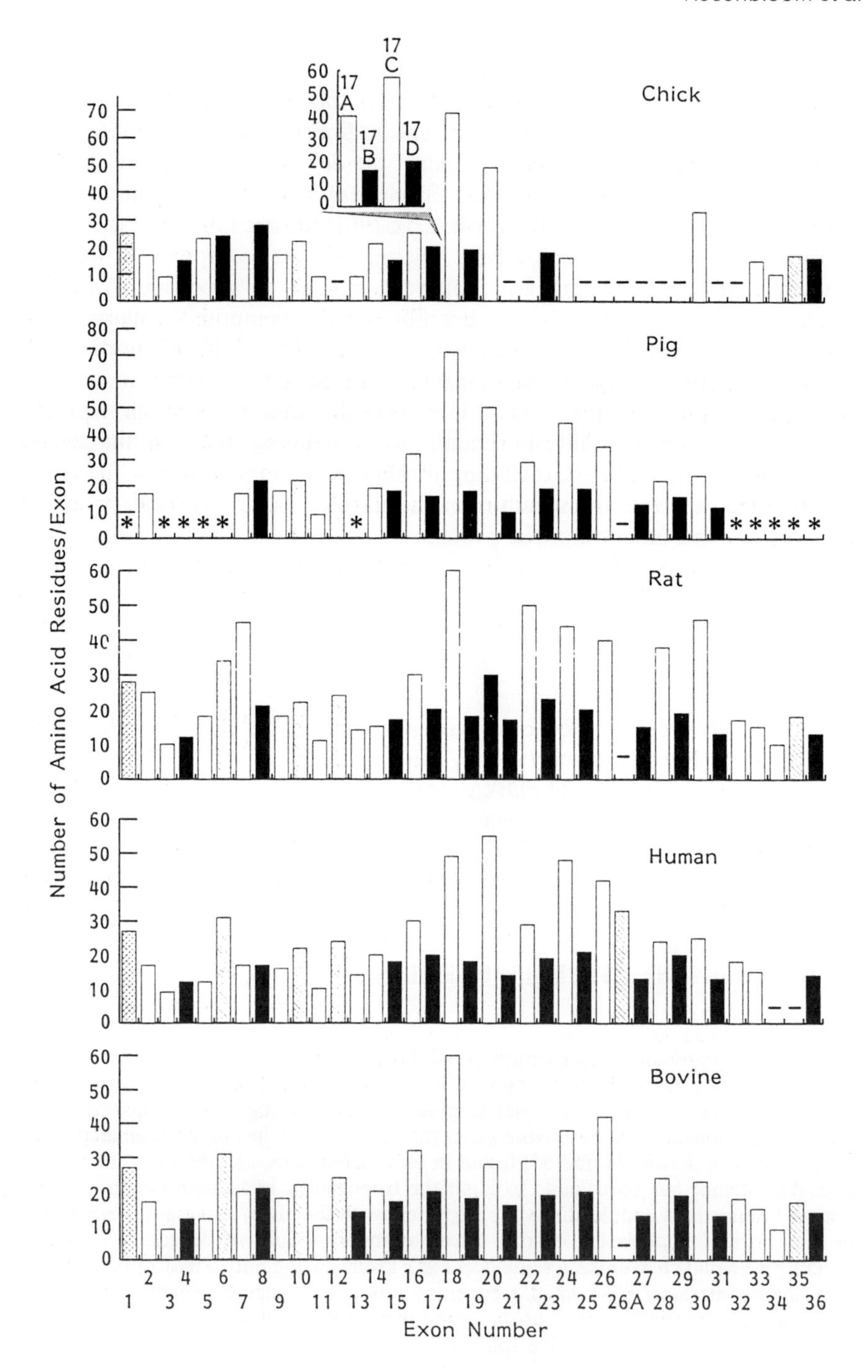
Chick
Pig
Rat
Human
Bovine
17 A
17 B
17 C
17 D
Number of Amino Acid Residues/Exon
Exon Number

a potential structure created by an internal disulfide bond, suggests that this portion of tropoelastin may interact strongly with acidic microfibrillar proteins. Such an interaction between tropoelastin and microfibril-associated glyprotein (MAGP) has recently been demonstrated (Bashir et al 1994).

The cDNA analyses demonstrated that tropoelastin consists predominantly of alternating hydrophobic and lysine-rich domains which project as relatively hydrophilic regions (Fig. 1). In these cross-link domains, the lysines usually occur in pairs. This arrangement supports the suggestion that a given desmosine/isodesmosine serves to join only two tropoelastin molecules, rather than the four that are theoretically possible. It is also apparent that these potential cross-linking sequences are not uniformly distributed and occur at shorter intervals in the first 200 residues, resulting in an asymmetry in the molecule. In addition, the potential cross-linking sequences in the first 200 residues frequently contain a proline or other residue between the lysines instead of the usual alanines. The conformation of tropoelastin in the cross-linking segments containing alanine residues is largely α-helical, which is undoubtedly important in the alignment and condensation of the lysine residues in desmosine formation. In contrast, the presence of proline residues clearly disrupts α-helix formation, and these segments may participate in cross-links other than the desmosines, such as the difunctional cross-link, lysinonorleucine. The 200-residue N-terminal segment of tropoelastin ends in a tyrosine-rich region the function of which is unknown, but this region may be involved in interaction with other matrix macromolecules or with the cell surface, mediating alignment of the molecules within the fibre. In two instances (exons 19 and 25), three lysines are found near one another and they may have a critical role in cross-linking the tropoelastin.

In general, there is good agreement at the nucleotide and encoded amino acid sequence levels among the mammalian elastins, which differ, however, in multiple segments from those of the chicken. Among mammalian elastins, most

---

FIG. 2. Comparison of exon sizes in elastin genes. Sizes for the entire bovine and human genes were determined by cDNA and genomic sequence analysis. Other exon sizes were estimated by comparison with the bovine gene and by using the hydropathy analyses illustrated in Fig. 1, making the assumption that sequences encoding hydrophobic and cross-linking domains are segregated in separate exons. Exons encoding hydrophobic sequences (□); exons encoding cross-linking sequences (■); exons having both a hydrophobic and a cross-linking character (⧅); exon 1 encoding most of the signal sequence (⊠); exon encoding the C-terminus (▩). Chick tropoelastin appears to contain a large insertion relative to mammalian tropoelastins and this insertion has been designated as exons 17A,B,C,D to preserve homologous numbering. The pig exon sizes were derived from peptide sequence data and a * indicates that no data was available for these segments. The human sequence contains the alternatively spliced exon 26A. A – indicates the corresponding exon is missing in that species.

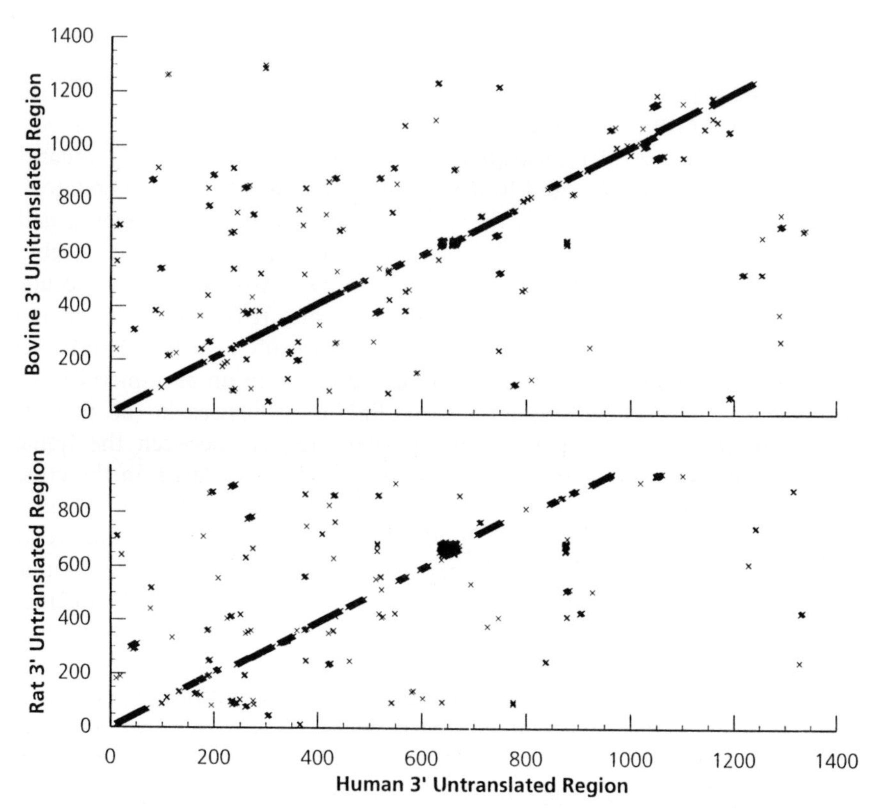


FIG. 3. Comparison of the 3′ untranslated regions of elastin cDNAs. The sequences of rat and cow were compared with the human sequence using the Wisconsin Sequence Analysis Package Program Compare, Version 8, September 1994, Genetics Computer Group. Note the striking regions of homology scored on the diagonal.

amino acid substitutions are of a conservative nature, but some significant differences do exist. For example, near the centre of bovine and porcine tropoelastins a pentapeptide, GVGVP, is repeated 11 times, but this repeat segment is considerably different and more irregular in human tropoelastin and is replaced in rat tropoelastin by GVGIP. Similarly, in human tropoelastin, a hexapeptide, GVGVAP, is repeated seven times, but only five times with conservative substitutions in bovine tropoelastin and it is absent altogether in rat tropoelastin. However, in the rat, several expansions of hydrophobic segments have occurred, significantly increasing the overall size of rat tropoelastin. These variations suggest that a particular number of amino acids and a precise sequence in a given hydrophobic region are not critical to the adequate

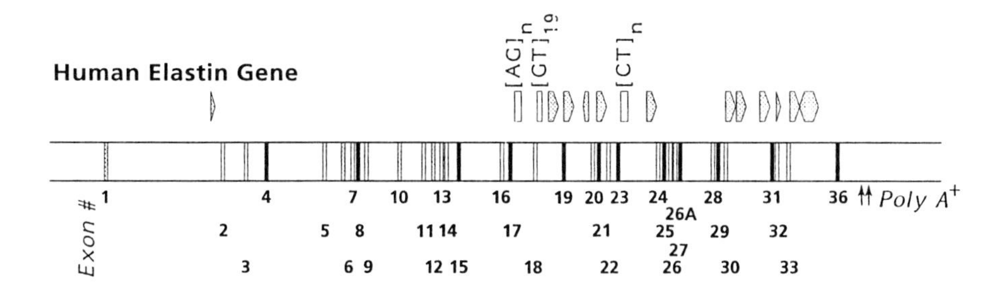


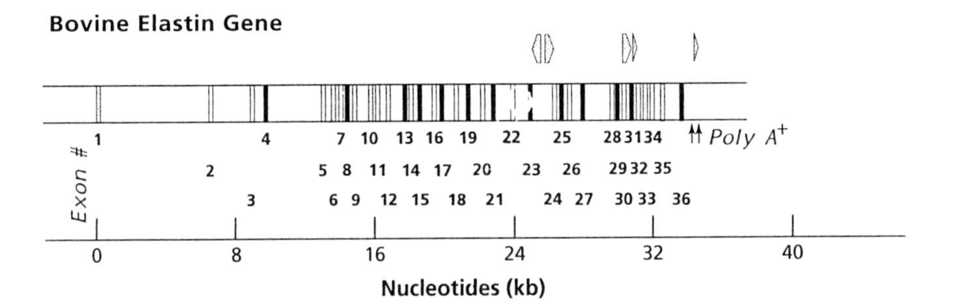


FIG. 4. Diagram of the human and bovine elastin genes. Exons are not drawn to scale. Exons encoding potential cross-linking domains (■); exons encoding hydrophobic domains (□); exons having both a hydrophobic and a cross-linking character (▧); exon encoding most of the signal sequence (⊠); exon encoding C-terminus (▩). Segments containing repetitive DNA (▨). Introns in the 5′ halves of the genes were not completely sequenced and they may contain additional repetitive sequences. Segments of dinucleotide repeats are also indicated.

functioning of the molecule. In contrast, the length of cross-linking segments is highly conserved, indicative of a strong functional requirement (Fig. 2). Four exons (numbers 6, 10, 12 and 35) have a dual hydrophobic and cross-linking character, suggesting that they may play a critical role in molecular assembly of the cross-linked fibre. The chicken sequence is quite homologous to the mammalian sequences for the first 302 residues and for the last 57 residues. In the central portion, although some segments are homologous, major differences exist which appear to be due to insertion, duplication and deletion events. The most striking of these differences is the occurrence in chicken tropoelastin of the repeating tripeptide $(GVP)_{12}$ (residues 374–409), which is not found in the mammalian elastins. The presence of this repeating tripeptide suggests that elastin may have a distant evolutionary relationship to collagen. All in all, there appears to be a strong tendency to conserve the size of the total polypeptide chain to 750–850 residues.

| *intron 3'* | | EXON | Bases Exon | | *5' intron* |
|---|---|---|---|---|---|
| ***Branch Point Consensus* yNyTuAy** yyyyyyyyyyNCAG | G | | | AG | GTuAG |
| 50 · 40 · 30 · 20 · 10 | | | | | |
| *Initiation* | | 1 | *81* | AG | GTAAGGACCC |
| ACCAGCCTAATAGTTGTGGCTCCTGGAGGACTGACTCTACCTGTTTCCTTTCAG | G | 2 | *50* | AG | GTAACGTAGA |
| GGCAAGCCTGGCAGAAGTACCGATGATCTCTCTTTCTCTTTCTCTCCCCCACAG | G | 3 | *27* | AG | GTGAGCTCAG |
| TAGTAGATGGATAAGCTGGGCCACCCCATT[illegible]ACTATCTTCTCTTCCCTCTGCAG | C | 4 | *36* | AG | GTAAGACCCA |
| CACTCTGGGCCTAGGAACACTGCCTACACTCCTGTCTCTGTTTCTTATACACAG | T | 5 | *36* | AG | GTGAGAGCTG |
| TAGGAGTCTTCATAGGTGTGGTAGCTCAGGACCTCACCCCATCCTCCCCTCGCA | G | 6 | *93* | TG | GTGAGTGGTG |
| CTTCTGCCTCCCCACTGTTCCTTACGCAATGCCTCACCTGTCCTGGCTCTGCAG | G | 7 | *51* | AG | GTACGATGGC |
| TCAAAGTTGCAGGCCTGGGTGGAGCCAACTGTGATGCAGCCCCTTCTGTGCCAG | G | 8 | *51* | GG | GTCAGTGCGG |
| GGGGTCCCTGGAGGCTGAGCTGCTGCTAGTAACTTTGCTTTCTTTTGGCCAGAG | G | 9 | *48* | AG | GTGAGAGCAA |
| CTGCAAGGCCTGCAAGGCCTGCCTTCCCACACTCACTGCTTTGTCCCCCGGCAG | G | 10 | *66* | AG | GTATGCAGCT |
| CTGGCCCAGTGTCCACAGTTCCAGGGCTGTAGTGACAGCTTTTTATCATTACAG | G | 11 | *30* | AG | GTGAGGCAAG |
| GGAGGGTTTGGAAGGGGGTGCTGGGACCTGAACTTGCTCTCTTTATTCCCACAG | G | 12 | *72* | TG | GTAAGTCAGA |
| GGGAGCCCAGCAAGGCATGGGGCAGCCCCTGAGTTTGCTCTGTCCTCTCTCCAG | G | 13 | *42* | TG | GTGAGTGAGA |
| GGGCAGCAGTGGTGATGTGTGCACAGATGACCATCAAGCCTCTCTGTTTTGCAG | G | 14 | *60* | AG | GTAAGGAAAG |
| TGGAGATACAGGAGCACTGTTTCAAGGTCTCTCCCCTCTGCTTCCTTCCCCCAG | G | 15 | *54* | CG | GTGAGTGCCC |
| CCTGTCCTCAGGAGGGTCCTTGGAAACTACATTGCACTGTCCCCATCTCAACAG | G | 16 | *90* | AG | GTAACATCTG |
| TGTCCTTGAGATGGCCACAGGCAAGGACCTCACCCTGTGTGGCTGTGTTTTCAG | G | 17 | *60* | TG | GTGAGTGCCT |
| CTAGCCCCTCTGAGGTTCCCATAGGTTAGGGGAACAATGCTTTTTCTTCCACAG | G | 18 | *147* | AG | GTGAGCTGGG |
| TGCTGCCTCCAATGCTGCTGCCTGAGCATGTTGTGTCCCTTTTGGTCTCTCCAG | G | 19 | *54* | CG | GTGAGTGCTA |
| GCCCAGCCTCTCTCACTGAGGCTTCTTTTCTACTTGGCTCCCTTCCCTCTGCAG | G | 20 | *165* | CC | GTGAGCCTTA |
| GAGGAGACCCAGGCACGGCTTCTGAGGGTCTCTATCTTTCTCGTTTCCTTGTAG | C | 21 | *42* | CG | GTAAGTGCCC |
| CCCCCAAAAAGTGAGTACTGGAGGGGCAAGGCTGAAAGTTCTCCACTCCCCGAG | G | 22 | *87* | AG | GTGAGCTGTG |
| AGCAGGGAGGGGTGTGAGAGATTACTCTCTCACCCCTTCTCTTCACACCTCCAG | G | 23 | *57* | TG | GTAAGTCCCC |
| TCTGTCCTCTTTGATCAGGTCTTGGTTAATGATCAGCTCTTCTCAATCTTGCAG | G | 24 | *144* | AG | GTGAGTTTCA |
| AGCCTCCATGGGCCCCGCCTCCATCTCTAATCCCCCTCTCTCTCCCTCCCTCAG | C | 25 | *63* | CC | GTGAGTGCCT |
| TCCTTAGGGGCATGCTCCCTGCCTGCTGTCGCCACCACTGCCCTCTGTCTGCAG | G | 26 | *126* | AG | GTGCAGATGA |
| | | 26A | *99* | GG | GTGCATAGTA |
| TCCCAGGCACAGAGCTCGGCTCCTGACCACTCCCCAACTTTTCTTTCTCCCCAG | T | 27 | *39* | TG | GTGAGTGCAC |
| AGGGAGACCCATCGTTCAGAAATGGAACACTCATTTTCCCTCCTCTCCCCGCAG | G | 28 | *72* | GG | GTGAGTTGAT |
| GGAGGGAATCTAACCAGTACAGAGTGCCTCCCTGAACTCGGTCTGTGTTCCCAG | G | 29 | *60* | TG | GTGAGCACTG |
| GCTTCAGTCCCACCTTTCTGACCAGCGGAGTCTAATGCTCAGCTGTCTCCACAG | G | 30 | *75* | AG | GTGAGAGTTG |
| GCCTGACCAGGTGGCATTGGCATTCCTGAGCCGTCATGTGCCTCATCTCCCCAG | G | 31 | *39* | CG | GTGAGTXCCC |
| AGGGCCTCTTCCCGATGGGGGTGTCTTATCCTGACCCCACCTGCCTCTTCTCAG | G | 32 | *54* | AG | GTAGGGGTGG |
| GCTGGAGTCAGTTTCCACCCCTACCAACCCACCAACCTGAAATCTCTCCTGCAG | G | 33 | *45* | AG | GTATGCCAGG |
| AGCCGAAACTGAGAGGGGCCGGACTCACAGTGATGTGCACCTCCTCCCGTCCAG | G | 36 | *42* | *Termination* | |
| 50 · 40 · 30 · 20 · 10 | | | | | |
| ***ELASTIN CONSENSUS SEQUENCE*** yyyyyyyyyyyyyyNCAG | G | | | NG | GTuAGy |

FIG. 5. Intron/exon borders of the human elastin gene. Possible branch point sequences with best fit to the consensus sequence yNyTuAy are indicated, where y = pyrimidine, u = purine and N = pyridimide or purine. Nucleotides that agree with the consensus sequence are solid; those which disagree are open faced. The bar identifies a non-consensus sequence at the 3′ border of intron 5.

Somewhat surprisingly, there is rather extensive (80%) nucleotide homology in the 3′ untranslated region, suggesting that this region may have a function either in regulating the stability of the mature mRNA or even in modulating translation (Fig. 3). Two polyadenylation signals are found 230 bp apart in the human and bovine genes, both of which are apparently utilized.

*Structure of the elastin gene*

Conclusive analyses have demonstrated that the elastin gene exists as a single copy, and the human gene has been localized to chromosome 7q11.1–21.1 (Fazio et al 1991). The entire bovine and human elastin genes have been isolated and extensively sequenced (Fig. 4) (Indik et al 1987, Yeh et al 1989, Bashir et al 1989). Rather small exons (27–165 bp in the human gene, Fig. 5) are interspersed between large introns resulting in a low coding ratio, about 1:20. Another important characteristic is that hydrophobic and cross-link domains of the protein are encoded in separate exons, so that the domain structure of the protein is a reflection of the exon organization of the gene. Although exons are all multiples of three nucleotides and glycine is usually found at the exon/intron junctions, the exons do not exhibit any regularity in size as is found in the fibrillar collagen genes (Fig. 5). Exon/intron borders always split codons in the same way in the elastin gene. Thus, the second and third nucleotides of a codon are included in the 5′ exon border, whereas the first nucleotide of a codon is found at the 3′ border of the previous exon. This consistent structure permits extensive alternative splicing of the primary transcript in a cassette-like fashion while maintaining the reading frame. Sequences homologous to exons 34 and 35 in other species have not been found in the human gene despite extensive searches. It is not presently known whether this difference has any functional significance.

*Intron characterization.* In the human genome, short interspersed elements of approximately 300 bp containing an *Alu*I restriction site (Alu repeat) constitute 3–6% of the total mass of DNA (Schmid & Jelinek 1982), but such sequences are found at a frequency of about four times the expected value in the introns of the elastin gene (Fig. 4). In addition to Alu repeats, rather long stretches composed of dinucleotide repeats occur. The function, if any, of these repetitive elements remains to be determined. In other human genes, deletions apparently mediated by recombination between repetitive sequences have occurred resulting in hereditary diseases. Evidence for genomic instability in regions of human DNA enriched in Alu repeat sequences exists (Calabretta et al 1982), but no mutations in the human elastin gene that might be mediated by similar mechanisms have been found. However, disruption of the elastin gene has been strongly linked to supravalvular aortic stenosis, an inherited disorder that causes haemodynamically significant narrowing of large arteries (Curran et al 1993).

Variability in the length of nucleotide repeats may result in informative polymorphisms in the population, but despite intensive searches no such polymorphisms have been identified in the elastin gene. However, several informative polymorphisms have been identified by PCR and restriction endonuclease analyses in the human elastin gene (Tromp et al 1991).

## Alternative splicing of elastin mRNA

Analysis of bovine, human and rat elastin cDNAs has clearly demonstrated alternative splicing of the primary transcripts (see Fig. 1 for identification of alternatively spliced exons) (Raju & Anwar 1987, Indik et al 1987, Yeh et al 1989, Pierce et al 1990). In most cases, splicing occurs in a cassette-like fashion in which an exon is either included or deleted, but occasionally a splicing event may divide an exon, as in the case of human exons 24 and 26. Both hydrophobic and cross-link domains are affected, so that two cross-link domains may be brought into apposition (e.g. deletion of exon 22) or the interval between cross-link domains may be increased (e.g. deletion of exon 23). It is not known whether these variations have any functional significance, although clearly a tighter or looser fibre network could be produced. S1 mapping experiments using elastin mRNA isolated from developing bovine and rat tissues have demonstrated that the alternative splicing of a few exons occurs frequently, but in the majority of cases it is infrequent (Yeh et al 1989, Heim et al 1991). These experiments also suggest that there is variation in exon use between tissues and that the alternative splicing pattern in the adult may differ significantly from that occurring during development. Translation of the alternatively spliced transcripts would result in significant primary sequence variation among individual tropoelastin molecules, and it is likely that such differences explain the finding of variant isoforms of tropoelastin in several species (Rich & Foster 1984, Wrenn et al 1987). It remains to be determined whether variation in splicing occurs in human disease situations. The possible variable expression of exon 26A in human elastin is particularly intriguing, since this segment, which is highly hydrophilic and atypical in amino acid sequence for elastin, appears to be rarely expressed under normal circumstances. Inclusion of this sequence might substantially alter the properties of the elastic fibre.

## Regulation of elastin gene expression

### *Sequence analysis of the promoter and transcription initiation*

DNA sequencing of the 5′ region flanking the start of the translation codon of the human and bovine genes revealed extremely strong conservation of portions of the sequence (94% homology from −1 to −192 and 86% from

**TABLE 1 Relative activity of elastin promoter–CAT constructs**

| *Elastin promoter fragment* | *% CAT activity* | |
|---|---|---|
| | *Control* | *IGF treated* |
| −2200 to +2 | 47.1 | 170.6 |
| −900 to +2 | 45.9 | 252.9 |
| −500 to +2 | 100.0 | 177.6 |
| −195 to +2 | 70.6 | 305.9 |
| −136 to +2 | 329.4 | 354.1 |
| −475 to −148 | 0 | |

Rat aortic smooth muscle cells were transiently transfected with plasmids containing various segments of the elastin promoter controlling expression of the CAT (chloramphenicol acetyltransferase) gene. Activities are expressed relative to that of the −500 to +2 fragment (taken as 100%). IGF, insulin-like growth factor.

−193 to −588), suggesting that these segments may have important functional roles (Yeh et al 1989, Bashir et al 1989). No canonical TATA boxes were found and although two CAAT sequences were found in the human gene, their locations (−57 to −61 and −599 to −603) suggest that they are not functionally significant. The 5′ flanking region is generally G+C rich (66%) with a high frequency of CpG dinucleotides. Many of these features have been previously associated with promoters of so called 'housekeeping genes', but more tissue-specific genes are being found to possess them as well, so that the distinction between these classes is now breaking down (Gardiner-Garden & Frommer 1987). As is becoming apparent for a number of genes, the elastin promoter contains a remarkable constellation of potential binding sites for transcription regulatory factors, indicative of complex regulation (Fig. 6A). These binding sites include multiple Sp1 and AP-2 binding sites, glucocorticoid-responsive elements, 12-*O*-tetradecanoylphorbol-13-acetate (TPA) and cAMP-responsive elements. There is also an extended sequence of alternating guanine and pyrimidine residues (−225 to −275), a type of sequence which may be associated with Z DNA and which may be involved in transcriptional regulation.

The absence of a TATA box in the putative promoter region suggested that there may be multiple sites of trancription initiation. S1 protection and primer extension analyses using human fetal aorta mRNA consistently identified three major clusters of transcription initiation centred around nucleotides −(7,8), −(15,16) and −(32,33). Five other minor initiation sites were observed between −45 and −195 (Bashir et al 1989). It remains to be determined whether a similar pattern is found in other tissues, whether the multiple initiation has any physiological significance, and whether there is any

**TABLE 2 Expression of human elastin promoter in transgenic mice**

| *Tissue* | *Relative CAT activity* |
|---|---|
| Skin | 1.0 |
| Brain | 11.0 |
| Kidney | 17.7 |
| Aorta | 28.7 |
| Lung | 98.0 |

Different tissues were dissected from a two-month-old transgenic mouse expressing CAT (chloramphenicol acetyltransferase) under the control of 5.2 kbp of the human elastin promoter. CAT activity was determined by the incubation of tissue extracts containing the same amount (100 μg) of protein with [$^{14}$C]chloramphenicol. The relative activities in different organs, in relation to the skin (1.0) are shown.

relationship between the position of transcript initiation and the pattern of alternative splicing.

### *Functional analysis of the human elastin promoter*

We used a panel of promoter/reporter gene constructs in transient transfection assays to determine the functional characteristics of the promoter (Kähäri et al 1990, Wolfe et al 1993). These experiments demonstrated the presence of multiple up- and down-regulatory elements within the 2.2 kbp 5′ flanking region tested (Table 1), and indicated that the core promoter necesssary for basal expression was contained within the region −148 to −1, since substantial levels of expression were obtained when it was retained and removal of this segment abolished all promoter activity. The positive regulatory and core promoter activity may be explained, at least in part, by the presence within this segment of multiple Sp1 and AP-2 binding sites, which may act as general transcriptional activating elements. This notion is supported by the observation that deletion of the segment −134 to −87 containing three putative Sp1 binding sites reduced the activity to 10–20% of the reference −475 to −1 construct.

Most of the detailed analyses were performed with rat aortic smooth muscle cells, but qualitatively similar results were obtained with NIH/3T3 cells, human skin fibroblasts, human HT-1080 fibrosarcoma cells and HeLa cells. The human skin fibroblasts and the HT-1080 cells express the endogenous elastin gene, albeit at a low level, but HeLa and NIH/3T3 cells do not. Gel mobility shift assays with segments of the elastin promoter (+2 to −195) have shown that nuclear extracts from elastin-producing cells (smooth muscle) give a different gel-shift pattern then extracts from cells which do not produce elastin (HeLa). Significantly, when tissues of transgenic mice expressing 5.2 kbp of the

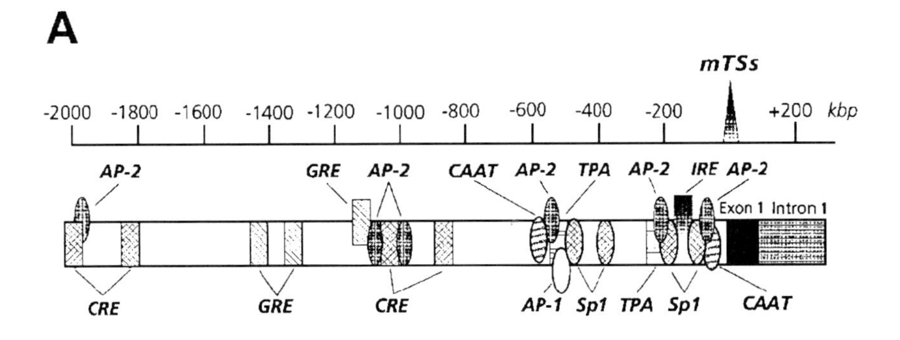


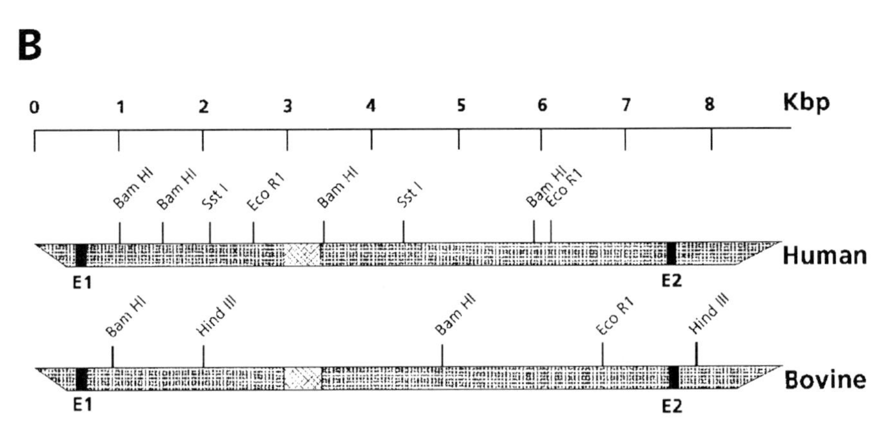


FIG. 6. (A) Cartoon diagram of the human elastin gene promoter. (B) Diagram of the first intron of the human and bovine elastin genes. Note the size similarity. The indicated segment within the intron (⊠) is a region of high sequence homology (77%) between the two genes which may be involved in the regulation of transcription. CRE, cAMP-responsive element; GRE, glucocorticoid-responsive element; IRE, insulin-like growth factor-responsive element; TPA, 12-*O*-tetradecanoylphorbol-13-acetate.

human elastin promoter were analysed, there was reasonable correspondence between expression of the endogenous gene and transgenes (Hsu-Wong et al 1994). Highest levels of expression of the transgene were found in the lung and aorta, whereas lower levels were detected in the kidneys, heart, brain and skin (Table 2). This distribution roughly parallels the accumulation of elastin in developing animals. Nevertheless, because of some inconsistencies between expression of transfected or transgenes and the endogenous gene, collectively these observations suggest that all elements for tissue- and development-specific expression of the elastin gene may not reside within the promoter regions tested. It has been observed in a number of genes, including three collagen genes, that the first intron contains segments which act as enhancer elements of promoter activity. Comparison of the intron sizes of the bovine and human elastin genes demonstrated that only in the case of the first intron was there strong conservation of length. Extensive sequence analysis identified a region of strong homology (77%) in the first intron (Fig. 6B), but enhancer activity of the homologous segment has yet to be conclusively demonstrated.

*Regulation of transcription by hormones and cytokines*

As noted above, analysis of the elastin promoter region identified a variety of potential regulatory elements, including those for glucocorticoids and cAMP, and earlier observations suggested that some of these may be functionally active. For example, injection of dexamethasone into developing chick embryos increased the rate of elastin accumulation in the aorta (Eichner & Rosenbloom 1979), and incubation of fetal bovine nuchal ligament fibroblasts with dexamethasone increased elastin gene expression (Mecham et al 1984), whereas cAMP abrogated the cGMP stimulation of elastin production by the ligament fibroblasts (Mecham et al 1985). However, only in a few cases is there compelling evidence that such effects are regulated at the transcriptional level. Insulin-like growth factor 1 has been shown to enhance elastin gene expression in the developing aorta, both *in vitro* (Rich et al 1992) and *in vivo* (Foster et al 1989). Detailed transfection analyses of rat aortic smooth muscle cells using elastin promoter/reporter gene constructs demonstrated that the insulin-like growth factor 1 was acting at the transcriptional level and that the responsive element was located between bases −195 and −136 (Table 1) (Wolfe et al 1993). In contrast, recombinant tumour necrosis factor α (TNF-α) markedly suppressed elastin mRNA levels in a time- and dose-dependent manner by up to 91% in cultured human skin fibroblasts and rat aortic smooth muscle cells (Kähäri et al 1992). TNF-α also suppressed the expression of elastin promoter/reporter constructs by up to 70% in transiently transfected cells, again indicating regulation at the transcriptional level. Detailed analyses of the mechanisms involved strongly suggested that the down-regulatory effect of TNF-α was mediated through Jun/Fos binding to an AP-1 site located at −223 to −229 in the elastin promoter. Conflicting evidence has been presented on the effects of interleukin 1β. One group reported that the cytokine inhibits elastin synthesis and decreases elastin mRNA levels in a particular subtype of neonatal rat lung fibroblasts, whereas another group reported that the cytokine increases elastin gene expression in cultured human skin fibroblasts apparently at the transcriptional level (Berk et al 1991, Mauviel et al 1993).

*Acknowledgement*

Supported by National Institutes of Health grants AR41474 and AR20553.

## References

Andersen SO, Florkin M, Stotz EH 1971 Resilin. In: Florkin M, Stotz EH (eds) Comprehensive biochemistry. Elsevier, Amsterdam, Vol 26C, p 633–645

Bashir MM, Abrams WR, Rosenbloom JC et al 1994 Microfibril-associated glycoprotein: characterization of the bovine gene and of the recombinantly expressed protein. Biochemistry 33:593–600

Bashir MM, Indik Z, Yeh H et al 1989 Characterization of the complete human elastin gene. Delineation of unusual features in the 5′-flanking region. J Biol Chem 264:8887–8891

Berk JL, Franzblau C, Goldstein RH 1991 Recombinant interleukin-1$\beta$ inhibits elastin formation by a neonatal rat lung fibroblast subtype. J Biol Chem 266:3192–3197

Bressan GM, Argos P, Stanley KK 1987 Repeating structure of chick tropoelastin revealed by complementary DNA cloning. Biochemistry 26:1497–1503

Burnett W, Finnigan-Burnick A, Yoon K, Rosenbloom J 1982 Analysis of elastin gene expression in the developing chick aorta using cloned elastin cDNA. J Biol Chem 257:1569–1572

Calabretta B, Robberson DL, Berrera-Saldana HA, Lambrou TP, Saunders GF 1982 Genome instability in a region of human DNA enriched in Alu repeat sequences. Nature 296:219–225

Cleary EG 1987 The microfibrillar component of the elastic fibres. Morphology and biochemistry. In: Uitto J, Perejda AJ (eds) Connective tissue disease. Molecular pathology of the extracellular matrix. Marcel Dekker, New York, p 55–81

Curran ME, Atkinson DL, Ewart AK, Morris CA, Leppert MF, Keating MT 1993 The elastin gene is disrupted by a translocation associated with supravalvular aortic stenosis. Cell 73:159–168

Eichner R, Rosenbloom J 1979 Collagen and elastin synthesis in the developing chick aorta. Arch Biochem Biophys 198:414–423

Fahrenbach WH, Sandberg LB, Cleary EG 1966 Ultrastructural studies on early elastogenesis. Anat Rec 155:563–576

Fazio MJ, Mattei M-G, Passage E et al 1991 Human elastin gene: new evidence for localization to the long arm of chromosome 7. Am J Hum Genet 48:696–703

Foster JA, Rich CB, Miller M, Benedict MR, Richman RA, Florini JR 1990 Effect of age and IGF-1 administration on elastin gene expression in rat aorta. J Gerontol 45:B113–B118

Gardiner-Garden M, Frommer M 1987 CpG islands in vertebrate genomes. J Mol Biol 196:261–282

Greenlee TK Jr, Ross R, Hartman JL 1966 The fine structure of elastic fibres. J Cell Biol 30:59–71

Heim RA, Pierce RA, Deak SB, Riley DJ, Boyd CD, Stolle CA 1991 Alternative splicing of rat tropoelastin messenger RNA is tissue-specific and developmentally regulated. Matrix 11:359–366

Hsu-Wong S, Katchman SD, Ledo I et al 1994 Tissue-specific and developmentally regulated expression of human elastin promoter activity in transgenic mice. J Biol Chem 269:18072–18075

Indik A, Yeh H, Ornstein-Goldstein N et al 1987 Alternative splicing of human elastin mRNA indicated by sequence analysis of cloned genomic and complementary DNA. Proc Natl Acad Sci USA 84:5680–5684

Kähäri V-M, Chen YQ, Bashir M, Rosenbloom J, Uitto J 1992 Tumor necrosis factor-α down-regulates human elastin gene expression. J Biol Chem 267:26134–26141

Kähäri V-M, Fazio MJ, Chen YQ, Bashir MM, Rosenbloom J, Uitto J 1990 Deletion analyses of 5′-flanking region of the human elastin gene: delineation of functional promoter and regulatory *cis*-elements. J Biol Chem 265:9485–9490

Karrer HE, Cox J 1961 Electronmicroscope study of developing chick embryo aorta. J Ultrastruct Res 4:420–454

Kelly RE, Rice RV 1967 Abductin: a rubber-like protein from the internal triangular hinge ligament of pecten. Science 155:208–210

Mauviel A, Chen YQ, Kähäri V-M et al 1993 Human recombinant interleukin-1β up-regulates elastin gene expression in dermal fibroblasts—evidence for transcriptional regulation *in vitro* and *in vivo*. J Biol Chem 268:6520–6524

Mecham RP, Levy BD, Morris SL, Madaras JG, Wrenn DS 1985 Increased cyclic GMP levels lead to a stimulation of elastin production in ligament fibroblasts that is reversed by cyclic AMP. J Biol Chem 260:3255–3261

Mecham RP, Morris SL, Levy BD, Wrenn DS 1984 Glucocorticoids stimulate elastin production in differentiated bovine ligament fibroblasts but do not induce elastin synthesis in undifferentiated cells. J Biol Chem 259:12414–12420

Pierce RA, Deak SB, Stolle CA, Boyd CO 1990 Heterogeneity of rat tropoelastin mRNA revealed by cDNA cloning. Biochemistry 29:9677–9683

Raju K, Anwar RA 1987 Primary structures of bovine elastin a, b, and c deduced from the sequences of cDNA clones. J Biol Chem 262:5755–5762

Rich CB, Foster J 1984 Isolation of tropoelastin from lathyritic chick aortae. Biochem J 217:581–584

Rich CB, Ewton DZ, Martin BM et al 1992 IGF-1 regulation of elastogenesis: comparison of aortic and lung cells. Am J Physiol 263:L276–L282

Sage H, Gray WR 1979 Studies on the evolution of elastin. I. Phylogenetic distribution. Comp Biochem Physiol B Comp Biochem 64:313–327

Sakai LY, Keene DR, Engvall EJ 1986 Fibrillin, a new 350-kD glycoprotein, is a component of extracellular microfibrils. Cell 103:2499–2509

Schmid CW, Jelinek WR 1982 The Alu family of dispersed repetitive sequences. Science 216:1065–1070

Shadwick RE, Gosline JM 1981 Elastic arteries in invertebrates: mechanics of the octopus aorta. Science 213:759–761

Tromp G, Christiano A, Goldstein N et al 1991 A polymorphism to G in ELN gene. Nucleic Acids Res 19:4314

Wolfe BL, Rich CB, Goud HD et al 1993 Insulin-like growth factor-1 regulates transcription of the elastin gene. J Biol Chem 268:12418–12426

Wrenn DS, Parks WC, Whitehouse LA et al 1987 Identification of multiple tropoelastins secreted by bovine cells. J Biol Chem 262:2244–2249

Yeh H, Anderson N, Ornstein-Goldstein N et al 1989 Structure of the bovine elastin gene and S1 nuclease analysis of alternative splicing of elastin messenger RNA in the bovine nuchal ligament. Biochemistry 28:2365–2370

Yoon K, May M, Goldstein N et al 1984 Characterization of a sheep elastin cDNA containing translated sequences. Biochem Biophys Res Commun 118:261–269

## DISCUSSION

*Keeley:* Has anyone found a serum response element in the elastin 5′ untranslated region?

*Rosenbloom:* No.

*Keeley:* The reason I ask is because the serum response element has been identified as a stretch-inducible element in both skeletal and cardiac muscle.

*Rosenbloom:* We're beginning to look at that. Dr Edward Macarack has performed some experiments on the effects of mechanical stress on elastin expression by smooth muscle cells. However, to date the results have been inconclusive.

*Pierce:* You mentioned alternative splicing and the sequences which surround the exons that are alternatively spliced in the elastin gene. I studied the sequences surrounding the exons in the rat elastin gene, and I found that in several cases there were differences that I felt might be significant (Pierce et al 1991). First of all, there were weak polypyrimidine tracks 5′ of those exons that were alternatively spliced. In a couple of other instances there were overlapping branch point sequences; these were only found in cases where there was alternative splicing. There was also a region of unusually strong secondary structure which involved the 5′ end of intron 15 and the actual coding sequence of exon 16. At that time we speculated that these changes had some effect on the alternative splicing. But in the absence of functional data, or data on the biological consequences of alternative splicing, we didn't pursue this any further. Is this a dead end, or do you think it will prove to be of interest?

*Rosenbloom:* There's still interest in alternative splicing; I don't think it's been thoroughly examined. *De novo* synthesis in either the adult situation or in some injury system would certainly be worth looking at. Alternative splicing has been described in hundreds of genes and, with rare exceptions, we don't know that much about its role.

*Davidson:* We know in collagens, because of the way the fibres pack, that the length of the molecule is ultra-critical to the proper arrangement of functional collagen (Yurchenko et al 1994). Is there any evidence that you can play with the length of the tropoelastin and alter its assembly in any way? Does variation in length play a positive role in elastic fibre organization?

*Rosenbloom:* That's a good question. There have not been any good experiments done on this. In the recombinant tropoelastins that we've made, we haven't attempted to make smaller molecules. One of the problems in looking at the assembly of the elastic fibre has been that we didn't have a good way of finding out whether the fibre was assembled correctly, i.e. we didn't have a particularly good endpoint. The only heritable disease of elastic fibres that we know about is supravalvular aortic stenosis (SVAS); in one case there is a translocation within exon 27 of the elastin gene. Individuals with SVAS have coarctation of the aorta, but we don't even know at this point whether that truncated allele is expressed. It is probably not expressed, just because you can't get accurate processing of the transcript, so it is likely a gene dosage effect.

*Davidson:* Alek, haven't you described a post-translational truncation of elastin that appears to have consequences on matrix assembly?

*Hinek:* Yes, we found that the ductus arteriosus, a fetal vessel which assembles elastin poorly in association with the formation of intimal cushions, produces abundant quantities of a 52 kDa tropoelastin which differs from the 68–72 kDa isoforms of tropoelastin in that it lacks the C-terminal 157 amino acids (Hinek & Rabinovitch 1993). The identification of the 52 kDa protein as tropoelastin was made on the basis of amino acid composition, amino acid sequence data and immunoreactivity with anti-tropoelastin antibodies (except the one that is specific to the tropoelastin C-terminus encoded by exon 36). *In vitro* translation showed, however, that the tropoelastin initially synthesized in ductus arteriosus smooth muscle cells has a molecular weight of 68–70 kDa, suggesting that the 52 kDa protein is a unique product of intracellular post-transcriptional modification of the full-length tropoelastin. We suggested that the 52 kDa tropoelastin might arise because ductus arteriosus smooth muscle cells produce a unique 68 kDa isoform particularly susceptible to enzymic or non-enzymic cleavage. We also speculated that post-translational cleavage of tropoelastin may be linked to the fact that ductus arteriosus smooth muscle cells lack the 67 kDa elastin-binding protein (Hinek et al 1991) which normally 'chaperones' the 68 kDa tropoelastin through the secretory pathways and protects it from proteolytic degradation (Hinek & Rabinovitch 1994).

Our pulse–chase experiments showed that the 52 kDa protein remained soluble and was not incorporated into elastic fibres. This clearly indicated that the intact C-terminal region of tropoelastin is necessary for the proper extracellular assembly of elastic fibres. This was later confirmed by Bob Mecham's group (Brown et al 1992) who showed that the cysteines of the C-terminus participate in anchoring tropoelastin to the microfibrillar scaffold.

*Uitto:* Joel, you alluded to the possible importance of the intron 1 as a regulatory element; in support of this is the fact that there is about 80% sequence conservation between human and bovine. However, do you really have any functional evidence for its regulatory role?

*Rosenbloom:* Not good evidence. We've made some promoter constructs, but we haven't tested that many. In addition, a critical feature of transfection experiments is that transcription of the transgene should reflect the activity of the endogenous gene; the problem has been that in this respect our test systems have not measured up very well. Our knowledge is limited—we didn't try this with the full promoter segment (5 kbp of flanking segment) coupled to the first intron, for instance. You really have to make a large number of constructs before you can start arguing one way or the other. You almost have to construct a minigene in which you do not perturb the promoter, the first exon and the whole first intron in order to start looking at this, otherwise you're going to be constantly worrying about whether you're changing distances or altering the relationship between any enhancer or negative elements in the first intron with the promoter.

*Keeley:* I would encourage you to express an elastin without the VGVAPG repeat for several reasons. First, I noted that this is exon 24, which is the exon adjacent to the cross-linking exon that you suggested might be of particular importance. Second, that is the sequence with which the elastin-binding protein interacts. Thirdly, this is also a domain which we believe may be of particular importance for self-assembly. I suspect that the effect of such a deletion might be very dramatic.

*Rosenbloom:* That's what I was going to ask Dan Urry: do you think that even though only a small percentage of the molecule is exact repeats, could these have a potential nucleating role with regard to assembly?

*Urry:* We've long believed that the repeating hexapeptide sequences have a role in assembly. The polyhexapeptides are soluble at low temperature in water, and if the temperature is raised, they go through the inverse temperature transition of hydrophobic folding and assembly. If you then lower the temperature, they don't disassemble—they don't re-dissolve. You can re-dissolve them in a solvent such as trifluoroethanol, but in water association is simply irreversible. This presents a nice opportunity to achieve an association which would then align the cross-linking regions.

Joel Rosenbloom, how many of the lysine residues in elastin are utilized during assembly in cross-link formation?

*Rosenbloom:* There are 35 lysines in bovine elastin and about 30 of them are cross-linked. About four or five are not.

*Urry:* That's an enormously improbable event if cross-linking were to occur on a random basis. In order for a pair of lysine residues in tropoelastin to form a cross-link, they must come together in space. With no more than 35 lysines in some 800 residues of the tropoelastin sequence, the chance on a random basis that contact residues between two protein molecules would be a pair of lysines is $(35/800)^2$; the chance that a second pair would occur is $(35/800)^2 \times (34/799)^2$. Consequently, the chance that 30 pairings of lysine would occur is in the order of 1 in $10^{100}$. This simply cannot happen on a random basis. Of course, when every repeating unit of polymer is a potential cross-link, as in the polymers that form rubbers, then a random network is possible. For elastin however, there has to be some aligning of these lysine residues to get them into position for cross-linking. This is the argument I used 20 years ago (Urry 1976): elastin can't be random. You can't take such a small number of lysines and have so many of them turn into cross-links without extensive aligning and arranging of them.

*Rosenbloom:* It is possible that one or more of the microfibrillar proteins may be playing a critical role in this.

*Mecham:* I agree with you. We have evidence that domains encoded by exons 19 and 25 (the only two cross-linking domains with three lysines) might act as nucleation sites for assembly of the polymer.

*Robert:* If I understand correctly, there is the possibility that splice variants could influence cross-linking.

*Rosenbloom:* In some cases they must. In human tropoelastin, I would like to point out that the sequence encoded in exon 22 is always spliced out. In this case, two cross-link domains are brought together, which probably alters cross-link formation. I have no explanation for why this occurs in human elastin; it doesn't happen in bovine elastin. In the human gene, exon 22 is present and I can't see anything different about the surrounding borders which would indicate why that exon should be spliced out so absolutely uniformly. But we have never found it in mature transcripts, and Jouni Uitto and I have both looked for it extensively.

*Robert:* Is there any difference in splicing and direction of cross-linking between fibres and the 2D sheet structures?

*Rosenbloom:* No one knows.

*Keeley:* This prompts another question: has anyone looked at the splicing pattern in ear cartilage elastin? This is the form of elastin which is architecturally most different from other elastins. It would be very interesting if it contained particular splice variants. I believe that one of the factors which may determine the final architecture of elastin is splicing, since this will alter the arrangements of cross-linking and hydrophobic exons and may determine which cross-linking exons align for formation of the desmosines. Has anyone ever looked at this?

*Parks:* Yes. A few years ago we assessed the pattern of alternative splicing by using splice-variant-specific oligomeric probes in an *in situ* hybridization assay (Parks et al 1992). We looked in a number of different tissues, including elastic cartilage, skin and pulmonary artery. The pattern of isoform expression within the tissue was identical. In other words, essentially every cell was producing the same alternative splice variants. We performed these studies to assess whether regional differences in the pattern of tropoelastin mRNA alternative splicing may be related to the distinct fibre organization among tissues. However, we detected no spatial heterogeneity. These findings suggest that alternative splicing may have another role, such as in the interactions of tropoelastin or elastin with other molecules. Another important point, and one that may give some clue as to its function, is that the pattern of alternative splicing changes with age.

*Robert:* How does alternative splicing change with age?

*Parks:* Although various exons are spliced in an age-specific pattern, exon 33 is omitted at a fairly high rate at all ages. In the adult, more exons are spliced, albeit at a low rate, thus producing a population of smaller tropoelastins. However, I'm not sure of the functional significance of this.

*Mariani:* You have indicated that in the bovine the number of tropoelastin isoforms produced is low during periods of high elastin production (neonate) and increases with age (Yeh et al 1989). Is this typical of other species as well?

It would seem to be important to focus on the level of alternative splicing taking place during periods of high elastic fibre deposition, as opposed to in the adult where elastin deposition is minimal.

*Parks:* That is not necessarily so. I know of no findings to indicate that the functional integrity of elastin produced late in development differs from the fibres made and deposited earlier in life. In essence, we can only guess as to the functional significance of tropoelastin alternative splicing. We have no assays for assessing function, and there are no data indicating, or even suggesting, a precise biological role for this extensive, age-specific heterogeneity.

*Hinek:* I think that the removal of internal sequences due to alternative splicing, or the simple modification of a certain domain that makes the tropoelastin isoform more susceptible to post-translational enzymic trimming, both lead to a similar final effect—lack of the proper assembly of elastic fibres. Although the absence of the C-terminal domain may prevent the initial interaction between tropoelastin and microfibrillar proteins, the deletion of internal sequences may prevent their subsequent cross-linking. The most exciting possibility is that the heterogeneity of tropoelastin isoforms, as well as their possible post-translational modifications, may be tissue specific or developmentally regulated. This would explain not only the different assembly patterns of elastic fibres and laminae in different tissues (vascular walls, skin, elastic cartilage), but may also provide a new perception of pathological conditions such as pulmonary hypertension and atherosclerosis, in which there is impaired assembly of elastic fibres despite the intensive synthesis of tropoelastin.

*Mecham:* But is it generally the case that there are fewer isoforms produced in the younger animal than in older animals? Is that consistent?

*Parks:* The data from the S1 mapping done in Joel Rosenbloom's lab (Yeh et al 1989) indicate that fewer isoforms are made in younger animals. During the most active periods of elastogenesis—that is, during late fetal and early neonatal life—the frequency of alternative splicing increases as well as the number of exons that are potentially spliced. Exon 33, which is omitted more often than any other exon, is spliced out from about 50% of transcripts at all ages. Indeed, the protein results that Bob Mecham and I published (Parks et al 1988) showed that a higher ratio of lower molecular weight isoforms were produced during peak periods of elastin production.

*Mecham:* It's interesting that when you look at the splicing patterns, the exons that are spliced out are actually hydrophilic domains. Exon 33, for example, contains the only arginine in the molecule that's not next to a helix or not at the C-terminus. Exons 13 and 14 contain lysines that don't fit the standard paradigm of a cross-linking sequence and are probably not oxidized by lysyl oxidase. If you disregard lysine residues that serve as cross-links, where are the hydrophilic amino acids? They're in exons 13, 14 and 33, and these are the exons that have the highest splice frequency. Just deleting these charged residues may be more important than changing the shape of the molecule.

## References

Brown PL, Mecham L, Tisdale C, Mecham RP 1992 The cysteine residues in the carboxy terminal domain of tropoelastin form an intrachain disulphide bond that stabilizes a loop structure and a positively charged pocket. Biochem Biophys Res Commun 186:549–555

Hinek A, Rabinovitch M 1993 Ductus arteriosus smooth muscle cells produce a 52 kD tropoelstin associated with impaired assembly of elastin laminae. J Biol Chem 268:1405–1413

Hinek A, Rabinovitch M 1994 67-kD elastin binding protein is a protective companion of extracellular insoluble elastin and intracellular soluble tropoelastin. J Cell Biol 126:563–574

Hinek A, Mecham RP, Keeley F, Rabinovitch M 1991 Impaired elastin fiber assembly is related to reduced 67 kD elastin binding protein in fetal lamb ductus arteriosus and in cultured aortic smooth muscle cells treated with chondroitin sulfate. J Clin Invest 88:2083–2094

Parks WC, Secrist H, Wu LC, Mecham RP 1988 Developmental regulation of tropoelastin isoforms. J Biol Chem 263:4416–4423

Parks WC, Roby JD, Wu LC, Grosso LE 1992 Cellular expression of tropoelastin mRNA splice variants. Matrix 12:156–162

Pierce RA, Alatani A, Deak SB, Boyd CD 1991 Elements of the rat tropoelastin gene associated with alternative splicing. Genomics 12:651–658

Urry DW 1976 Molecular mechanisms of elastin coacervation and calcification. Faraday Dis 161:205

Yeh H, Anderson N, Ornstein-Goldstein N et al 1989 Structure of the bovine elastin gene and S1 nuclease analysis of alternative splicing of elastin mRNA in the bovine nuchal ligament. Biochemistry 28:2365–2370

Yurchenko PD, Birk DE, Mecham RP (eds) 1994 Extracellular matrix assembly and structure. Academic Press, San Diego, CA

# Regulation of elastin synthesis in pathological states

Jeffrey M. Davidson, Man-Cong Zang, Ornella Zoia and Maria Gabriella Giro

*Department of Pathology, Vanderbilt University School of Medicine and Research Service, Department of Veterans Affairs Medical Center, Nashville, TN 37232-2561, USA*

*Abstract.* Elastin is rapidly deposited during late gestation in resilient tissues such as the arteries, lungs and skin owing to increased concentration of its mRNA. Pathological states can arise from congenital insufficiency or disorganization of elastin (cutis laxa). Other elastin deficiencies may be due to excess elastolysis or gene dosage effects. In the former, high turnover rates can be assessed by measurements of elastin degradation products in urine. Excess elastin accumulation by skin fibroblasts is characteristic of genetic diseases such as Buschke–Ollendorff syndrome, Hutchinson–Gilford progeria and keloid. Elastin expression is modulated by peptide growth factors, steroid hormones and phorbol esters, among which transforming growth factor $\beta$ (TGF-$\beta$) is an especially potent up-regulator, acting largely through stabilization of mRNA. Recent evidence indicates cutis laxa fibroblasts that express little or no elastin have normal transcriptional activity but abnormal rates of elastin mRNA degradation. This defect is substantially reversed by TGF-$\beta$ through mRNA stabilization. Current studies explore the hypothesis that stability determinants lie within the 3′ untranslated region of elastin mRNA. Post-transcriptional control of elastin expression appears to be a major regulatory mechanism.

*1995 The molecular biology and pathology of elastic tissues. Wiley, Chichester (Ciba Foundation Symposium 192) p 81–99*

Elastin is the biological rubber of the vertebrates. In concert with microfibrillar glycoproteins, which constitute about 10% of the elastic fibre, elastin provides resiliency to a wide variety of connective tissue structures (Parks et al 1993). The proper function of elastin has great physiological significance in pulmonary, vascular, cutaneous and laryngeal tissues, among others. Convergent lines of evidence suggest that the bulk of elastin is deposited in elastic fibres during the latter stages of development of each of these tissues. Unlike other fibrous matrix proteins such as the collagens, elastin does not appear to undergo extensive, post-developmental remodelling. Neosynthesis of elastin does occur after injury to some of these tissues, however; arterial walls subject to hypertension or neointimal hyperplasia are probably the most extreme examples of neosynthesis of elastin in response to physiological stress. In contrast, destruction or injury to the elastic tissue architecture of the lung and skin often fails to result in an orderly, new deposition of elastic fibres, although net concentration of elastin may approach pre-injury values (Davidson et al 1992). In severely damaged dermis, elastin is absent from scar tissue for many years.

A number of human genetic diseases are associated with altered elastic tissue (Uitto et al 1991), although only one genetic disease, supravalvular aortic stenosis, has been positively linked to the elastin gene locus (Curren et al 1993). This disease, which is principally associated with deranged elastic fibre organization in the aortic trunk, appears to be due to large deletions or truncations at the 3′ terminus of one elastin allele. Cutis laxa, in its most severe, perinatal form, results in the near absence of detectable elastic fibres in the skin and the internal organs, leading to very early demise of these patients (Uitto et al 1993). Cutis laxa is quite heterogeneous in severity, however, and no form of the disease, which is reported to have both autosomal dominant and recessive patterns of inheritance, has, as yet, been convincingly linked to the elastin gene. Cutis laxa also exists in so-called ‘acquisita’ forms, in which loss of cutaneous elastic fibres can be very dramatic as a result of localized or generalized inflammatory events. Excessive, distorted and frequently calcified elastic fibres are typical of the pathology of pseudoxanthoma elasticum, an autosomal dominant disease which affects most particularly the cutaneous areas of highest flexion and extension such as the axilla, and produces internal pathology through calcification of an elastic structure in the eye, Bruch’s membrane, and blood vessels (Neldner 1993). Specific over-accumulation of elastin is also noted in the cutaneous, autosomal dominant disorder, Buschke–Ollendorff syndrome (BOS), in which papules containing dense collections of elastic fibres are found on the extremities (Giro et al 1992).

## Regulation of elastin

There is ample evidence of elastin regulation in a wide variety of systems (Parks et al 1993). The two organs which most abundantly express the molecule are the aorta and, in grazing animals, the nuchal ligament. Both of these organs show rapid increases in elastin synthesis and concentration during the perinatal period; this amplification is certainly due to increased steady-state concentrations of elastin mRNA ($mRNA_E$). Indeed, there are no reports at this time which contradict the concept that the production of elastin by cells is a direct consequence of the steady-state concentration of $mRNA_E$. Elastin synthesis, however, is not equivalent to elastic fibre formation. Early ultrastructural studies and more-sophisticated cell culture studies have made it apparent that the elastic fibre arises as the result of condensation of the soluble precursor of elastin, tropoelastin, upon a microfibrillar network of glycoproteins, the best characterized of which are the large glycoproteins, fibrillins 1 and 2, and microfibril-associated glycoprotein (MAGP), a 31 kDa, cysteine-rich glycoprotein. Since elastic fibre microfibrils have a complex ultrastructure, it might be reasonable to expect that multiple protein components comprise these elements. In addition, elastic fibre assembly appears to be dependent upon interaction with an elastin/laminin binding protein of molecular mass 67 kDa. This molecule, which appears to be closely associated with the cell surface as a multimeric complex, also has lectin properties, and binding of galactosyl sugars to the molecule results in loss of affinity for elastin and disruption of elastic fibre assembly *in vitro*. One group has identified the 67 kDa component as an alternate form (S-) of $\beta$-galactosidase (Hinek et al 1993). There are both clinical and experimental situations in which elastin may be synthesized but very inefficiently incorporated into its characteristic, highly insoluble, polymerized structure. The net result is elevated rates of excretion of biomarkers of elastin turnover (Davidson 1990).

Other factors in elastic fibre durability include lysyl oxidase and prolyl hydroxylase. In addition, glycosaminoglycans such as hyaluronan and proteoglycans such as decorin can be found associated with elastic fibres, suggesting that these charged polysaccharides also have a role in normal or pathological elastin fibrillogenesis (Pasquali-Ronchetti et al 1984). Regulation of elastic fibre formation is therefore potentially affected by expression of many different gene products.

Elastin gene expression is readily observed in the developmental systems mentioned above and in explants and cell cultures derived from these tissues at varying stages of development (Davidson & Giro 1986, Parks et al 1993). Cultured vascular smooth muscle cells (SMCs) from aortic tissue, cultured fibroblasts from nuchal ligament, skin and vocal fold, and chondrocytes from elastic cartilage maintain the elastogenic phenotype in culture, reflecting the *in vivo* state. The elastin phenotype in cell culture, however, is often unstable (Ruckman et al 1994), suggesting that the culture environment, which includes a 2D growth pattern and exposure to moderate concentrations of serum in a growth medium akin to wound fluid, is not conducive to the long-term stability of an elastogenic phenotype.

There are several examples of elastin regulation at the level of gene transcription, but post-transcriptional events may play a critical role in determining the overall level of expression of elastin during later development and tissue repair (Parks et al 1993). Significant physiological modulators include both steroid hormones and soluble peptide factors. The glucocorticoid effect on elastin production is cell-type and context dependent, in that glucocorticoids can either up- or down-regulate elastin production in cultured cells, depending on developmental stage and tissue of origin (Russell et al 1995). 1,25-dihydroxyvitamin $D_3$ down-regulates elastin

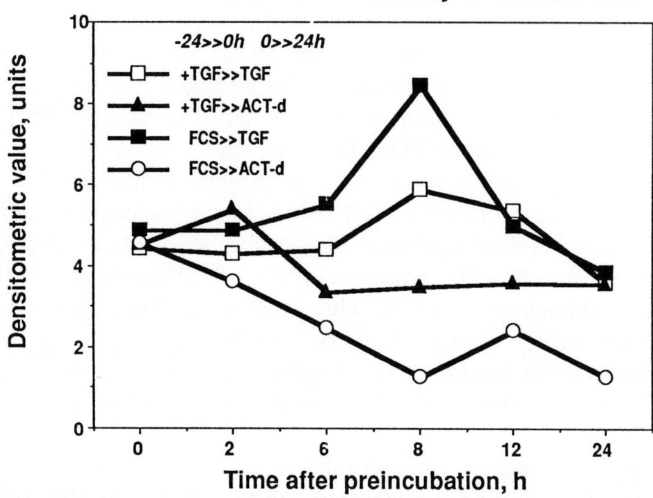


FIG. 1. Stabilization of elastin mRNA by TGF-$\beta$ in vascular smooth muscle cells. Confluent cultures of porcine aortic smooth muscle cells, maintained in 10% fetal calf serum, were treated with either 10 ng/ml recombinant human TGF-$\beta$1 (TGF; Genentech) or fetal calf serum only (10%; FCS) for 24 h. At the end of 24 h (0 h), culture media were changed, and cultures were supplemented either with TGF-$\beta$1 alone, TGF-$\beta$1 in combination with actinomycin D (ACT-d; 5 $\mu$g/ml), or actinomycin D in standard culture medium as indicated in the legend. RNA was isolated from sister cultures at the indicated times after preincubation, and elastin mRNA was quantified via Northern blot hybridization, autoradiography and densitometry with normalization to a constitutively expressed gene, cyclophilin. Hybridization signal strengths for all samples were arbitrarily normalized to the 0 h value. Strong induction of $mRNA_E$ was seen in the cultures switched from FCS to FCS plus TGF-$\beta$1 (closed squares), with milder induction by a second addition of TGF-$\beta$1 (open squares). Addition of actinomycin D to TGF-$\beta$-treated cultures showed a slight decline in $mRNA_E$ over 24 h (closed triangles), whereas serum-only cultures showed a characteristic rate of degradation with a half-life of about 8 h (open circles).

production in cultured cells by reducing $mRNA_E$ stability (Pierce et al 1992). Elastin production is up-regulated by insulin-like growth factor 1 (IGF-1) (Foster et al 1987, Davidson et al 1993a) and transforming growth factor (TGF)-$\beta$ (Liu & Davidson 1988, Davidson et al 1993a). On the basis of promoter fusion constructs, IGF-1 appears to act at the transcriptional level by inactivating *cis*-acting repressor element (Wolfe et al 1993). The predominant effect of TGF-$\beta$ is to increase elastin production by stabilization of transcripts in fibroblasts (Kähäri et al 1992a) and SMCs (Fig. 1), although there are reports of a minimal promoter region upstream of the transcription start site in the human elastin promoter that exhibits a response to TGF-$\beta$ in a cell-dependent fashion (Marigo et al 1994). Tumour necrosis factor (TNF)-α acts negatively at the transcriptional level, as demonstrated by elastin promoter constructs (Kähäri et al 1992b). As discussed below, both basic fibroblast growth factor (bFGF) and TGF-α can exert negative effects on elastin production, also antagonizing TGF-β (Davidson et al 1993a). Interleukin 1 (IL-1) is another regulator of elastin production, found to be either inhibitory in SMCs (Mauviel et al 1993) or stimulatory in fibroblasts (Berk et al 1991). Phorbol esters that mimic the activation of protein kinase C pathways by cytokines and other stimuli, also destabilize $mRNA_E$ (Parks et al 1992). How these modulating agents affect elastin transcription *in vivo* remains to be seen. The elastin promoter region characterized thus far (5.2 kb) appears to confer partially tissue-specific expression of elastin (Rosenbloom et al 1993).

In summary, there are very many pathologies or developmental problems that involve abnormal regulation of the accumulation of elastic tissue. Although the elastic fibre can be considered to be physiologically inert during adult life, a wide range of insults to elastic tissue can result in either its chronic loss or excess, dysfunctional accumulation. Means by which one might regulate elastin production are uncertain. We lack specific knowledge on detailed mechanisms of control of transcription, control of post-transcriptional processing, control of elastic fibre fibrillogenesis and regulation of elastin degradation. Although all four levels must be considered, this paper will focus on the second of these potential levels of control of elastin accumulation.

We have evaluated the relative impact of several soluble peptide factors on elastin production in SMCs and fibroblasts (Liu & Davidson 1988, Davidson et al 1993a). TGF-β is the most potent stimulator of elastin production in these cells, and cytokines such as bFGF and TGF-α can have contrary effects on elastin metabolism. Stimulation of cell division with mitogens that probably act through protein kinase C may selectively diminish elastin production. This observation is also concordant with previous evidence that elastin production is quite low during the logarithmic phase of cell growth. Although IGF-1 is reported to have a substantial activity in removing a negative transcription element from the human elastin promoter (Rich et al 1993), we found that, at least in the SMCs we examined, TGF-β was six to 10 times more potent than IGF-1 in stimulating elastin production (Fig. 2) (Davidson et al 1993a).

Regarding *in vivo* regulation, we have shown by *in situ* hybridization that topical TGF-β1 was able to induce $mRNA_E$ in porcine and murine wounds (Quaglino et al 1990). Unpublished histological and ultrastructural findings in these wounds confirm that there is little if any elastin protein found in the scar tissue of porcine wounds as long as 45 days after repair has begun (unpublished findings). In contrast, rat wounds were observed to contain very small amounts of amorphous elastin 28 days after injury (Davidson et al 1992). These latter findings suggest that elastic fibre assembly is uncoupled from elastin expression

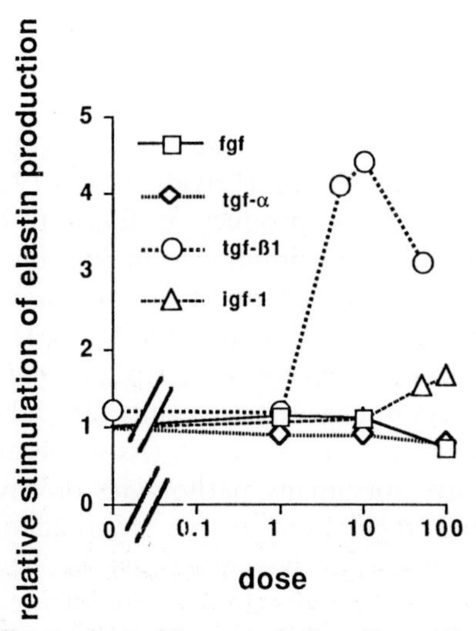


FIG. 2. Differential effects of peptide growth factors on elastin expression. Porcine vascular smooth muscle cells were treated for 48 h with the indicated concentrations (dose, ng/ml) of peptide growth factor and analysed for elastin production by enzyme-linked immunoassay. fgf, basic fibroblast growth factor (Synergen); tgf-$\alpha$, transforming growth factor $\alpha$; tgf-$\beta$1, transforming growth factor $\beta$1 (Genentech); igf-1, insulin-like growth factor 1 (UBI). Data were normalized to a unity value at zero doses.

at the wound site, or that levels of elastic fibre component expression are not sufficient to bring about significant accumulation of the fibre (Davidson et al 1993b). It is speculated that the oxidative wound environment may be unfavourable for expression of elastin.

### *Regulation in disease states: biochemical basis for altered elastin accumulation in Buschke–Ollendorff syndrome, keloid, cutis laxa and Hutchinson–Gilford progeria*

BOS presents as a complex, autosomal dominant syndrome, with elevated accumulation of elastin in localized areas of the dermis. Fibroblasts from patients with BOS produced exceptionally high levels of elastin under standard culture conditions, and it was difficult to stimulate further elastin production in these patients using TGF-$\beta$ as a challenge (Giro et al 1992). Thus, overproduction of elastin by skin cells is diagnostic of this condition, and it represents a stable phenotype in this dominant disorder. Although some linkage analyses have been performed (unpublished work), there is no definitive linkage or exclusion of BOS at the elastin gene (*ELN*) locus. We recently performed an evaluation of elastin regulation in keloids (Russell et al 1995). These studies have revealed that the keloid fibroblast, unlike its normal

**TABLE 1 Comparison of elastin biosynthesis in normal and cutis laxa fibroblast strains: effect of TGF-β[a]**

| | *Elastin production (mol eq/cell·h × 10⁴)* | | *mRNA level* | | *mRNA half-life (h)* | | *TGF-β decay (d)* |
|---|---|---|---|---|---|---|---|
| *TGF-β* | − | + | − | + | − | + | |
| Strain | | | | | | | |
| Control | 30 | 120 (4) | 1 | 2–3 (3) | 8.0 | 16 (2) | 3.3 |
| Cutis laxa | <1 | 11 (50) | 0–0.1 | 1–2 (16.5) | 0.25 | 2.5 (18) | 0.5 |
| Ratio | ⩾10 | 10.9 | ⩾10 | 2–3 | 32 | 6.4 | 6.6 |

[a]Fibroblast cultures from a control fibroblast strain (GM4390) and a cutis laxa strain (KT; Sephel et al 1989) were evaluated for elastin production, $mRNA_E$ level, half-life of $mRNA_E$ after arrest of RNA synthesis with 5,6-dichlorobenzimidazole (60 μM), and half-time to return to basal mRNA values after TGF-β withdrawal. Cultures were maintained in the presence (+) or absence (−) of 10 ng/ml recombinant human TGF-β2 (a gift of Celltrix). Numbers in parentheses indicate the ratio between stimulated and unstimulated cultures. Data are taken from Zhang et al 1995.

counterpart, is resistant to down-regulation of elastin (and collagen) production by glucocorticoids. Such findings suggest that a common regulatory pathway for the two proteins must be deranged in this benign tumour. This is in contrast to our unpublished findings, which show that collagen and elastin are oppositely regulated by ascorbate, that the effect depends on $pO_2$, and that mRNA levels for elastin are reduced by a mechanism that does not involve decreased transcript stability. Keloid fibroblasts produced more elastin than their normal counterparts, on average. As has been observed with collagen production, over-expression of elastin was confined to the deep zone of the keloid. These observations supported previous biochemical evidence of elastin accumulation in what is often presumed to be a purely collagenous tumour. Lack of glucocorticoid down-regulation of elastin in keloid fibroblasts is similar to the behaviour of fetal skin cells. However, fetal skin fibroblasts actually express very low levels of elastin transcript (Sephel et al 1987), and glucocorticoid down-regulation appears to accompany differentiation of elastin expression in the skin at 17–19 weeks of gestation.

Cutis laxa, in its most severe, autosomal recessive form, is characterized by low production of elastin and low levels of $mRNA_E$ (Olsen et al 1988, Sephel et al 1989). Fibroblast populations analysed in this laboratory illustrated greater heterogeneity in terms of elastin production and fibre organization, leading us to the conclusion that multiple abnormalities of elastic fibres can give rise to the cutis laxa phenotype (Sephel et al 1989). We have investigated the mechanisms for low production in two cutis laxa fibroblast strains (KT and KL) that expressed little or no elastin transcript. Surprisingly, TGF-β was a potent stimulus for elastin production in these 'non-producers'. Elastin production in both strains rose from virtually undetectable levels to 25–75% of normal levels, representing more than a 50-fold increase over baseline (Zhang et al 1995). We also established that cutis laxa strains had virtually normal rates of elastin gene transcription through nuclear run-on assays and that TGF-$\beta$ had no measurable effect on transcription in cutis laxa or normal fibroblasts. Expression stimulated by TGF-$\beta$2 was largely reversed by bFGF (100 ng/ml). *In situ* hybridization analysis of cells from the KT strain showed a clear, dose-dependent relationship of accumulation of $mRNA_E$ to cytokine level, and this was further established by quantitative Northern hybridization. $mRNA_E$ and cell-free translation products of mRNA from TGF-$\beta$-induced cutis laxa cells appeared to be full-length. Since our own evidence and that of others indicated TGF-$\beta$ could be expected to act largely at the level of transcript stability, we performed several mRNA half-life studies. Elastin transcripts were 20–30 times less stable in cutis laxa than in normal strains, and TGF-$\beta$ treatment increased stability more than 15-fold (Table 1). COL1A2 mRNA stability was equivalent in normal and cutis laxa strains. Although $mRNA_E$ was strongly induced in KT cells, TGF-$\beta$ withdrawal led to an extremely rapid decay that was partly overcome by cycloheximide. Thus, the mechanism for elastin deficiency in these cutis laxa strains appears to be a rate of $mRNA_E$ degradation that vastly exceeds its rate of transcription. We are currently evaluating the molecular basis for instability of $mRNA_E$, whether it be a defect in the transcript, its translation or RNA catabolism.

Not all forms of cutis laxa show a genetic basis. The so-called 'acquisita' form of the disease is associated with many types of inflammation as well as malignancy. Recent observations on one such case remarkably implicate the over-expression of the serine proteinase, cathespin G, in fibroblasts (Fornieri et al 1994). Elastin production by these cells was normal, whereas elastin (desmosine) content of the skin was markedly reduced. Diagnosis and therapy of cutis laxa acquisita can be assisted by use of an assay we developed for elastin degradation products (EDPs) present in plasma and urine in many forms of cutis laxa, including ones brought on by multiple myeloma (McCarty et al 1995) and localized cutaneous inflammation. EDP assays have also proven useful for studying the effect of anti-elastases on septic shock (Gossage et al 1993), elastin turnover during emphysema and pulmonary hypertension (Davidson et al 1993b), neutrophil activation by dialysis membranes and the progression of bronchopulmonary dysplasia in neonates. EDP levels correlate with desmosine levels, at least in the case of a hypertension model.

Recent studies on Hutchinson–Gilford progeria have shown that skin fibroblasts, which express very high levels of elastin and type IV collagen transcripts (Sephel et al 1988, Colige et al 1991), are unresponsive to further TGF-$\beta$ stimulation (Giro & Davidson 1993). Another aspect of abnormal regulation of elastin that we have explored has been the concept that over-production may result from autocrine expression of stimulatory cytokines. In Hutchinson–Gilford progeria, but not Buschke–Ollendorff syndrome (Giro et al 1992), conditioned media appeared to contain elastogenic activity (Giro & Davidson 1993). Studies are underway to determine whether or not this activity can be neutralized by anti-TGF-$\beta$ or anti-IGF-1 antibodies.

*Relationship between structure and function in the elastin gene*

It remains firmly established that elastin synthesis is closely coupled to the steady-state level of its mRNA (Davidson & Giro 1986, Davidson & Sephel 1987, Parks et al 1993), although elastin accumulation may not necessarily be directly coupled to elastin synthesis (Johnson et al 1993), leading to the concept that there may be situations, usually sequelae to inflammation, in which elastin turnover is remarkably high (Davidson 1990, Davidson et al 1993b). Many investigators have sought to understand how different *cis*-acting elements within the elastin gene control its expression (Kähäri et al 1990, 1992b, Rich et al 1993, Rosenbloom et al 1993, Marigo et al 1994). Ongoing studies by others led us to believe that the elastin promoter, at least as currently defined, may not be a dominant regulatory element in all circumstances, and post-transcriptional control seemed to be an important area to consider for regulation.

There is considerable evidence for post-transcriptional regulation of elastin expression in cells and tissues. Modulators such as TGF-$\beta$, 1,25-dihydroxyvitamin $D_3$ and phorbol esters can down-regulate production of elastin by decreasing the stability of elastin transcripts (Kähäri et al 1992a, Parks et al 1992, 1993, Pierce et al 1992). In many other biosynthetic systems, increased or decreased transcript stability arises as a result of interaction of *cis* elements in transcripts and *trans*-acting factors (Belasco & Brawerman 1993). Interestingly, phorbol esters have been shown to alter (stabilize) TGF-$\beta$ mRNA through a *trans*-acting factor (Wager & Assoian 1990). Although the cutis laxa fibroblasts we have studied offer an opportunity to analyse certain mechanisms of RNA stabilization, there is as yet no evidence that the post-transcriptional regulatory systems will be precisely the same in normal and mutant elastin-producing cells.

The 3′ untranslated region (UTR) of the *ELN* gene is a highly conserved, 1.3 kb element encoded in the last exon, together with a small, unique, hydrophilic region of the C-terminal of tropoelastin of most mammalian transcripts (Indik et al 1990, Parks et al 1993, Rosenbloom et al 1993). By analogy with several other mRNAs, the 3′ UTR represents a good candidate region of the molecule for regulation of RNA stability. We have subcloned this region, including some sequences beyond the transcription termination sites, from the human elastin gene as a *Hin*dIII-*Pst*I fragment into a pBR322 derivative and thence into a pGEM-4Z vector. This entire region or

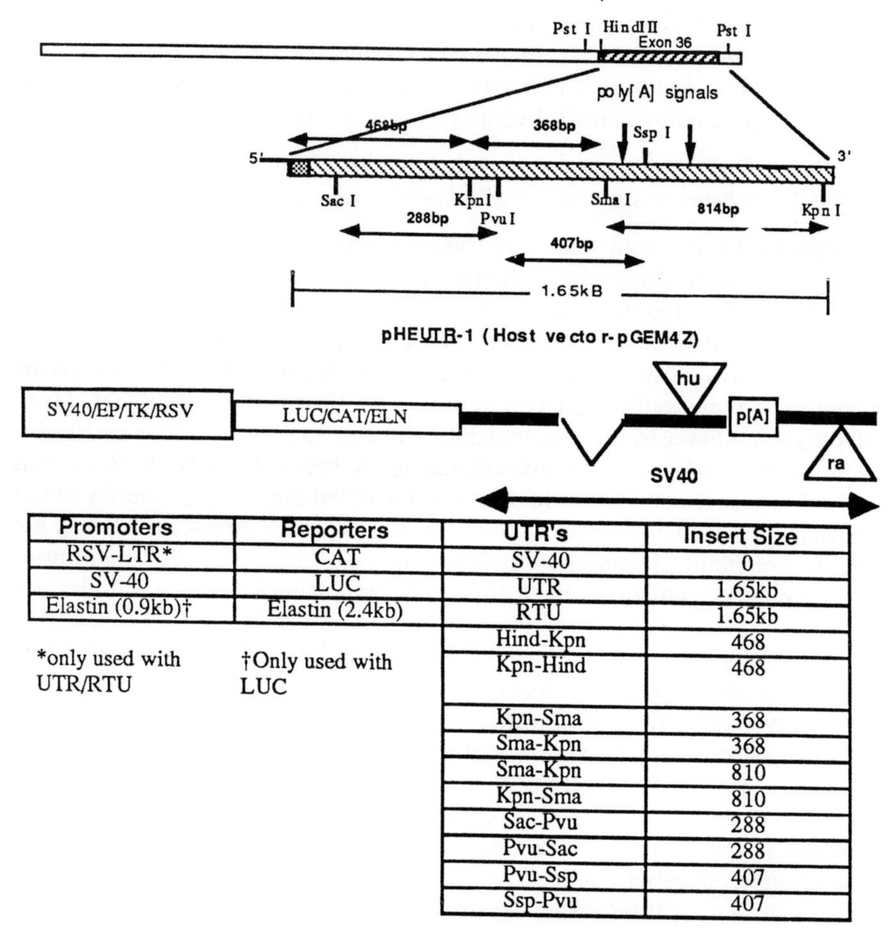


| Promoters | Reporters | UTR's | Insert Size |
|---|---|---|---|
| RSV-LTR* | CAT | SV-40 | 0 |
| SV-40 | LUC | UTR | 1.65kb |
| Elastin (0.9kb)† | Elastin (2.4kb) | RTU | 1.65kb |
| | | Hind-Kpn | 468 |
| | | Kpn-Hind | 468 |
| | | Kpn-Sma | 368 |
| | | Sma-Kpn | 368 |
| | | Sma-Kpn | 810 |
| | | Kpn-Sma | 810 |
| | | Sac-Pvu | 288 |
| | | Pvu-Sac | 288 |
| | | Pvu-Ssp | 407 |
| | | Ssp-Pvu | 407 |

*only used with UTR/RTU

†Only used with LUC

FIG. 3. Construction of reporter plasmids for RNA stability effects. The upper part of the figure illustrates the 3′ region of the human elastin gene containing exon 36 (line 1, hatched area), which encodes the C-terminus of the protein and the entire 3′ untranslated region (UTR). This region was excised from a genomic clone as a *Hin*dIII-*Pst*I fragment, pHEUTR-1. In the enlargement of the exon (line 2), the two vertical arrows indicate the two potential polyadenylation sites within $mRNA_E$. The double-headed arrows indicate various subregions of the 3′ UTR that have been excised and ligated into the *Hpa*I site of an SV40-based expression vector (line 3) at the site indicated by hu, just upstream of the SV40 polyadenylation signal (p[A]). This portion of the figure also indicates the relative positions of the various promoter and reporter regions. Thymidine kinase constructs (generated by Drs Pierce and Parks at Washington University in St Louis) have used a downstream site designated ra in this figure. The table beneath indicates the various fragment sizes in bp. Each of these fragments has been inserted in both orientations, using an intermediate transcription vector to produce RNA products for gel shift analysis.

subfragments therefrom have been incorporated by blunt-end ligation in both orientations, into the *Hpa*I site within the SV40 polyadenylation/splicing region of chloramphenicol acetyltransferase (CAT) and luciferase (LUC) reporter plasmids. Figure 3 indicates the various combinations of promoters, reporter segments and UTR subregions that have been constructed. In addition to using a strong viral promoter, we have modified the LUC plasmids by substituting the human elastin promoter (EP, 0.9 kb) (Marigo et al 1993) for a viral promoter in order to make LUC expression more sensitive to changes in transcript stability. This EP-LUC-UTR/RTU construct shows much higher expression than an irrelevant DNA insert, and LUC expression is upregulated by TGF-$\beta$2 in both 3T3 fibroblasts and aortic SMCs. We have also tested a set of plasmids built by a slightly different strategy: R. Pierce and W. Parks at Washington University in St Louis have developed a plasmid expression vector based on a thymidine kinase promoter, LUC reporter and an SV40 polyadenylation region into which has been introduced a multiple cloning site.

Initial studies have used transient and stable transformation of HeLa cells with SV40-CAT and EP-LUC constructs. Although both promoters were active, the addition of the UTR in either orientation had little effect on CAT activity, and TGF-$\beta$ response was insignificant. On the basis of more recent experiments, we feel that HeLa cells show inappropriate regulation of these constructs. After transient transfection into canine SMCs of Rous sarcoma virus promoter (RSV)-CAT combined with the human UTR, both UTR orientations increased CAT activity, but only the RTU orientation was enhanced by TGF-$\beta$2. Stable transformants of 3T3 cells containing RSV-LUC-UTR/RTU showed positive regulation by TGF-$\beta$, although again somewhat greater with the RTU orientation. Orientation-independent behaviour would support the concept that the regulatory determinant in this region may be in RNA secondary structure. Because the strong, viral promoters may override stability regulation, we have performed transient transfection in porcine SMCs with the EP-LUC-UTR/RTU constructs. These transient assays showed that transcripts from both UTR orientations were more actively expressed than a control sequence, both the UTR and the RTU orientations were responsive to TGF-$\beta$2, and the UTR orientation was more responsive. Half-life experiments are in progress to confirm regulation at the level of mRNA stability. Thus, we are currently exploring in fine detail the properties of the 3′ UTR, believing that mRNA stability may play an important role in regulation of elastin accumulation in health and disease.

*Acknowledgements*

The work described here was supported by National Institutes of Health grants GM37387 and AG06528 and the Department of Veterans Affairs.

**References**

Belasco JG, Brawerman G (eds) 1993 Control of messenger RNA stability. Academic Press, New York, p 517

Berk JL, Franzblau C, Goldstein RH 1991 Recombinant interleukin-1$\beta$ inhibits elastin formation by a neonatal rat lung fibroblast subtype. J Biol Chem 266:3192–3197

Colige A, Nusgens B, Lapiere CM 1991 Altered response of progeria fibroblasts to epidermal growth factor. J Cell Sci 100:649–655

Curren ME, Atkinson DL, Ewart AK, Morris CA, Leppart MF, Keating MT 1993 The elastin gene is disrupted by a translocation associated with supravalvular aorta stenosis. Cell 73:159–168

Davidson JM 1990 Biochemistry and turnover of lung interstitium. Eur Resp J 3: 1048–1068

Davidson JM, Giro MG 1986 Control of elastin synthesis: molecular and cellular aspects. In: Mecham RP (ed) Regulation of matrix accumulation. Academic Press, New York, p 177–216

Davidson JM, Sephel GC 1987 Regulation of elastin synthesis in organ and cell culture. Methods Enzymol 144:214–232

Davidson JM, Quaglino JD, Giro MG 1992 Elastin repair. In: Cohen IK, Diegelmann RE, Lindblad WJ (eds) Wound healing: biochemical and clinical aspects. W. B. Saunders, Philadelphia, PA, p 223–236

Davidson JM, Zoia O, Liu JM 1993a Modulation of transforming growth factor beta-1 stimulated elastin and collagen production and proliferation in porcine vascular smooth muscle cells and skin fibroblasts by basic fibroblast growth factor, transforming growth factor-$\alpha$, and insulin-like growth factor-I. J Cell Physiol 155:149–156

Davidson JM, Perkett EA, Gossage JR, Christman JW, Meyrick B 1993b Elastin metabolism in the vessel wall during pulmonary hypertension. Eur Resp Rev 3:581–590

Fornieri C, Quaglino DJ, Lungarella G et al 1994 Elastin production and degradation in cutis laxa acquisita. J Invest Dermatol 103:583–588

Foster JA, Rich CB, Florini JR 1987 Insulin-like growth factor 1, somatodemin c, induces the synthesis of tropoelastin in aortic tissue. Collagen Relat Res 7:161–169

Giro MG, Davidson JM 1993 Familial co-segregation of the elastin phenotype in skin fibroblasts from Hutchinson–Gilford progeria. Mech Ageing Dev 70:163–176

Giro MG, Duvic M, Smith LT et al 1992 Buschke–Ollendorff syndrome associated with elevated elastin production by affected skin fibroblasts in culture. J Invest Dermatol 99:129–137

Gossage JR, Kuratomi Y, Davidson JM, Lefferts PL, Snapper JR 1993 The effects of elastase inhibitors, SC-37698 and SC-39026, on endotoxin-induced lung dysfuntion in awake sheep. Am Rev Respir Dis 147:1371–1379

Hinek A, Rabinovitch M, Keeley F, Okamuroaho Y, Callahan J 1993 The 67-kD elastin/laminin-binding protein is related to an enzymatically inactive, alternatively spliced $\beta$-galactosidase. J Clin Invest 91:1198–1205

Indik Z, Yeh H, Ornstein-Goldstein N, Rosenbloom J 1990 Structure of the elastin gene and alternative splicing of elastin mRNA. In: Sandell LJ, Boyd CD (eds) Extracellular matrix genes. Academic Press, San Diego, CA, p 221–250

Johnson DJ, LaBourene J, Rabinovitch M, Keeley FW 1993 Relative efficiency of incorporation of newly synthesized elastin and collagen into aorta, pulmonary artery and pulmonary vein of growing pigs. Connect Tissue Res 29:213–221

Kähäri V-M, Fazio MJ, Chen YQ, Bashir MM, Rosenbloom J, Uitto J 1990 Deletion analyses of 5′-flanking region of the human elastin gene: evidence for a functional promoter and regulatory *cis*-elements. J Biol Chem 265:9485–9490
Kähäri V-M, Olsen DR, Rhudy DW, Carrillo P, Chen YQ, Uitto J 1992a Transforming growth factor-β upregulates elastin gene expression in human skin fibroblasts: evidence for post-transcriptional modulation. Lab Invest 66:580–588
Kähäri V-M, Chen YQ, Bashir MM, Rosenbloom J, Uitto J 1992b Tumor necrosis factor-α downregulates human elastin gene expression. Evidence for the role of AP-1 in the suppression of promoter activity. J Biol Chem 267:26134–26141
Liu JM, Davidson JM 1988 The elastogenic effect of recombinant transforming growth factor-beta on porcine aortic smooth muscle cells. Biochem Biophys Res Commun 154:895–901
Marigo V, Volpin D, Bressan GM 1993 Regulation of the human elastin promoter in chick embryo cells—tissue-specific effect of TGF-beta. Biochim Biophys Acta 1172:31–36
Marigo V, Volpin D, Vitale G, Bressan GM 1994 Identification of a TGF-beta responsive element in the human elastin promoter. Biochem Biophys Res Commun 199:1049–1056
Mauviel A, Chen YQ, Kähäri V-M et al 1993 Human recombinant interleukin-1β upregulates elastin gene expression in dermal fibroblasts—evidence for transcriptional regulation *in vitro* and *in vivo*. J Biol Chem 268:6520–6524
McCarty MJ, Davidson JM, Cardone JS, Anderson LL 1995 Cutis laxa acquisita associated with multiple myeloma: a case report of the literature. Arch Dermatol, submitted
Neldner K 1993 Pseudoxanthoma elasticum. In: Royce PM, Steinmann B (eds) Connective tissue and its heritable disorders: molecular, genetic, and medical aspects. Wiley-Liss, New York, p 425–436
Olsen DR, Fazio MJ, Shamban AT, Rosenbloom J, Uitto J 1988 Cutis laxa: reduced elastin gene expression in skin fibroblast cultures as determined by hybridizations with a homologous cDNA and an exon 1-specific oligonucleotide. J Biol Chem 263: 6465–6467
Parks WC, Kolodziej ME, Pierce RA 1992 Phorbol ester-mediated downregulation of tropoelastin expression is controlled by a posttranscriptional mechanism. Biochemistry 31:6639–6645
Parks WC, Pierce RA, Lee KA, Mecham RP 1993 Elastin. Adv Mol Cell Biol 6:133–182
Pasquali-Ronchetti I, Bressan GM, Fornieri C, Baccarani-Contri M, Castellani I, Volpin D 1984 Elastin fiber-associated glycosaminoglycans in beta-aminopropionitrile-induced lathyrism. Exp Mol Pathol 40:235–245
Pierce RA, Kolodziej ME, Parks WC 1992 1,25-dihydroxyvitamin D3 represses tropoelastin expression by a posttranscriptional mechanism. J Biol Chem 267:11593–11599
Quaglino D, Nanney LB, Kennedy R, Davidson JM 1990 Localized effects of transforming growth factor-beta on extracellular matrix gene expression during wound healing. 1. Excisional wound model. Lab Invest 63:307–319
Rich CB, Goud HD, Bashir M, Rosenbloom J, Foster JA 1993 Developmental regulation of aortic elastin gene expression involves disruption of an IGF-I sensitive repressor complex. Biochem Biophys Res Commun 196:1316–1322
Rosenbloom J, Abrams WR, Mecham R 1993 Extracellular matrix. 4. The elastic fiber. FASEB J 7:1208–1218
Ruckman JL, Luvalle PA, Hill KE, Giro MG, Davidson JM 1994 Phenotypic stability and variation in cells of the porcine aorta: collagen and elastin production. Matrix Biol 14:135–145

Russell SB, Trupin JS, Kennedy RZ, Russell JR, Davidson JM 1995 Glucocorticoid regulation of elastin synthesis in human fibroblasts: down-regulation in fibroblasts from normal dermis but not from keloids. J Invest Dermatol 104:241–245

Sephel GC, Buckley A, Davidson JM 1987 Developmental initiation of elastin gene expression in human skin fibroblasts. J Invest Dermatol 88:732–735

Sephel GC, Buckley A, Giro MG, Davdison JM 1988 Increased elastin production by progeria skin fibroblasts is controlled by the steady-state levels of elastin mRNA. J Invest Dermatol 90:643–647

Sephel GC, Byers PH, Holbrook KA, Davidson JM 1989 Heterogeneity of elastin expression in cutis laxa fibroblast strains. J Invest Dermatol 93:147–153

Uitto J, Fazio MJ, Bashir M, Rosenbloom J 1991 Elastic fibers of the connective tissue. In: Goldsmith LA (ed) Physiology, biochemistry, and molecular biology of the skin, 2nd edn. Oxford University Press, New York, p 530–557

Uitto J, Fazio MJ, Christiano AM 1993 Cutis laxa and premature aging syndromes. In: Royce PM, Steinmann B (eds) Connective tissue and its heritable disorders: molecular, genetic, and medical aspects. Wiley-Liss, New York, p 409–423

Wager RE, Assoian RK 1990 A phorbol ester-regulated ribonuclease system controlling transforming growth factor β1 gene expression in hematopoetic cells. Moll Cell Biol 10:5983–5990

Wolfe BL, Rich CB, Goud HD et al 1993 Insulin-like growth factor-I regulates transcription of the elastin gene. J Biol Chem 268:12418–12426

Zhang MC, Giro MG, Quaglino D Jr, Davidson JM 1995 TGF-β reverses a posttranscriptional defect of elastin synthesis in a cutis laxa skin fibroblast strain. J Clin Invest 95:986–994

## DISCUSSION

*Uitto:* You highlighted TGF-β primarily as a stabilizer of the mRNA. However, there clearly are TGF-β-responsive elements within the elastin promoter region. We have recent data from transgenic mice which express 5.2 kb of the human elastin promoter, linked to the CAT reporter gene in a tissue-specific and developmentally regulated manner (Hsu-Wong et al 1994). If we inject as little as 100 ng of TGF-β1 subcutaneously to these mice, the CAT activity goes up about 10-fold (Katchman et al 1994).

*Davidson:* I am aware of the two examples of a TGF-β-responsive element in portions of an elastin promoter linked to a reporter gene. However, I don't know what the situation is *in vivo* for the activity of the endogenous elastin gene under the same circumstances. The data I showed were all from cells in culture; in these conditions using our transcription assay we couldn't detect a difference. I'm aware of your data, and G. Bressan has also reported that there is a TGF-β-responsive element present when one isolates the promoter from the other surrounding elements (Marigo et al 1994).

*Keeley:* We have some recent unpublished evidence that indicates that elastin mRNA stability is important in development. Elastin mRNA from day-old chick aorta has a half-life of about 25–30 h. However, from a six-week-old chick the half-life has fallen to about 6 h. Therefore, it seems that destabilization of elastin mRNA may make an important contribution to the decline in elastin production with age. We also find that the half-life of elastin mRNA in second-passage smooth muscle cells cultured from one-day-old chick aorta has decreased to less than 10 h. This raises the possibility that in cell culture studies we are actually working with a system in which the elastin message is chronically destabilized.

*Parks:* Some recent results of ours do indeed demonstrate that mRNA stability contributes to the regulation of tropoelastin expression during development (Swee et al 1995). The onset of elastin expression and the progressive increase that occurs in fetal and neonatal development are controlled at the level of gene transcription. We determined this *in vivo* by assessing the expression of tropoelastin pre-mRNA as an indicator of ongoing transcription. Interestingly, and quite surprisingly, the level of tropoelastin transcription is not reduced in adult tissues, even though the steady state mRNA levels drop more than 70-fold. It seems to be a very peculiarly regulated gene.

*Keeley:* We became interested in message stability in the first place because we observed that over a 24 h incubation period in organ culture, the ability of aortic tissue to make elastin falls several-fold, and there is a corresponding decrease in the steady-state mRNA levels for elastin. This is selective for elastin, since these aortic tissues continue to produce collagen perfectly well, and mRNA levels for collagen do not fall. Over the same incubation period, mRNA levels for glyceraldehyde phosphate dehydrogenase increase several-fold, which might suggest some problem with oxygenation of the tissue. However, there is certainly a high enough level of oxygen to ensure hydroxylation of collagen, so it is not clear to us whether there is a connection between the fall in elastin mRNA and the rising glyceraldehyde phosphate dehydrogenase mRNA.

You showed that glucocorticoids are stimulators of elastin production. This certainly seems to be true in the fetus. However, some time ago we showed that although glucocorticoids stimulate elastin production in aortic tissues of chicken embryos, similar doses of these agents in 14-day-old chickens suppress aortic elastin synthesis (Keeley & Johnson 1987). Therefore, the effects of glucocorticoids are particularly interesting because they appear to be developmentally modulated.

*Davidson:* We have followed developmental regulation of elastin in human skin fibroblast cultures. In fetal human skin, glucocorticoid treatment causes slight up-regulation or no change in elastin production until 19 weeks, after which time elastin expression is suppressed (Russell et al 1995). Conversely, if we look at continuously or repeatedly cultured vascular smooth muscle cells, or fibroblasts in the vocal fold, elastin production is stimulated by glucocorticoid (Broadley et al 1994). So there's a real heterogeneity of glucocorticoid response in development and in different tissues, and we don't know at what level that's operating.

*Starcher:* I was very interested in your high-oxygen tissue culture studies. Bob Rucker's group published a paper on the lack of effects of high ascorbic acid *in vivo* on elastin synthesis (Critchfield et al 1985). We have also looked at whether this could cause emphysema in neonatal mice pups, since several studies have shown that application of ascorbic acid to cells in culture lowers elastin synthesis (Faris et al 1984, Bergethon et al 1989). We fed a 10% vitamin C diet to pregnant mice and found no evidence of problems in elastin synthesis in the lungs of the mouse pups. Perhaps what we were seeing is that in circumstances of normal (*in vivo*) oxygen tension you won't find an effect on elastin synthesis. It is an abnormal situation when you look at cells in culture because of the high oxygen levels.

*Mungai:* Apart from the genetic diseases that you have shown us, would you be in a position to explain the increased synthesis of elastin in repairing bone? When a bone is fractured, the temporary bone which rivets the two fractured ends, callus, has a scaffold of elastin fibres (Murakami & Emery 1966, 1967). Would that fall under this kind of explanation?

*Davidson:* I wasn't aware that there was an elastic scaffold in callus. There are two sides to elastin and tissue repair. In vascular and pulmonary disease there can be extremely rapid and intense reactivation of elastogenesis and accumulation of elastic fibres, and so there's a nice architecture deposited very quickly. In contrast, in cutaneous injury, elastin is the last thing to appear after all the other architecture is established. It's a very tissue-dependent system. I've mentioned our ascorbate/oxygen observations, because I'm beginning to think that the oxidative environment may have much to do with how efficiently the elastin is deposited under circumstances of inflammation and repair. This is not only through biosynthetic mechanisms, but perhaps also for reasons that inflammation may alter the structure of the protein through other kinds of modifications.

*Kimani:* You touched very briefly on ascorbate and how it down-regulates production of elastin. But let's look at oxygen tension in a more general way as it relates to altitude and hypoxia. I have seen a study suggesting that elastin biosynthesis in the pulmonary vasculature is up-regulated in animals and humans living at high altitude (Walker et al 1984). Would you comment on this?

*Parks:* Stated simply, elastin production is stimulated in response to high altitude. This increase is confined to the pulmonary vasculature where, after all, haemodynamic changes occur in response to altitude. Interestingly, the response to hypoxia differs with age. In persistent pulmonary hypertension in the neonate, increased elastin deposition is confined to areas of pre-existing elastin, namely the media. In the adult, however, primary pulmonary hypertension is characterized by formation of a neointimal lesion that contains a fair bit of elastin. We don't know how these age-specific responses are regulated. However, working with Kurt Stenmark at the

University of Colorado, we have recently shown that the regulatory defect in neonatal pulmonary hypertension is a failure to shut off tropoelastin from its high levels of expression in the newborn, rather than a stimulation in response to hypoxia (Stenmark et al 1994).

*Davidson:* I brought up the oxygen/ascorbate interaction in my paper because there is evidence published that ascorbate modulates collagen production through lipid peroxide intermediates (Chojkier et al 1989, Geesin et al 1991). Because we have had data that production of the protein falls and production of message falls under stimulation with ascorbate for some time, we wanted to try and dissect that mechanism out more accurately. It's still not a very clear system, because ascorbate can also act to modify the primary structure of the protein through prolyl hydroxylation.

*Robert:* It's a pity we couldn't get any information on Linus Pauling's aorta; for decades he consumed huge amounts of ascorbic acid every day.

I remember from my early years when I worked with ascorbic acid, that as soon as you dissolve it, it is immediately oxidized. Did you use deoxy ascorbate as well as ascorbate in the first experiment you just cited? Is the redox state of ascorbic acid of importance in the experiments you described?

*Davidson:* We've done experiments with oxidized ascorbate and dehydroascorbic acid; they have no effects. Because ascorbate is rapidly oxidized, the way we have to treat cell culture systems to get reproducible effects is to add it to cultures 2–3 times a day. Apparently, if you do not maintain a steady level of reduced ascorbate you do not see consistent effects.

There's also a dose-dependent time delay for the ascorbate effect. That is, you can see repression of elastin production early with high doses of ascorbate, whereas it requires many days of exposure to see inhibition at low ascorbate levels.

*Robert:* So it is not a question of the reduced or oxidized form?

*Davidson:* No, I think it's the reduced form of the molecule that initially has the activity. The problem is that ascorbate can also act as an oxidizing agent in the presence of high levels of $O_2$, so we're not certain that it's acting in the way we would like it to act under artificial culture conditions where the media are equilibrated with room air.

*Kagan:* As you know, lysyl oxidase is a copper-dependent enzyme, and you might expect that ascorbate could shift the balance of $Cu^{2+}$ to $Cu^{+}$. The functional form of copper in lysyl oxidase is $Cu^{2+}$. Perhaps excess ascorbate might influence the cross-linking process by affecting lysyl oxidase in this model.

*Starcher:* If that's true wouldn't you expect, with the high levels of ascorbic acid that Bob Rucker gave his rats, that he would almost be giving a lathyritic diet?

*Kagan:* But Dr Davidson has coupled high concentrations of ascorbate with low oxygen in this model, which overall could reduce the oxidation potential of the system to the degree that the $Cu^{2+} : Cu^{+}$ ratio is low.

*Davidson:* The assay system we use does not assess elastin fibrillogenesis. Very little cross-linking occurs in either the skin fibroblast cultures or in the smooth muscle cells that we use in short-term experiments. There may well be effects on the copper-dependent cross-linking system.

*Kagan:* Among all the transcription regulatory sites that occur in the elastin promoter, is there a metalloregulatory element? This is another possible connection.

*Davidson:* I'm not aware of any metal response elements.

*Boyd:* Certainly in rats there aren't any.

*Pasquali-Ronchetti:* I don't know whether this will confuse the issue further, but several years ago we did experiments giving a diet with 10% ascorbic acid to chickens (Quaglino et al 1991). After two months we looked at the aorta and measured the lysyl oxidase activity of the tissue; this was lower than normal. The distribution of the elastin fibres in the tissue was different from normal.

## References

Bergethon PR, Mogayzel PJ, Franzblau C 1989 Effect of the reducing environment on the accumulation of elastin and collagen in cultured smooth muscle cells. Biochem J 258:279–284

Broadley C, Gonzalez DA, Nair R, Koriwchak M, Ossoff RH, Davidson JM 1994 A tissue culture model for the study of canine vocal fold fibroblasts. Laryngoscope 104:1–5

Chojkier M, Houglum K, Solis HJ, Brenner DA 1989 Stimulation of collagen gene expression by ascorbic acid in cultured human fibroblasts. A role for lipid peroxidation? J Biol Chem 264:16957–16962

Critchfield JW, Dubick M, Last J, Cross C, Rucker RB 1985 Changes in response to ascorbic acid administered orally to rat pups: lung collagen, elastin and protein synthesis. J Nutr 115:70–78

Faris B, Salcedo LL, Cook V, Johnson L, Foster JA, Franzblau C 1984 Effect of varying amounts of ascorbate on collagen, elastin and lysyl oxidase synthesis in aortic smooth muscle cell cultures. Biochim Biophys Acta 797:71–75

Geesin J, Hendricks L, Falkenstein P, Gordon J, Berg R 1991 Regulation of collagen synthesis by ascorbic acid: characterization of the role of ascorbate-stimulated lipid peroxidation. Arch Biochem Biophys 290:127–132

Hsu-Wong S, Katchman S, Ledo I et al 1994 Tissue-specific and developmentally regulated expression of human elastin promoter activity in transgenic mice. J Biol Chem 269:18072–18075

Katchman SD, Hsu-Wong S, Ledo I, Wu M, Uitto J 1994 Transforming growth factor-$\beta$ up-regulates human elastin promoter activity in transgenic mice. Biochem Biophys Res Commun 203:485–490

Keeley FW, Johnson DJ 1987 Age differences in the effect of hydrocortisone on the synthesis of insoluble elastin in aortic tissue of growing chicks. Connect Tissue Res 16:259–268

Marigo V, Volpin D, Vitale G, Bressan GM 1994 Identification of a TGF-$\beta$ responsive element in the human elastin promoter. Biochem Biophys Res Commun 199: 1049–1056

Murakami H, Emery MA 1966 Elastic fibers in the periosteum in fracture healing. Surg Forum 17:452–454

Murakami H, Emery MA 1967 The role of elastic fibers in the periosteum in fracture healing in guinea pigs. I. Histological studies of the elastic fibers in the periosteum and the possible relationship between the osteogenic cells and the cells that form elastic fibers. Can J Surg 10:359–370

Quaglino D, Fornieri C, Botti B, Davidson JM, Pasquali-Ronchetti I 1991 The opposing effects of ascorbate on collagen and elastin deposition in the neonatal rat aorta. Eur J Cell Biol 54:18–26

Russell SB, Trupin JS, Kennedy RZ, Russell JR, Davidson JM 1995 Glucocorticoid regulation of elastin synthesis in human fibroblasts: down-regulation in fibroblasts form normal dermis but not from keloids. J Invest Dermatol 104:241–245

Stenmark KR, Durmowicz AG, Roby JD, Mecham RP, Parks WC 1994 Persistence of the fetal pattern of tropoelastin expression in neonatal pulmonary hypertension. J Clin Invest 93:1234–1242

Swee M, Parks WC, Pierce RA 1995 Developmental regulation of elastin production. Expression of tropoelastin pre-mRNA persists after downregulation of steady-state mRNA levels. J Biol Chem, in press

Walker BR, Berend N, Voelkel N 1984 Comparison of muscular pulmonary arteries in low and high altitude hamsters and rats. Respir Physiol 56:45–50

# Catalytic properties and structural components of lysyl oxidase

H. M. Kagan*, V. B. Reddy*, N. Narasimhan* and K. Csiszar†

**Department of Biochemistry, Boston University School of Medicine, Boston, MA 02118 and †Department of Surgery, UMDNJ-Robert Wood Johnson Medical School, New Brunswick, NJ 08903, USA*

*Abstract.* Key aspects of the biosynthesis and catalytic specificity of lysyl oxidase (LO) have been explored. Oxidation of peptidyl lysine in synthetic oligopeptides is markedly sensitive to the presence of vicinal dicarboxylic ami\no acid residues. Optimal activity is obtained with the -Glu-Lys- sequence within a polyglycine 11-mer, whereas the -Lys-Glu- sequence is much less efficiently oxidized. The -Asp-Glu-Lys- sequence is a very poor substrate, although this sequence is oxidized in type I collagen fibrils. These results are considered in the light of a model requiring collagen to be assembled as fibrils prior to oxidation by LO. An *in vitro* system for the expression of catalytically active LO has been devised. Deletion or inclusion of the cDNA coding for the propeptide region in the expressed construct results in apparently identical, catalytically active enzyme products, indicating the lack of essentiality of this region for active enzyme production. These effects are considered with respect to the conservation of the amino acid sequence of LO produced by different species.

*1995 The molecular biology and pathology of elastic tissues. Wiley, Chichester (Ciba Foundation Symposium 192) p 100–121*

Lysyl oxidase (LO) is the catalyst found in connective tissues that oxidatively converts peptidyl lysine in collagen and elastin substrates to peptidyl $\alpha$-aminoadipic-$\delta$-semialdehyde. This chemical change is critical to the formation of mechanically sound, properly functioning connective tissues, since the peptidyl aldehyde spontaneously undergoes condensations with other peptidyl aldehydes and amino groups to give rise to the variety of inter- and intramolecular covalent cross-linkages which stabilize and insolubilize fibres of collagen and elastin (Kagan 1986).

The enzyme purified from bovine aorta functions by an aminotransferase-type mechanism involving a substituted (aminated) enzyme intermediate (Williamson & Kagan 1986) and requires a tightly bound copper cofactor (Gacheru et al 1990) as well as covalently bound carbonyl cofactor (Williamson et al 1986), the latter likely to be a quinone moiety. In this paper, we describe a possible basis for the unique specificity of LO for selected lysines in collagen and offer a rationale for its ability to utilize both collagen and elastin substrates. In addition, recent advances in the understanding of its pathway of biosynthesis, explored in part with a newly developed mammalian cell expression system, will be described. Progress in the analysis of the expression and structure of the lysyl oxidase gene will also be considered.

## Methods

Purified LO was prepared from bovine aorta (Williams & Kagan 1985). Peptides were synthesized by solid-phase methodology, purified and characterized as described (Nagan & Kagan 1994). Lysyl oxidase activity against simple organic amines and synthetic peptides was assayed by a continuous fluorescence method (Trackman et al 1981) and against recombinant human tropoelastin prepared and utilized as described (Bedell-Hogan et al 1993).

## Results and discussion

### *Specificity of lysyl oxidase for sequences vicinal to peptidyl lysine*

The ability of LO to utilize both elastin and collagen substrates has long been an enigma, given the remarkably diverse natures of these two proteins. Thus, oxidized lysines in elastin commonly appear in $\mathrm{Ala_m}$-Lys-$\mathrm{Ala_p}$ or $\mathrm{Ala_m}$-Lys-$\mathrm{Ala_n}$-Lys-$\mathrm{Ala_p}$ sequences, where $m$ and $p$ are variable and $n$ is 2 or 3. In contrast, the susceptible sites of oxidation in types I, II and III collagens are restricted to the single lysines in the N- and C-telopeptide, non-triple-helical sequences. In the N-terminal telopeptides, the apparent consensus sequence is -X-Asp-J-Lys-Z where X is Tyr or Phe, J is variable [although most commonly J is Glu in the $\alpha$ 1(I) and $\alpha$ 1(II) chains] and Z is Ser or Gly. Similarly, the -Glu-Lys-Z sequence, where Z is Ala, Ser or Gly, is common to the $\alpha$ 1 chains of types I and III collagens. The apparent proximity of the substrate lysines in these collagens to anionic residues contrasts with the deficiency of anionic residues near susceptible lysines in elastin, and with the observations that oxidation is highly favoured by the net cationic charge and strongly inhibited by anionic charge in elastin (Kagan et al 1981) and globular proteins which can be oxidized by LO *in vitro* (Kagan et al 1984). To further explore these effects, we have synthesized lysine-containing oligopeptides and characterized the kinetic constants for their oxidation by LO.

Lysine and other polar residues were incorporated within polyglycine backbones, whose N- and C-terminal glycines were acetylated and amidated, respectively. Initial results revealed that the $k_{cat}/K_m$ parameter increased with increasing chain length in Ac-$\mathrm{Gly_m}$-Lys-$\mathrm{Gly_m}$-$\mathrm{NH_n}$ ($m$ and $n$=1 to 5), consistent with an extended substrate binding site. Each oligopeptide subsequently analysed was 11 residues in length, unless otherwise indicated, with the single lysine residue as centrally contained within the sequence as possible.

*Effect of vicinal glutamate on peptidyl lysine oxidation.* As shown in Fig. 1A, Glu immediately N-terminal ($-1$ position) to Lys decreases the $K_m$ and marginally increases the $k_{cat}$, thus markedly increasing the overall catalytic efficiency, as indexed by the $k_{cat}/K_m$ parameter (Fig. 1B). The decrease in $K_m$ is also seen with a Glu at the $-2$ position, but then increases significantly above that of the control with Glu at the $-3$ position (Fig. 1A), whereas the $k_{cat}/K_m$ decreases dramatically with the placement of Glu at the $-2$ or $-3$ positions (Fig. 1B). In contrast to the favourable effects seen with the -EK- sequence, Glu immediately C-terminal to Lys ($+1$ position; -KE-) dramatically increases the $K_m$ while increasing the $k_{cat}$ to a lesser degree. Notably, the $K_m$ for -KE- exceeded that for -EK- 24-fold while the $k_{cat}/K_m$ is reduced from the control lacking a Glu residue (Fig. 1B). The $+1$ position appears uniquely sensitive to a Glu residue, since $K_m$ and $k_{cat}/K_m$ return approximately to control values with Glu at the $+2$ or $+3$ positions (Fig. 1A,B).

*Effect of vicinal aspartate on peptidyl lysine oxidation.* Introduction of an Asp immediately N-terminal to Lys decreases $K_m$ to a degree similar to that seen with Glu at this position (Fig. 2A). The $k_{cat}$ is decreased 6.5-fold with the -DK- sequence, however, when compared with the control. Thus, the marked increase in $k_{cat}/K_m$ seen with the - EK- sequence (Fig. 1B) does not occur with the -DK- sequence (Fig. 2B). The changes in kinetic parameters seen by introduction of Asp at selected positions in the Ac-Gly$_m$-Lys-Gly$_n$-$NH_2$ control 11-mer overall are considerably less than those seen with substitution with Glu (compare Figs 1 and 2). Together, these results reveal that LO exhibits clear sequence preferences for single, vicinal anionic residues, with the -EK- sequence acting as the most favourable and the -KE- sequence as the least favourable arrangement. Moreover, the efficiency of oxidation is clearly sensitive to the one-carbon difference in side chain length between aspartate and glutamate residues when these are immediately N- or C-terminal to lysine.

*Effect of collagen-like sequences on peptidyl lysine oxidation.* As noted, lysine becomes oxidized in the -YDEKS- sequence of the N-terminal telopeptide of the $\alpha 1$(I) chain which contains two potentially anionic residues. Remarkably, the 11-mer containing the -DEK- sequence (Table 1, peptide 2) was not oxidized, although the -DGK- or -GEK- sequences are productive substrates for LO (Figs 1 and 2). Similarly, a 12-mer sequence closely matching that of the N-terminal telopeptide of the $\alpha 1$(I) chain was a productive although very inefficient substrate for LO (Table 1, peptide 3),

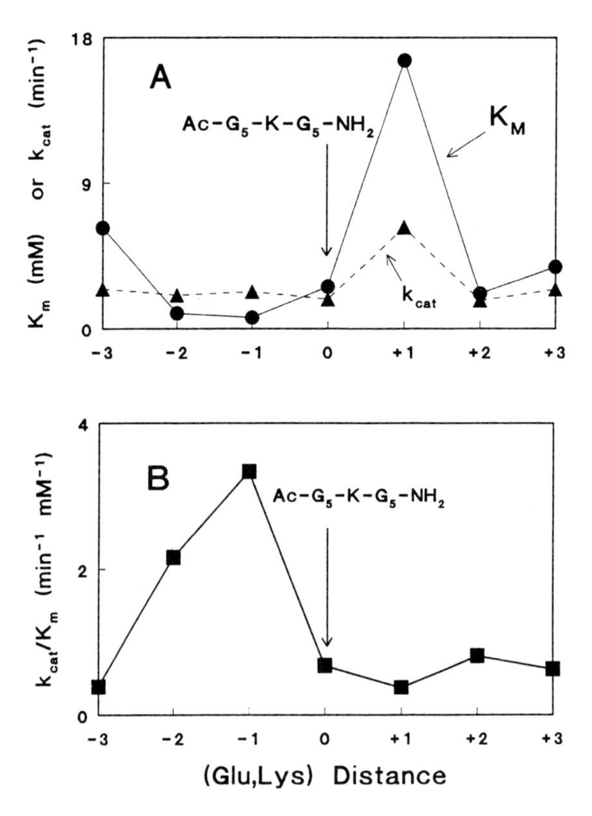


FIG. 1. Effect of glutamic acid on the oxidation of peptidyl lysine. Residue distances indicated on the $x$-axis: 0, no glutamic acid; (+) and (−) values, residues C-terminal or N-terminal to peptidyl lysine, respectively. (A) Variation of $K_m$ (solid line, circles) or $k_{cat}$ (dashed line, triangles) with glutamate position. (B) Variation of $k_{cat}/K_m$ with glutamate position.

with a $k_{cat}/K_m$ 5.6-fold less than that of the control 11-mer (Table 1, peptide 1) and 28-fold less than that of the favourable -EK- sequence. Replacing Glu in the -DEK- sequence with a glutamine residue (-DQK-, Table 1, peptide 4), results in a large increase in $K_m$ accompanied by a large increase in $k_{cat}$, with the resulting $k_{cat}/K_m$ indicating that this is a reasonably efficient substrate, relative to the control 11-mer. Replacing the glutamate in the -DEK- sequence with a proline (Table 1, peptide 5) markedly lowers the $K_m$ while increasing the $k_{cat}$, thus increasing the catalytic efficiency of oxidation relative to the control 11-mer. Similarly, replacing the aspartate in the -DEK- with a glutamine (-QEK-, Table 1, peptide 6) restores favourable substrate properties to the 11-mer oligopeptide. These results reveal that the kinetic effects of a favourably positioned glutamate (-EK-) or aspartate (-DGK-) residue are not additive, since

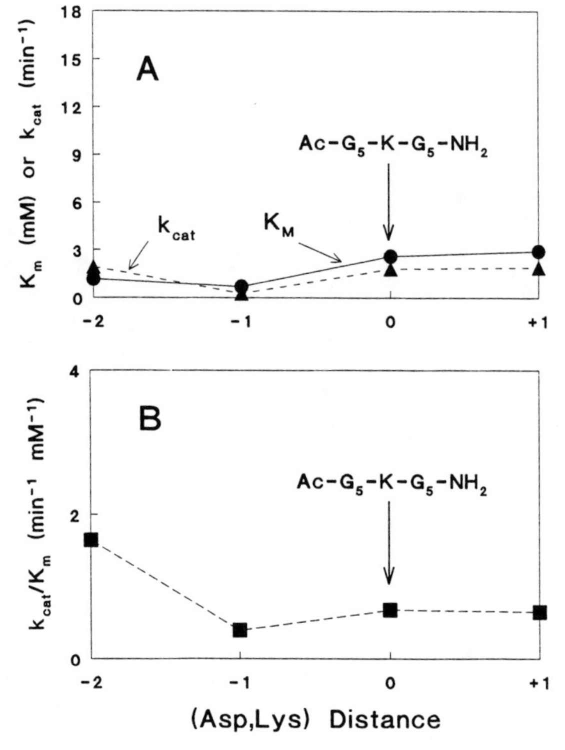


FIG. 2. Effect of aspartic acid on the oxidation of peptidyl lysine. Residue distance indicated on the $x$-axis: 0, no aspartic acid; (+) and (−) values, residues C-terminal or N-terminal to peptidyl lysine, respectively. (A) Variation of $K_m$ (solid line, circles) or $k_{cat}$ (dashed line, triangles) with aspartate position. (B) Variation of $k_{cat}/K_m$ with aspartate position.

**TABLE 1 Effect of collagen-like sequences on peptidyl lysine oxidation**

| *Peptide number* | *Sequence* | $K_m$ *(mM)* | $k_{cat}$ *(min*$^{-1}$*)* | $k_{cat}/K_m$ *(min*$^{-1}$ *mM*$^{-1}$*)* |
|---|---|---|---|---|
| 1 | $G_5$-K-$G_5$ | 2.61 | 1.82 | 0.68 |
| 2 | $G_4$-DEK-$G_4$[a] | — | — | — |
| 3 | ELSYGYDEKSTG[b] | 3.64 | 0.44 | 0.12 |
| 4 | $G_3$-DQK-$G_5$ | 20.00 | 22.13 | 1.11 |
| 5 | $G_3$-DPK-$G_5$ | 0.33 | 2.17 | 6.58 |
| 6 | $G_3$-QEK-$G_5$ | 0.83 | 1.15 | 1.39 |

[a]No apparent rate of oxidation.
[b]α1(I) collagen telopeptide sequence with N-terminal *N*-acetylglutamate replacing N-terminal pyroglutamate.

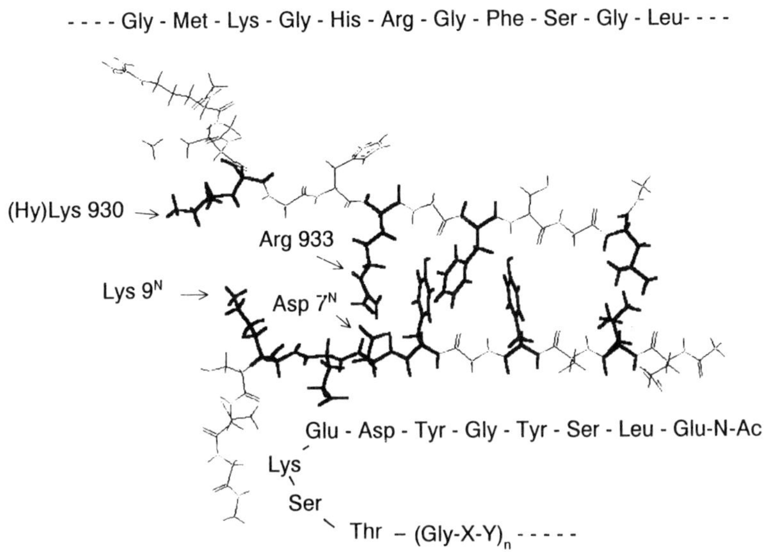

- - - - Gly - Met - Lys - Gly - His - Arg - Gly - Phe - Ser - Gly - Leu- - - -

(Hy)Lys 930

Arg 933

Lys 9N

Asp 7N

Glu - Asp - Tyr - Gly - Tyr - Ser - Leu - Glu-N-Ac

Lys

Ser

Thr – (Gly-X-Y)n - - - - -

FIG. 3. Representation of interactions between N-telopeptide and triple-helical sequences within type I collagen fibrils. The conserved, cationic, triple-helical sequence (residues Gly-928 to Leu-938) shown at the top of the interpeptide complex can interact with the N-telopeptide sequence (*N*-Acetyl-Glu-$1^N$ to Thr-$11^N$), shown at the bottom of the complex. The N-telopeptide, as represented, contains a $\beta$-turn (Helseth et al 1979). Lys-$9^N$ is oxidized by lysyl oxidase. Lys-930 of the triple helix can exist as a hydroxylysine residue.

their combined presence in the -DEK- sequence, as in type I collagen, presents a highly unfavourable substrate to LO. Moreover, the data obtained with the peptides containing glutamine argue that it is the anionic nature of the dicarboxylic amino acids in the -DEK- sequence that suppresses the efficiency of oxidation. Since there is a fourfold increase in $k_{cat}/K_m$ of the -DPK- sequence in comparison with that for -DGK-, the specific orientation of the aspartate two residues from lysine also appears to be critical to the catalytic processing, in view of the different conformation imposed on the 11-mer by the proline residue.

These results were considered in the light of the presence of anionic residues near susceptible lysines in collagen; the fact that the type I collagen N-telopeptide sequence, -DEK-, is a very poor substrate; and in view of the evidence that type I collagen monomers are not oxidized by LO *in vitro* unless they are first assembled into the quarter-staggered fibril arrangement (Siegel 1974). Within such fibrils, a conserved sequence, -Gly-(Met, Ile)-Lys-Gly-His-Arg-Gly-Phe-, found at both the N- and C-terminal regions of the triple helices of one molecule, aligns in close proximity to the N- and C-terminal telopeptide lysines of adjacent molecules. Figure 3 illustrates potential intermolecular interactions between these peptide regions, including hydrophobic bonding between leucines, stacking between aromatic side chains and, of particular interest, an ionic bond between Arg-933 of the triple helix and Asp-$7^N$ of the telopeptide. The Arg-to-Asp interaction would thus neutralize the anionic nature of Asp-$7^N$, thus presenting Lys-$9^N$ in a local environment in which the only full negative charge would stem from Glu in the -Glu-Lys sequence, consistent with the favourable effects seen here with the -GEK- sequence. We predict that this interaction underlies the requirement for native fibril formation prior to Lys-$9^N$ oxidation. It is also of interest that a lysyl oxidase molecule approaching Lys-$9^N$, upon diffusing into the hole regions between linearly adjacent head-to-tail aligned tropocollagen units within native fibrils, would confront Lys-933 of the triple helix as well as Lys-$9^N$, while the side chain of Glu-$8^N$ would be pointing away from that of Lys-$9^N$. Thus, the enzyme would approach a region with a net positive charge, consistent with the preference seen with elastin substrates *in vitro* (Kagan et al 1981). This microenvironment of substrate lysine in type I collagen would then more closely approximate those in which lysine is oxidized in elastin in that cationic lysine residues tend to be grouped in an otherwise ionically deficient environment.

*Expression and processing of lysyl oxidase in Chinese hamster ovary cells*

As demonstrated in arterial smooth muscle cell cultures, LO is synthesized as a 46 kDa preproprotein. After signal peptide cleavage, the resulting proprotein becomes *N*-glycosylated and then is secreted as a 50 kDa species which is proteolytically processed to the active 32 kDa species in the medium by a secreted metalloenzyme (Trackman et al 1992). We developed an expression system to assess the effect of deletion of the bulk of the propeptide region on enzyme production in chinese hamster ovary (CHO) cells which otherwise do not express LO. Among several clones isolated, we characterized two in detail. One of these, LOD-06, was developed from CHO cells which had been transfected with a plasmid containing a 1.6 kb segment of rat LO cDNA coding for the full coding region of the preproprotein. A truncated version of the enzyme expressed in clone LOD32-2 was derived from CHO cells transfected with a plasmid containing a 1.2 kb cDNA construct coding for the signal peptide sequence (residues 1–21) placed in frame and continuous with the sequence coding for residues 134–411 of the 411-residue preproprotein, the latter region representing the mature enzyme portion of the preproenzyme. Thus, LOD32-2 lacked the bulk of the sequence coding for the propeptide region of prolysyl oxidase. The full-length and truncated LO cDNA constructs were incorporated in plasmids pMLD and pMLD32, respectively, each of which contained an SV40 early promoter, mouse dihydrofolate reductase (DHFR) cDNA, SV40 polyadenylation signal, and mouse metallothionein promoter.

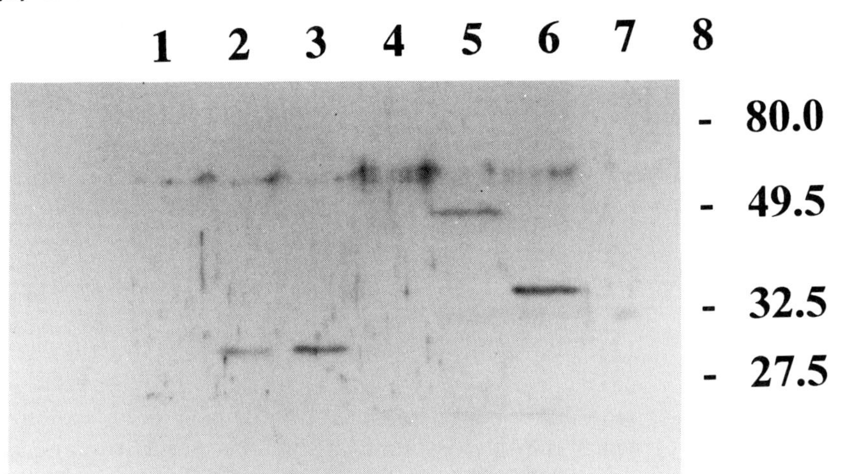


FIG. 4. Western blot analysis of lysyl oxidase expressed in CHO cells. Lanes 1 and 4, medium and cell lysate, respectively, from control CHO cells transfected with plasmid lacking lysyl oxidase cDNA; lanes 2 and 3, media from LOD-06 and LOD32-2 cell lines, respectively; lanes 5 and 6, cell lysates from LOD-06 and LOD32-2, respectively; lane 7, lysyl oxidase purified from bovine aorta; and lane 8, molecular weight markers.

Lysyl oxidase transcription would be driven by metallothionein promoter and polyadenylated by SV40 polyadenylation signal in these expression vectors. These plasmid DNAs were transfected into DHFR-deficient CHO cells, which were then grown and selected in alfa-modified Eagle's medium supplemented with 10% dialysed fetal bovine serum and 0.02 $\mu$M methotrexate (Kagan et al 1995).

*Western blotting.* As shown in Fig. 4, 29 kDa bands appear in the Western blots of the conditioned medium of LOD-06 (lane 2) and LOD32-2 (lane 3). In addition, a band is seen at approximately 50 kDa derived from the cell lysate of LOD-06 (lane 5), whereas a band occurs at 35 kDa derived from the lysate of LOD32-2 (lane 6). Bands corresponding to mature or precursor forms of LO were not seen in control cells transfected with plasmids lacking LO cDNA (lanes 1 and 4). Purified bovine aorta LO migrates as a 32 kDa protein (lane 7). The bands at 50 kDa seen here are consistent with the expected size of *N*-glycosylated prolysyl oxidase (Trackman et al 1992) and those at 29 kDa in the medium are consistent with the mature forms of this enzyme isolated from connective tissues reported to vary between 28 and 32 kDa (Kagan 1986). The calculated molecular weight of the protein derived from the truncated cDNA is 32 533 Da after signal peptide cleavage between residues 21 and 22 (Trackman et al 1990). The observed molecular weight of 35 kDa suggests

**TABLE 2 Activity of recombinant forms of lysyl oxidase against tropoelastin**[a]

| *Enzyme source* | *[$^2H$]$H_2O$ (cpm/2 h)* |
|---|---|
| LOD-06 (760 μl) | 969 ± 64 |
| LOD32-2 (760 μl) | 987 ± 73 |
| Bovine aorta lysyl oxidase (2 μg) | 987 ± 1 |
| CHO cells (760 μl) | 33 ± 29 |

[a]Data are presented as the means ± SD of assays in triplicate.

that this product may have undergone post-translational modification, a probable form of which is *N*-glycosylation at the -Asn-Arg-Thr- consensus sequence persisting at the site equivalent to residues 138–140 in the full rat proenzyme sequence (Trackman et al 1990).

*Catalytic properties.* Aliquots of the conditioned medium of clones LOD-06 and LOD32-2 both catalysed the oxidative deamination of lysine in a recombinant human tropoelastin substrate. In contrast, negligible activity was found in the conditioned medium of CHO cells transfected only with a DHFR gene but not with LO cDNA (Table 2). The specific activities of the two clones were estimated to be similar to each other, on the basis of the relative densities of the mature forms seen in Western blots of the conditioned media and on the relative activities against tropoelastin.

*Inhibition profiles.* The enzyme activity secreted into the conditioned media of each clone was inhibited by agents known to inhibit purified LO. As shown in Table 3, the preparations of expressed enzymes are inhibited irreversibly and to similar degrees by phenylhydrazine (Williamson et al 1986) and ethylenediamine (Gacheru et al 1989), consistent with a functional carbonyl cofactor; by α,α′-dipyridyl and diethyldithiocarbamate, consistent with a copper cofactor (Gacheru et al 1990); and irreversibly by β-aminopropionitrile (BAPN), a mechanism-based, irreversible inhibitor of LO (Tang et al 1983). The sensitivities to BAPN of the enzyme activity of both clones were essentially the same, with $IC_{50}$ values of 1–2 μM, in excellent agreement with the $IC_{50}$ values of 2.6 and 1.4 μM reported for the 32 kDa LO enzymes purified from bovine lung and aorta, respectively (Cronlund & Kagan 1986).

*Messenger RNA expression.* Total RNA was prepared from CHO cells transfected with the full-length or truncated forms of LO cDNA. As shown, cells transfected with plasmids lacking LO cDNA constructs do not yield a visibly hybridizing band on Northern blots probed with $^{32}$P-labelled, full-length

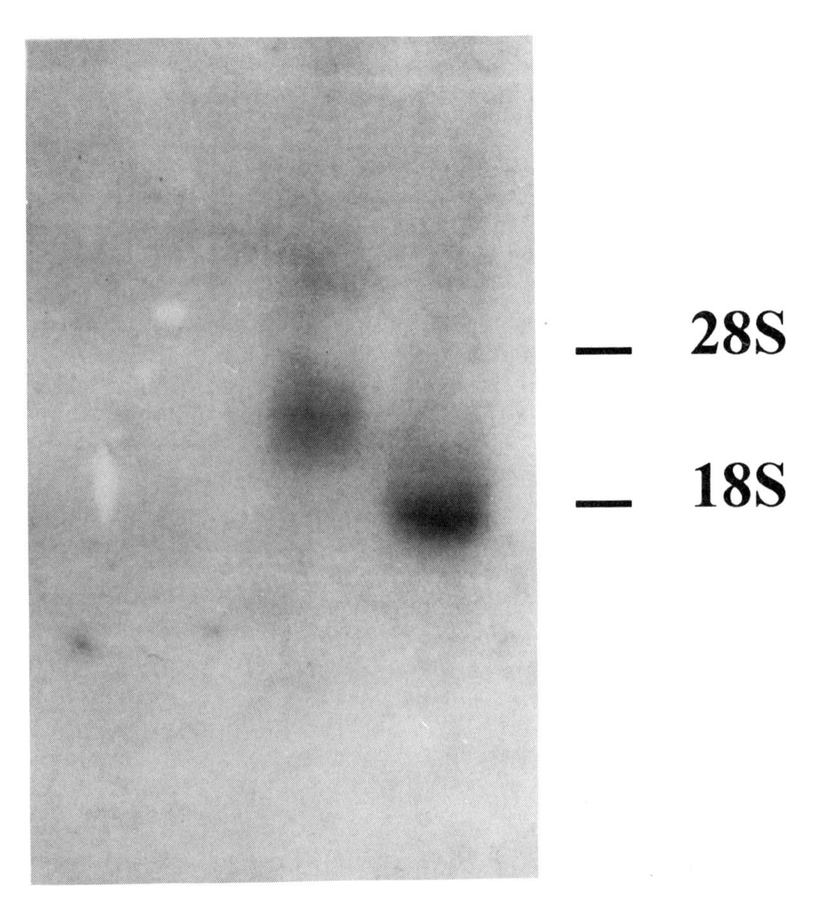


FIG. 5. Northern blot analysis of RNAs from lysyl oxidase-producing cell lines. Lane 1, RNA from CHO cells transfected with plasmid lacking LO cDNA; lane 2, RNA from LOD-06 cell line; and lane 3, RNA from LOD32-2 cell line. The approximate positions of 18S and 28S ribosomal RNAs are indicated by horizontal bars.

LO cDNA (Fig. 5). In contrast, total RNA isolated from LOD-06 displays a band at ca. 2 kb, whereas that from clone LOD32-2 displays a band at ca. 1.6 kb. These sizes are consistent with those expected of the LO mRNA species derived from the two LO cDNA constructs.

Notably, functional enzymes with apparently the same molecular weights and catalytic properties appeared in the medium of both clones. The protein derived from the truncated construct was intended to include a dibasic Arg-Arg sequence (residues 134 and 135) of the rat proenzyme as a possible proenzyme proteolytic processing site. Cleavage at this site predicts the release of a 32 kDa enzyme derived from the C-terminal sequence of the proenzyme (Trackman et al 1990). Since LO species of 29 kDa were produced in the CHO cell medium, cleavage likely occurred further downstream in the sequence. The N-terminus of the four ionic variants of the enzyme isolated from bovine aorta was identified as aspartic acid in each case (Sullivan & Kagan 1982). Conceivably, consecutive cleavages of the proenzyme, possibly at the dibasic site and ultimately including cleavage within the -Gly-Asp-Asp- sequence at residues 162–164 (Trackman et al 1990), would then yield a 29 kDa product. This possibility is currently under investigation.

The generation of apparently identical catalysts from both the full-length and truncated proenzymes indicates that the full propeptide sequence is not necessary for the generation of a processed, functional catalyst. Thus, information required for the proper folding of the protein to generate a functionaly active site appears to be contained largely (and possibly entirely) within the N-terminal sequence of which the mature catalyst is composed. It is of interest in this regard that the predicted amino acid sequences of the rat, human, mouse and chick proproteins corresponding to the propeptide region are not nearly as highly conserved as the sequences corresponding to the mature enzyme region of the proprotein (see below). The present results contrast with effects seen with other proproteins such as the precursor form of transforming growth factor $\beta$ in which the pro-region appears to act as a chaperone essential to the proper folding and secretion of the mature protein segment (Lopez et al 1992). It is notable that each of the 10 cysteine residues of rat prolysyl oxidase are contained within the segment corresponding to the mature catalyst, and at least six of these exist as disulfide bonds (Williams & Kagan 1985). This feature is doubtlessly critical to the stabilization of the functional conformation of the mature enzyme and to the unusual thermostability of lysyl oxidase (Trackman et al 1981).

*Protein sequence and gene structure of lysyl oxidase*

*Derived amino acid sequences from different species.* Sequence alignment of translated LO cDNAs from different species, including human (Hamalainen et al 1991, Mariani et al 1992), rat (Trackman et al 1990), mouse (Kenyon et al 1991) and chicken (Wu et al 1992), identified both conserved and divergent sequence elements (Fig. 6). The N-terminus (150 amino acid residues) revealed the greatest divergence in sequence. The first 21 amino acid residues that constitute the signal peptide were 70–72% identical among human, rat and mouse. There was only a 38% identity between the chick and rat sequences. The majority of deletions and insertions occurred in the propeptide domain (amino acids 22–150), leading to slight size differences in the open reading frame. The similarity in this region is only 60–67%. In contrast, a high level of

**TABLE 3 Inhibition of lysyl oxidase activity against tropoelastin[a]**

| | *LOD-06* | | *LOD32-2* | |
|---|---|---|---|---|
| *Inhibitor* | *[$^3$H]H$_2$O release (cpm/2 h)* | *% control* | *[$^3$H]H$_2$O release (cpm/2 h)* | *% control* |
| Control | 551±26 | 100±5 | 441±30 | 100±6.8 |
| Phenylhydrazine (1 mM) | 71±0.4 | 12.9±0.1 | 71±0.4 | 12.9±0.1 |
| Ethylenediamine (1 mM) | 24.3±0.2 | 4.4±0 | 26±2.9 | 4.7±0.5 |
| α,α″-Dipyridyl (5 mM) | 105±8.6 | 19±1.6 | 100±7.5 | 18.2±1.4 |
| Diethyldithiocarbamate (1 mM) | 183±16.3 | 33.3±3 | 125±4.7 | 22.6±0.8 |
| BAPN (0.5 mM) | 95±5 | 17.3±0.8 | 95±4.5 | 17.3±0.8 |

[a]Data are presented as the means±SD of assays in triplicate.

conservation (90–95% identity) existed within the C-terminal region of the molecule (amino acids 150–417 in human and 144–411 in rat), supporting the view that the conserved C-terminus is the catalytic part of the enzyme and the less-conserved propeptide N-terminal region may provide a non-enzymic function, possibly suppressing activity in the intracellular compartment or having a cellular targeting function.

The potential Asn-linked glycosylation sites NAS and NQT are present only in the human protein. The *N*-glycosylation site NRT is conserved in human, rat and mouse, and the glycosylation site NPS is present in all four species (Hamalainen et al 1991). A copper-binding domain is conserved in LO from all four species and four histidine residues present within this conserved sequence are believed to be involved in the copper-binding coordination complex. There are two additional putative metal-binding consensus sequence domains present in all four species (Trackman et al 1990, Krebs & Krawetz 1993). As noted, the ten cysteine residues of the proenzyme are restricted to the functional enzyme region and are conserved in all LO proteins.

*Regulation of expression of the lysyl oxidase gene.* Lysyl oxidase mRNA was detected in total RNA preparations obtained from heart, placenta, lung, liver, skeletal muscle, kidney and pancreatic tissue but no expression was evident in brain (Kim et al 1995). The single-copy LO gene codes for several messages, two major transcripts of 4.8 and 3.8 kb and a less abundant 2.0 kb message in human (Mariani et al 1992), rat (Trackman et al 1990) and mouse cells (Contente et al 1993). The chicken gene seems to be an exception with a larger, 6.0 kb single mRNA species (Wu et al 1992). There are tissue and developmental stage-specific differences in the levels of expression of the individual transcripts that arise through differential use of multiple polyadenylation signals in the 5′

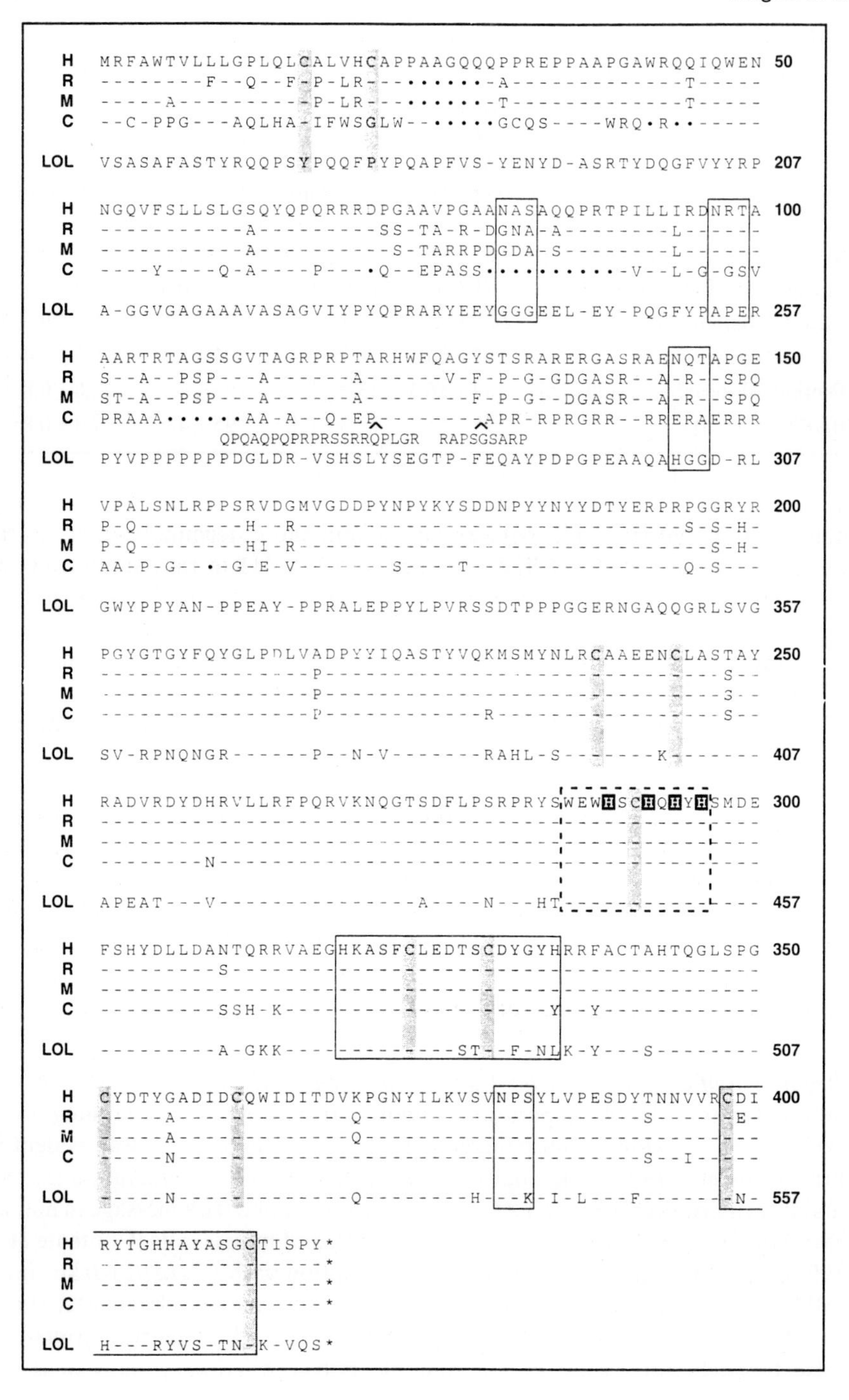
H MRFAWTVLLLGPLQLCALVHCAPPAAGQQQPPREPPAAPGAWRQQIQWEN 50
LOL VSASAFASTYRQQPSYPQQFPYPQAPFVS-YENYD-ASRTYDQGFVYYRP 207
H NGQVFSLLSLGSQYQPQRRRDPGAAVPGAANASAQQPRTPILLIRDNRTA 100
LOL A-GGVGAGAAAVASAGVIYPYQPRARYEEYGGGEEL-EY-PQGFYPAPER 257
H AARTRTAGSSGVTAGRPRPTARHWFQAGYSTSRARERGASRAENQTAPGE 150
QPQAQPQPRPRSSRRQPLGR RAPSGSARP
LOL PYVPPPPPPPDGLDR-VSHSLYSEGTP-FEQAYPDPGPEAAQAHGGD-RL 307
H VPALSNLRPPSRVDGMVGDDPYNPYKYSDDNPYYNYYDTYERPRPGGRYR 200
LOL GWYPPYAN-PPEAY-PPRALEPPYLPVRSSDTPPPGGERNGAQQGRLSVG 357
H PGYGTGYFQYGLPDLVADPYYIQASTYVQKMSMYNLRCAAEENCLASTAY 250
LOL SV-RPNQNGR------P--N-V-------RAHL-S------K------- 407
H RADVRDYDHRVLLRFPQRVKNQGTSDFLPSRPRYSWEWHSCHQHYHSMDE 300
LOL APEAT---V---------------A----N---HT--------------- 457
H FSHYDLLDANTQRRVAEGHKASFCLEDTSCDYGYHRRFACTAHTQGLSPG 350
LOL ---------A-GKK--------------ST--F-NLK-Y---S-------- 507
H CYDTYGADIDCQWIDITDVKPGNYILKVSVNPSYLVPESDYTNNVVRCDI 400
LOL -----N-------------Q--------H---K-I-L---F-------N- 557
H RYTGHHAYASGCTISPY*
LOL H---RYVS-TN-K-VQS*

untranslated region. This mechanism, in addition to transcriptional regulation, may contribute to the post-translational modulation of LO gene expression (C. D. Boyd, T. J. Mariani, Y. Kim & K. Csiszar, unpublished results).

*Structure of the lysyl oxidase gene.* The LO gene was mapped to human chromosome 5q23 (Hamalainen et al 1991, Mariani et al 1992) and to mouse chromosome 18 (Mock et al 1992, Lossie et al 1993, Chang et al 1993). Seven exons and six introns constitute the human and mouse genes alike. There is a considerable conservation of exon and intron sizes and exon–intron junctions between the two genes. All seven exons and the first five introns of the mouse gene agree in size with the human LO gene. There is a significant size difference in intron six only. This conservation confirms a strong evolutionary pressure to maintain both the structure and function of the LO gene (Csiszar et al 1993, Hamalainen et al 1993).

*A lysyl oxidase-like gene.* A new gene on human chromosome 15 was recently isolated and characterized that exhibited significant sequence and structural similarity to the gene encoding LO. The 2.4 kb LO-like mRNA sequence is encoded within seven exons, distributed through 25 kb of genomic DNA. Exons 2–6 encode the region of greatest homology to LO. The sizes of these exons, moreover, are exactly the same as the sizes of corresponding exons within the LO gene. The conservation of sequence and exon and intron sizes indicates that at least exons 2–6 within the LO-like gene and exons 2–6 in the gene encoding LO arose from a common ancestral gene by gene duplication (Kim et al 1994).

The cDNA-derived amino acid sequence alignment of a putative LO-like protein with the human LO sequence shows 84% identity (Kenyon et al 1993) and a similar pattern of conserved domains (Fig. 6). The N-terminal end has little homology to LO but the C-terminus, encoding a putative catalytic domain, is conserved. Though glycosylation sites are absent in the LO-like protein, the entire copper-binding domain, including the four histidines, is strictly conserved. Similarly, ten of the twelve cysteine residues in LO are positionally conserved in the derived LO-like sequence. The presence of these cysteine residues in the LO-like protein suggests that the tertiary structure of this protein is similar to that of LO and that it may fulfil a similar enzymatic role, as well.

---

FIG. 6. Amino acid comparison of the human (H), rat (R), mouse (M) and chicken (C) lysyl oxidase (LO) and the human LO-like proteins (LOL). Dashes denote identical amino acids, dots indicate deletions, ∧ indicate insertions in the chick protein, asterisks identify a stop codon. Numbers are based on the amino acid sequence of human LO. The amino acid residues of the LO-like protein are numbered according to Kenyon et al (1993). *N*-glycosylation sites NAS, NGT and NTR are boxed; conserved NPS is in a shaded box. Conserved cysteine residues are in shaded boxes; shaded boxes also identify copper and putative metal-binding sequences. The four copper-coordinating histidine residues within the copper-binding domain are indicated in reverse font.

*Acknowledgements*

Research summarized here was supported by National Institutes of Health grants R37 AR 18880 and HL 13262 (to H. M. K.) and HL 37438 (to K. C.)

## References


Bedell-Hogan D, Trackman P, Abrams W, Rosenbloom J, Kagan HM 1993 Oxidation, crosslinking and insolubilization of recombinant tropoelastin by purified lysyl oxidase. J Biol Chem 268:10345–10350

Chang YS, Svinarich DM, Yang TP, Krawetz SA 1993 The mouse lysyl oxidase gene (*Lox*) resides on chromosome 18. Cytogenet Cell Genet 63:47–49

Contente S, Csiszar K, Kenyon K, Friedman RM 1993 Structure of the mouse lysyl oxidase gene. Genomics 16:395–400

Csiszar K, Mariani TJ, Gosin JS, Deak SB, Boyd CD 1993 A restriction fragment length polymorphism results in a nonconservative amino acid substitution encoded within the 1st exon of the human lysyl oxidase gene. Genomics 16:401–406.

Cronlund AL, Kagan HM 1986 Comparison of lysyl oxidase of bovine aorta and lung. Connect Tissue Res 15:173–185

Gacheru SN, Trackman PC, Calaman SD, Greenaway FT, Kagan HM 1989 Vicinal diamines as PQQ-directed, irreversible inhibitors of lysyl oxidase. J Biol Chem 264:12963–12969

Gacheru SN, Trackman PC, Shah MA et al 1990 Structural and catalytic properties of copper in lysyl oxidase. J Biol Chem 265:19022–19027

Hamalainen RR, Jones TA, Sheer D, Taskinen K, Pihlajaniemi T, Kivirikko KI 1991 Molecular cloning of human lysyl oxidase and assignment of the gene to chromosome 5q23.3-q31.2. Genomics 11:508–516

Hamalainen RR, Kemppainen R, Pihlajaniemi T, Kivirikko KI 1993 Structure of the human lysyl oxidase gene. Genomics 17:544–548

Helseth DL, Lechner JH, Veis A 1979 Role of the amino-terminal extrahelical region of type-I collagen in directing the 4D overlap in fibrillogenesis. Biopolymers 18: 3005–3014

Kagan HM 1986 Characterization and regulation of lysyl oxidase. In: Mecham RP (ed) Biology of the extracellular matrix, vol 1: Regulation of matrix accumulation. Academic Press, Orlando, FL, p 321–398

Kagan HM, Tseng L, Simpson DE 1981 Control of elastin metabolism by elastin ligands: reciprocal effects on lysyl oxidase activity. J Biol Chem 256:5417–5421

Kagan HM, Williams MA, Williamson PR, Anderson JM 1984 Influence of sequence and charge on the specificity of lysyl oxidase toward protein and synthetic peptide substrates. J Biol Chem 259:11203–11207

Kagan HM, Reddy VB, Panchenko M et al 1995 Expression of lysyl oxidase from cDNA constructs in mammalian cells: the propeptide region is not essential to the folding and secretion of the functional enzyme. J Cell Biochem, in press

Kenyon K, Contente S, Trackman PC, Tang J, Kagan HM, Friedman RM 1991 lysyl oxidase and *rrg* messenger RNA. Science 253:802

Kenyon K, Modi WS, Contente S, Friedman RM 1993 A novel human cDNA with a predicted protein similar to lysyl oxidase maps to chromosome 15q24–q25. J Biol Chem 268:18435–18437

Kim Y, Boyd CD, Csiszar K 1995 A new gene with sequence and structural similarity to the gene encoding human lysyl oxidase. J Biol Chem, in press

Krebs CJ, Krawetz SA 1993 Lysyl oxidase copper-talon complex: a model. Biochim Biophys Acta 1202:7–12
Lossie AC, Buckwalter MS, Camper SA 1993 Lysyl oxidase maps between Grl-4 and Adrb-2 on mouse chromosome 18. Mamm Genome 4:177–178
Mariani TJ, Trackman PC, Kagan HM et al 1992 The complete derived amino acid sequence of human lysyl oxidase and assignment of the gene to chromosome 5. Matrix 12:242–248
Mock BA, Contente S, Kenyon K, Friedman RM, Kozak CA 1992 The gene for lysyl oxidase maps to mouse chromosome 18. Genomics 14:822–823
Lopez AR, Cook J, Deininger PL, Derynck R 1992 Dominant negative mutants of transforming growth factor-beta 1 inhibit the secretion of different transforming growth factor-beta isoforms. Mol Cell Biol 12:1674–1679
Nagan N, Kagan HM 1994 Modulation of lysyl oxidase activity toward peptidyl lysine by vicinal dicarboxylic amino acid residues. Implications for collagen crosslinking. J Biol Chem 269:22366–22371
Siegel RC 1974 Biosynthesis of collagen crosslinks: increased activity of purified lysyl oxidase with reconstituted collagen fibrils. Proc Natl Acad Sci USA 71:4826–4830
Sullivan KA, Kagan HM 1982 Evidence for structural similarities in the multiple forms of aortic and cartilage lysyl oxidase and a catalytically quiescent aortic protein. J Biol Chem 257:13520–13526
Tang SS, Trackman PC, Kagan HM 1983 Reaction of aortic lysyl oxidase with $\beta$-aminopropionitrile. J Biol Chem 258:4331–4338
Trackman PC, Bedell-Hogan D, Tang J, Kagan HM 1992 Post-translational glycosylation and proteolytic processing of a lysyl oxidase precursor. J Biol Chem 267:8666–8671
Trackman PC, Pratt AM, Wolanski A et al 1990 Cloning of rat aorta lysyl oxidase cDNA: complete codons and predicted amino acid sequence. Biochemistry 29: 4863–4870
Trackman PC, Zoski CG, Kagan HM 1981 Development of a peroxidase-coupled fluorometric assay for lysyl oxidase. Anal Biochem 113:336–342
Williams MA, Kagan HM 1985 Assessment of lysyl oxidase variants by urea gel electrophoresis: evidence against disulfide isomers as bases of enzyme heterogeneity. Anal Biochem 149:430–437
Williams PR, Kagan HM 1986 Reaction pathway of bovine aortic lysyl oxidase. J Biol Chem 261:9477–9482
Williamson PR, Kittler JM, Thanassi JW, Kagan HM 1986 Reactivity of a functional carbonyl moiety in bovine lysyl oxidase. Evidence against pyridoxal 5′-phosphate. Biochem J 235:597–605
Wu Y, Rich CB, Lincecum J, Trackman P, Kagan HM, Foster JA 1992 Characterization and developmental expression of chick lysyl oxidase. J Biol Chem 267:24199–24206

## DISCUSSION

*Bailey:* As an extension on Herb Kagan's talk on lysyl oxidase I would like to make a brief comment on the cross-linking of elastin. Little work has been done on the cross-links since the desmosines were characterized by Partridge in the 1960s. However, there is a discrepancy in the number of lysine residues involved in the cross-linking mechanism. The initial lysine content for the non-cross-linked tropoelastin is 35 residues, but the 22 involved in known cross-linking and the five residual lysines in mature elastin only total 27 lysines, leaving approximately eight lysine residues unaccounted for. Additional cross-links have been identified, but their yields are too low to account for the missing lysines. For example, Nagai (1983) identified a pyridinium compound he termed neodesmosine, and Starcher et al (1987) identified a pyridinium ring structure similar to desmosine in which the aldol participated in place of a lysine aldehyde. The compound was therefore derived from five lysines and was given the trivial name pentasine. Suyama & Nakamura (1990) identified a similar compound derived from the reduced aldol, termed allodesmosine, but again the yield was only about one tenth that of desmosine. Suyama & Nakamura (1991) also identified a dipyridine structure, termed aldosine. However, this was derived from elastin in low yield as an acid hydrolysis product of the aldol and cannot be considered a native cross-link. Halme et al (1985) identified a new cyclic amino acid that increased with age, but this appeared to be derived from reaction of the aldol with an adjacent α-amino group, presumably arising from peptide bond cleavage during ageing, and similarly cannot be considered a native cross-link.

We have recently been studying the role of pyrroles as cross-links in collagen and have applied the techniques to elastin. We found that the pyrroles were destroyed by acid and alkali purification techniques, but were retained in autoclave-purified elastin. Such a pyrrole could be formed by reaction of lysine aldehyde with the aldol or with δ-lysinonorleucine. Each pyrrole would therefore represent three lysine equivalents. At the present time we have not been able to characterize the particular pyrrole and hence determine its yield, but in view of its structure it could act as an additional cross-link and possibly account for the missing lysines in elastin.

*Kagan:* It would be interesting to see what the sequences surrounding those lysines are.

*Bailey:* Unfortunately, as soon as we get a peptide fragment small enough to sequence, the pyrrole breaks down. I don't know why.

*Keeley:* Herb Kagan, I was intrigued to see that the 50 kDa prolysyl oxidase is processed into the active form by a secreted metalloproteinase. Is this a $Ca^{2+}$-dependent enzyme?

*Kagan:* We know that it can be inhibited by EDTA, and we can overcome this inhibition with $Ca^{2+}$.

*Keeley:* The reason for asking is that addition of EGTA to our aortic organ culture system appears to inhibit assembly of insoluble elastin but not synthesis of tropoelastin. Perhaps EGTA might be acting in this way by blocking processing of lysyl oxidase.

*Kagan:* It could be. We now have some evidence for a candidate metalloproteinase that appears to activate prolysyl oxidase. Surprisingly, we have found that matrix metalloproteinase 2 (type IV collagenase) can selectively cleave the 50 kDa proenzyme to the 32 kDa form.

*Robert:* No more than 10% of the activity of the metalloendopeptidase we described in mesenchymal cells is ever secreted; it is a membrane-bound enzyme (Archilla-Marcos et al 1993). Is the one you mentioned secreted?

*Kagan:* Yes; the activity appears in the conditioned medium of aortic smooth muscle cell cultures.

*Robert:* It might that be just enough of this enzyme is secreted to perform the processing you mentioned, or it may be a different enzyme.

*Kagan:* It might be the same; there might be some partitioning between the medium and the cell. We haven't actually tested for a cell-bound protease.

*Pasquali-Ronchetti:* You showed that lysyl oxidase is more efficient when you work with collagen fibres than with collagen molecules in solution. What happens with elastin? Is lysyl oxidase activity higher when you work with coacervates of tropoelastin?

*Kagan:* Some years ago, Dan Urry and I studied the effects of lysyl oxidase on peptidyl lysine contained within elastin-like peptides containing polypentapeptide or polyhexapeptide repeat sequences. There was considerably more lysine oxidation and cross-linkage formation if these synthetic peptides were maintained at temperatures favouring coacervation (Kagan et al 1980). Recently, together with Joel Rosenbloom, we found that human recombinant tropoelastin is also an efficient substrate for lysyl oxidase *in vitro* (Bedell-Hogan et al 1993). In fact, a cross-linked, hot-alkali-insoluble product eventually formed. In this case, however, it is difficult to assess the degree to which coacervation may contribute to the substrate potential for lysyl oxidase, since one must incubate at temperatures which allow lysyl oxidase to function and which may also induce coacervation. Thus, it is certainly possible that tropoelastin monomers are oxidised by lysyl oxidase and this would be consistent with the fact that, intrinsically, tropoelastin is cationic, owing to its lysine and arginine content prior to lysine oxidation. It becomes much less cationic as oxidation proceeds. However, coacervation or polymerization of monomers may be important for the appropriate intermolecular alignment of lysines which cross-link immediately after oxidation.

*Hinek:* Do you know how lysyl oxidase is immobilized in the framework of the elastic fibres? Despite the fact that it catalyses the oxidation of lysines it is not an elastin-binding protein, so it has to be strategically located by microfibrils.

*Kagan:* Some years ago, we found by immunoelactron microscopy that the enzyme localized predominantly at the interface between the microfibrils and the amorphous elastin fibres in the extracellular matrices of bovine and rat aortas (Kagan et al 1986). If, in fact, elastic fibres grow in the extracellular space by accretion of tropoelastin units on their surface, this is a reasonable site for the enzyme where it may oxidise and cross-link new tropoelastin units as they come along. We saw negligible quantities of lysyl oxidase inside the amorphous fibre, and significant amounts along the periphery between the microfibrils and elastin.

*Hinek:* Was there any periodicity?

*Kagan:* We didn't take that into account.

*Davidson:* Is processing of the precursor required for activity?

*Kagan:* That's clearly a key question. As yet, we have not isolated enough of the pure 50 kDa prolysyl oxidase to be able to test that unequivocally. However, certain transformed cell lines that only express the 50 kDa species have no measurable lysyl oxidase activity. In a related (unpublished) study, we pulsed chick embryo aortas in organ culture with [$^{14}$C]BAPN. Since BAPN is a mechanism-based inhibitor of lysyl oxidase, it should label only the catalytically functional enzyme. We detected radioactive bands only at positions equivalent to 29–30 kDa in SDS gels of extracts of the pulsed aortas. These results are consistent with the conclusion that the 50 kDa enzyme is not active, whereas the 32 kDa species is.

*Davidson:* Are either of these forms active when expressed in bacteria?

*Kagan:* We have expressed them in bacteria, but the protein products are not active. That is likely a consequence of a problem in folding.

*Boyd:* But in your CHO cells you do get enzyme activity on an elastin substrate from the sequence of the recombinant protein that contains the propeptide.

*Kagan:* What we tested here was the conditioned medium. Thus, the medium contained both the 29 kDa species, and the 50 kDa precursor or the truncated 35 kDa precursor, depending on the construct we used. We haven't shown unequivocally that activity comes from the 29 kDa species but, given the results I've just described, I would say that this is highly likely.

*Boyd:* David Hulmes (personal communication) has indicated that the processing site in human lysyl oxidase that he's been looking at differs from the predicted arginine sites that you have in the propeptide sequence. Does your cleavage site match the site that David is talking about?

*Kagan:* I'm not aware of his results; suffice to say that the specific cleavage site in prolysyl oxidase has yet to be unequivocally established.

*Boyd:* Several people interested in 3′ untranslated region (UTR) sequences have been quite excited by an observation made by Phil Trackman and Herb Kagan a few years ago. They showed that there was a sequence in the 3′ UTR that contained a 200 bp region very similar to a domain in the rat tropoelastin message (Trackman et al 1990). This suggests some coordinate regulation (possibly post-transcriptional) of expression of these two proteins. Tom Mariani sequenced the entire 3′ UTR from human lysyl oxidase mRNA (exon 7 of the gene), and could not show homology to any sequence within the human or the rat tropoelastin 3′ UTR (Boyd et al 1995). K. Kivirikko showed the same sort of results (Hamalainen et al 1993). If there is a consensus sequence in the 3′ UTR of lysyl oxidase and tropoelastin mRNAs, it is certainly a species-specific phenomenon because it doesn't occur in human.

*Parks:* When this putative shared element was first reported, it suggested an intriguing regulatory mechanism to co-ordinately control matrix production.

But since elastin is not the only substrate for lysyl oxidase, it is unlikely that the regulation of this enzyme would be tied so specifically to just one of its extracellular matrix substrates. Since you don't always find collagen where elastin is, and vice versa, I would expect that the regulation of lysyl oxidase would be distinct from that of elastin.

*Starcher:* In the same context, is lysyl oxidase production developmentally regulated in parallel with elastin? As you see a shut down in elastin synthesis in later stages of development, does lysyl oxidase shut down in the same pattern?

*Kagan:* Judy Foster showed that in chick, lysyl oxidase expression did increase in embryogenesis coincident with the increase in collagen and elastin mRNAs, but its production started to shut down while the expression of those substrates was continuing to increase. There was a peak of lysyl oxidase at about Day 12 while elastin synthesis was still increasing. Nevertheless, lysyl oxidase mRNA continued to be expressed, although at reduced levels, throughout the time that collagen and elastin were being synthesized (Wu et al 1992).

*Pasquali-Ronchetti:* Lysyl oxidase from bovine aorta is localized by immunocytochemistry between microfibrils and elastin fibres. I had the opportunity to work with polyclonal antibodies raised against lysyl oxidase isolated from human placenta, which doesn't contain elastin but does contain collagen. These antibodies stained collagen beautifully, but didn't stain elastin at all, in both human aorta and skin (Baccarani-Contri et al 1989). Is there any evidence for different isoforms of lysyl oxidase in different tissues? Probably *in vitro* both isoforms work on substrates like collagen and tropoelastin, but what happens *in vivo*?

*Kagan:* I am not aware of any demonstration that different isomorphs exist in different tissues. However, *in vitro*, both collagen and elastin will be oxidized by each of the four isomorphs of bovine aorta lysyl oxidase, but the situation might be different *in vivo*.

*Mariani:* When I was in Charles Boyd's laboratory, we addressed the potential role of alternative splicing in production of the different lysyl oxidase isoforms. We used a very sensitive PCR-based assay for the detection of splice variants in total RNA from a number of different human tissues. We were unable to find any evidence of alternative splicing for the human lysyl oxidase gene (unpublished observations), suggesting this is not a mechanism of isoform production.

I know the polyarginine sites in prolysyl oxidase are pretty well conserved. Are the GDD sites also well conserved between species?

*Kagan:* They're conserved in human, rat, mouse and chick, whereas the Arg-Arg appears at approximately the same sequence position in the rat, mouse and chick, whereas the human sequence contains only a single Arg at this site.

*Mariani:* Does anyone want to comment on the possible functionality of the lysyl oxidase-related protein, and how this could relate to the differences in immunostaining of collagen and elastin?

*Kagan:* The important question is whether it possesses lysyl oxidase activity: we don't know this yet. But certainly, it appears to be highly homologous to the other lysyl oxidase species, even though its gene occurs on a different chromosome (Kenyon et al 1993). It seems likely that it plays an important although yet unknown role.

*Boyd:* The expression studies that Katalin Csiszar is presently doing with Herb Kagan to see whether this lysyl oxidase-related protein has lysyl oxidase activity, will directly address whether a separate gene may encode another lysyl oxidase. If this is the case, it may explain Ivonne Pasquali-Ronchetti's observation of tissue-specific distribution of lysyl oxidases that don't cross-react with polyclonal antibodies to the known lysyl oxidases.

*Robert:* Is there any way a cell can allocate its lysyl oxidase to collagen or elastin if it makes both fibrous proteins simultaneously? Does it always make an excess of lysyl oxidase, or does it 'decide' which will be cross-linked and to what extent?

*Kagan:* A crude estimate made several years ago (unpublished) of the molar ratio of lysyl oxidase to collage present in calvarial tissue was approximatley 1:1. This certainly suggests that the catalytic turnover of lysyl oxidase is limited.

*Robert:* Can ascorbic acid reduce copper to inactivate the enzyme?

*Kagan:* The possibility that ascorbate can reduce divalent copper tightly bound at the active site of lysyl oxidase has not been assessed, to my knowledge.

## References


Archilla-Marcos M, Robert L 1993 Control of the biosynthesis and excretion of the elastase-type protease of human skin fibroblasts by the elastin receptor. Clin Physiol Biochem 10:86–91

Baccarani-Contri M, Vincenzi D, Quaglino D Jr, Pasquali-Ronchetti I 1989 Immunolocalization of placental lysyl oxidase in human placenta, skin and aorta. Matrix 9:428–436

Bedell-Hogan D, Trackman P, Abrams W, Rosenbloom J, Kagan HM 1993 Oxidation, crosslinking and insolublilization of recombinant tropoelastin by purified lysyl oxidase. J Biol Chem 268:10345–10350

Boyd CD, Mariani TJ, Kim Y, Csiszar K 1995 The size heterogeneity of human lysyl oxidase RNA is due to alternate polyadenylation site and not alternate exon usage. Mol Biol Rep, in press

Halme T, Jutila M, Vihersaari T, Oksman P, Light ND, Penttinen R 1985 The borhydride-reducible compounds of human aortic elastin. Biochem J 232:169–175

Hamalainen RR, Kemppainen R, Pihlajaniemi T, Kivirikko KI 1993 Structure of the human lysyl oxidase gene. Genomics 17:544–548

Kagan HM, Tseng L, Trackman PC, Okamoto K, Rapaka RS, Urry DW 1980 Repeat polypeptide models of elastin as substrates for lysyl oxidase. J Biol Chem 255:3656–3659

Kagan HM, Vaccaro CA, Bronson RE, Tang SS, Brody JS 1986 Ultrastructural localization of lysyl oxidase in vascular connective tissue. J Cell Biol 103:1121–1128

Kenyon K, Modi WS, Contente S, Friedman RM 1993 A novel human cDNA with a predicted protein similar to lysyl oxidase maps to chromosome 15q24-q25. J Biol Chem 268:18435–18437

Nagai Y 1983 Neodesmosine, a new elastin crosslink. Connect Tissue (Japan) 14: 112–113

Starcher B, Cook G, Gallop P, Henson E, Shoulders B 1987 Isolation and characterization of a pentameric amino acid from elastin. Connective Tissue Res 16:15–25

Suyama K, Nakamura F 1990 Isolation and characterization of a new cross-linking amino acid ‘allodesmosine’ from acid hydrolysates of elastin. Biochem Biophys Res Commun 170:713–718

Suyama K, Nakamura F 1991 Aldosine, an acid hydrolysis product of the aldol crosslink of bovine elastin and collagen. Agric Biol Chem 55:3147–3149

Trackman PC, Pratt AM, Wolanski A et al 1990 Cloning of rat aorta lysyl oxidase cDNA: complete codons and predicted amino acid sequence. Biochemistry 29: 4863–4870

Wu Y, Rich CB, Lincecum J, Trackman P, Kagan HM, Foster JA 1992 Characterization and developmental expression of chick aortic lysyl oxidase. J Biol Chem 267:24199–24206

# General discussion II

## Evolutionary aspects

*Mungai:* I was intrigued by Leslie Robert's mention of the evolution of elastin in his introductory remarks. I wanted to add that in terrestrial animals it appears the effect of gravity has determined the pattern of evolution of elastin in the form of ligaments. This is best demonstrated by the way the heads of quadrupeds hang in front of the body, with the ligamentum nuchae acting as a cantilever from the vertebral column to the head. In the case of the elephant, I've had the opportunity of dissecting out the ligamentum nuchae. It consists of two large cords, which are inserted to the back of the very large head. This would appear to be part of the evolution of terrestrial animals with the effect of gravity. But the other aspect of evolution is demonstrated by humans and other primates in the assumption of the upright posture. Two of our colleagues have studied two aspects of this. First, Mike Rose, who was with us about 20 years ago, showed that up to 75% of a monkey's time is spent in a postural position, either resting or eating, in contrast to locomotion (Rose 1974). Later, Fortunato Fasana (1976) studied the distribution and organization of vertebral ligaments in the monkey, and found (as we expected) that these are very carefully distributed according to the demands made on the vertebral column as the monkey uses its arms in a postural activity. The assumption of the upright posture by humans and the sitting upright posture by monkeys has therefore had an extra effect on the distribution of elastic fibres in vertebral ligaments. This is also shown by the fact that with the development of cervical curvature in humans, which is concave posteriorly, the ligamentum nuchae that is so important in the quadruped has more or less disappeared into an insignificant median raphe between the muscles.

*Robert:* This is complementary to what I said on evolution in the introduction; the evolution of elastin was crucial for the development of the respiratory and the circulatory systems, but obviously it was also vital in the assumption of the upright posture by primates. Anthropologists and comparative anatomists claim that assumption of the upright position was crucial for the development of the brain, so here we have a link, for the first time, between elastin and brain development!

*Starcher:* Before we leave this discussion on evolution, could you share the interesting story you told me earlier about why the elephant has to have elastin in its toes?

*Mungai:* The average African elephant weighs 5–6 tons, and it walks on the tips of its toes. In the spaces between the toes are large numbers of loculae of fat, surrounded by capsules of elastic fibres. When the elephant steps on its toes it dissipates the force of its weight on these loculae of fat, by 'rolling' on them.

*Kimani:* This is not unique to the elephant. If you look at the weight-bearing areas of the human foot, you will find similar loculae of fat in the subcutaneous tissue, encapsulated by elastic fibres. A while back, we published a paper showing the distribution of elastic fibres in the human foot (Kimani 1984).

*Pierce:* I have some observations regarding repeating peptide sequences in tropoelastin. When I cloned the rat tropoelastin cDNA, I found remarkable nucleotide conservation in the repeating regions. As an example, in exon 30, there is a 30 bp perfect repeat in the rat tropoelastin cDNA. In exon 22, there are three identical 24 bp repeats. This suggests to me that the repetitive nucleotide sequences arose long before the repetitive amino acid sequences. A particular peptide sequence, for instance a hexamer, can be encoded by many different nucleotide sequences. The likelihood of large perfect repeats at the nucleotide level occurring by chance is astronomically small.

*Rosenbloom:* Generally speaking, there is some conservation in codon usage. Charles Boyd has looked at this more carefully than I have. I don't know about rat, but in human there is certainly a heavy bias in the usage of glycine codons, which is GGX (where X is T or C). I haven't generalized this; I don't know whether the rat codon usage is the same as human or not. I don't know what role those highly conserved nucleotide sequences play.

*Boyd:* Rich Pierce and I found that the codon usage in the rat sequence is similar to human. But Rich has brought up a bigger issue; this relates to the evolution of a coding sequence for a protein that is essentially modular in its various domains. Some of Fred Keeley's recent work on lamprin (Robson et al 1993) has put together rather nicely some of the predictions which many of us have been worrying about concerning elastin sequences. Helene Sage was the first to suggest that elastin (defined here as a protein that would stain with an elastic fibre stain), first made its appearance in fish. The last time we shared a common ancestor with fish was about 400 million years ago; consequently, this protein first made its appearance in vertebrate phylogeny at about that time. However, the protein itself is made up of cross-linked and hydrophobic sequences, which (particularly the hydrophobic sequences) will occur in many different proteins. An earlier paper on hydrophobic domains in spider dragline silk emphasizes this point (Xu & Lewis 1990). The same sorts of hydrophobic sequences, very conserved in sequence, have been shown to be present in the elastic protein in lamprey cartilage. Because we last shared a common ancestor with lampreys about 500 million years ago, which predates our common ancestry with teleost fish, it is clear that exon sequences encoding hyrophobic domains have been around in evolution for a very long time. The ancient phylogeny of those sequences is reflected in that rather conserved sequence Rich Pierce has referred to.

*Keeley:* I just want to comment on a couple of things Charles Boyd has said. The lamprey cartilage in which the protein with these tandem repeats is found is not extensible; it behaves like other cartilage tissue in that respect. Secondly, we have to be careful about distinguishing between convergent and divergent evolution. It is certainly possible that similar motifs were independently reinvented more than once. These hydrophobic repeat sequences seem to occur in proteins that are predominantly $\beta$-sheet, hydrophobic, extracellular proteins which self-assemble into a matrix. I would agree that this is an ancient motif which is used for assembling hydrophobic matrix proteins.

There is also a rather striking similarity between sequences in spider dragline silk and in elastin. This spider silk contains repeated domains which are rich in alanine (and have been proposed to be $\alpha$-helical in structure) interspersed with hydrophobic domains, some of which contain the GGLGY sequence. In this respect, the protein looks like elastin without the lysines in the cross-linking domains. I don't know whether this is an example of convergent or divergent evolution, or whether the similarities are merely figments of my imagination.

*Tamburro:* May one hypothesize divergent evolution? Paradoxically, one could suggest that ancestral elastin was inelastic polyglycine which, after divergent evolution, gave rise to different elastomeric proteins.

*Keeley:* One of the reasons that we thought it might not be divergent evolution is that elastin contains those kinds of sequences in both forward and backward orientations. It's hard to explain this in terms of divergent evolution.

*Urry:* Richard Pierce, are these exact base sequence repeats? Is there no sequence variation in these repeats that would reflect multiple codon usage for a given amino acid?

*Pierce:* They're exact repeats and they differ both at the nucleotide level and at the coding level from say bovine and human. This suggests that it's not a divergent evolution, but a gene conversion event.

*Urry:* How many of these exact base sequence repeats are there?

*Pierce:* In exon 30, for example, there are two perfect 30 bp repeats; in exon 24, there are three 24 bp repeats.

*Urry:* With Dave McPherson, Casey Morrow and Jie Xu, we have been using gene construction and expression methods to prepare repeating sequences of elastin and related peptides. It appears best to randomize the multiple triplet codons available for each amino acid and to weight the codon usage for the gene construction in a manner consistent with the organism's preferred codon usage. In this way the gene for $(GVGVP)_{10}$ was constructed without repeating nucleotide sequences. This basic gene is then concatemerized (polymerized) to obtain, for example, $(GVGVP)_{250}$. The repeating base sequence was made 150 nucleotides in length in order to minimize homologous recombination or looping out with deletion of sequence. The result has been a stable high level of expression. Others, using shorter base sequence repeats, have not reported comparable success. It would seem in the natural gene, where there are fewer repeats in a continuous sequence, that randomization of codon usage is not necessary.

*Davidson:* On the basis of several amino acid analyses, there are several forbidden amino acid residues, at least in higher vertebrate elastins. Certainly one could conceive that the appearance of some residues would have drastic effects on elastin organization, but what would the functional consequences of a tryptophan or histidine in elastin be in your type of analysis?

*Urry:* In our hydrophobicity scale, tryptophan is the most hydrophobic amino acid. It would drive hydrophobic association perhaps more irreversibly than you would want. For example, some of the alternative splicing involves the removal of charged sequences. When we try to see how the charged sequences can be compatible with hydrophobic assembly, noting that there is this apolar–polar repulsive free energy (Urry 1993) between the hydrations of apolar and polar species, you look to a balance for the expression of hydrophobicity. If there is too much hydrophobicity, the protein can collapse to form a metastable state, due to hydrophobic interaction. We would argue, in general, that this could be an important reason for a chaperone, to tend the more hydrophobic regions until the proper moment for completion of folding and assembly.

While we say that elastin is a very hydrophobic protein, it's not that hydrophobic a protein if you consider the intensity of the hydrophobicity. It has a very high proportion of hydrophobic residues, but it doesn't have the most hydrophobic residues like tryptophan, and not so many tyrosines and phenylalanines, which would really make it hydrophobic. The polypentapeptide is hydrophobically balanced: it is soluble and dissociated below 25 °C, and as the temperature is raised to 37 °C it assembles nicely. If you had a sequence that was much more hydrophobic, it would collapse hydrophobically. It would not be free enough conformationally for fibre assembly to occur.

*Keeley:* One of the interesting things about other proteins containing the GGLGY tandem repeats (lamprey cartilage protein, silk moth chorion proteins) is that, at least in the chorion proteins, the tyrosine residue is apparently involved in hydrogen bonding. These proteins are able to be stabilized to a remarkable degree without the need for covalent cross-links. Thus, these sequences may be important not only for alignment, but also for stabilization. In the case of elastin, as Dan Urry has just said, you want alignment but stabilization will come from the desmosine cross-links. In elastin, the tyrosines in these motifs are replaced by valine or alanine, so that stabilization by hydrogen bonding is not possible.

*Urry:* The periodicity of hydrophobicity is also very important. Fred Keeley has made the analogy to the silk proteins. There you get the hydrophobic periodicity, but with that particular periodicity the backbone can fall into stable, hydrogen-bonded cross-$\beta$ structures, for example. Hydrogen bonding ties up the backbone rigidly and removes the elasticity. The way one gets ideal elasticity is to have a hydrophobic construct that leaves the backbone with a lot of freedom. On extending, the internal chain dynamics are damped and one gets entropic elasticity.

*Robert:* Allen Bailey, years ago you produced a beautiful scheme showing how the lysine-based cross-links increase in development then go down again, to be replaced by different cross-links. If I understood correctly, there would be no lysyl oxidase present which could catalyse this transformation—so why do the lysine-based interchain cross-links decrease during development?

*Bailey:* You are quite correct, you do not need lysyl oxidase to form the mature cross-links, because the molecules are packed so precisely in the fibre that the reactive groups are in precise apposition and consequently react spontaneously. Initially, the lysyl oxidase is believed to bind to the sequence Hyl-Gly-His-Arg of one molecule and oxidises the telopeptide lysine of an adjacent molecule to lysyl aldehyde. That's all you need. The divalent Schiff base cross-link forms between the lysyl aldehyde and the hydroxylysine in the above sequence. This cross-link is still chemically reactive and as the fibre matures it may react with the lysyl aldehyde of another fibre. If you think in terms of pentafibrils, the molecules are quarter-staggered relative to each other and only head-to-tail cross-linking can occur. However, during maturation these fibrils pack closely together, but this time with the molecules in register such that the cross-link can react with a telopeptide lysyl aldehyde (Bailey et al 1980). No enzymes are needed for this second maturation reaction. It's not like waiting for the reactants to get together in solution. In the fibre there is a very high concentration of the reactants because of the close apposition of the reactive groups within the fibre.

*Robert:* Yes, but I thought the sequential replacement of different types of cross-links changes with age.

*Bailey:* Certainly, as turnover slows down, then there is a decrease in the new divalent cross-links since they now have time to interact to form the mature cross-links, which consequently increase as there is a slow conversion from one to the other. It's a question of the decreasing rate of turnover with age.

## References

Bailey AJ, Light ND, Atkins EDT 1980 Chemical cross-linking restrictions on models for the molecular organization of the collagen fibre. Nature 288:408–410

Fasana FV 1976 Microscopic organization of the ligamenta flava in Cercopithecinae. Acta Anat 94:127–142

Kimani JK 1984 The structural and functional organisation of the connective tissue in the human foot with reference to the histomorphology of the elastic fibre system. Acta Morphol Neerl Scand 22:313–323

Robson P, Wright GM, Sitar ZE et al 1993 Characterization of lamprin, an unusual matrix protein from lamprey cartilage: implications for evolution, structure, and assembly of elastin and other fibrillar proteins. J Biol Chem 268:1440–1447

Rose MD 1974 Postural adaptations in New and Old World monkeys. In: Jenkins FA (ed) Primate locomotion. Academic Press, New York, p 201–222

Urry DW 1993 Molecular machines: how motion and other functions in living organisms can result from reversible chemical changes. Angew Chem Int Ed Engl 32:819–841

Xu M, Lewis RV 1990 Structure of a protein superfibre: spider dragline silk. Proc Natl Acad Sci USA 87:7120–7124

# The structure and function of fibrillin

Dieter P. Reinhardt*, Steve C. Chalberg* and Lynn Y. Sakai*†

*Shriners Hospital for Crippled Children and †Department of Biochemstry and Molecular Biology, Oregon Health Sciences University, 3101 SW Sam Jackson Park Road, Portland, OR 97201, USA

*Abstract.* Fibrillin is a very large molecule whose primary structure is now known from the cloning and sequencing of 10 kb of cDNA. Immunohistochemical results suggest that one of the functions of fibrillin molecules is to contribute to the structure of the microfibril. The importance of fibrillin as a structural macromolecule has been demonstrated by the identification of the gene for fibrillin (*FBN1*) as the disease-causing gene in Marfan's syndrome. While it is clear that fibrillin contributes to the structure of the microfibril, it is not known whether fibrillin molecules self-assemble or whether fibrillin interacts with other molecules in order to form microfibrils. In order to investigate whether particular domains of fibrillin are important to the assembly of the microfibril and to specify domains that participate in interactions with other proteins, we have produced recombinant fibrillin 1 peptides in human cells and used them in studies described here. Additionally, new information regarding the 5′ end of *FBN1* has been obtained from studies investigating promoter activity, and potential proteolytic cleavage sites have been identified in the N- and C-terminal domains.

*1995 The molecular biology and pathology of elastic tissues. Wiley, Chichester (Ciba Foundation Symposium 192) p 128–147*

Two morphological components—a microfibrillar component and an amorphous component—can be identified in all elastic fibres. While it has been known for some time that elastin makes the major contribution to the amorphous component, progress in identifying the protein constituents of the microfibrillar component has only recently been made. It is now well established that fibrillin, a large non-collagenous glycoprotein, is the major protein component of microfibrils (Sakai et al 1986, 1991, Maddox et al 1989, Keene et al 1991). Immunolocalization of fibrillin demonstrated that the microfibrils of elastic fibres and other morphologically similar microfibrils, which exist in many different tissues independent of an amorphous component, are composed of the same major structural macromolecule (Sakai et al 1986). Why and how some microfibrils and not others are associated with an amorphous component is unknown. The molecular basis for the association of the microfibrillar component with the amorphous component may be determined by interactions between fibrillin and elastin and/or between these components and other molecules also associated with elastic fibres.

A number of other molecules have been immunolocalized to microfibrils. These include mircrofibril-associated glycoprotein (Gibson et al 1989, 1991), associated microfibril glycoprotein (Horrigan et al 1992), 36 kDa microfibril-associated glycoprotein (Kobayashi et al 1989), amyloid P (Breathnach et al 1981), vitronectin (Dahlbäck et al 1989), and versican (Zimmermann et al 1994). Lysyl oxidase, necessary for cross-linking elastin molecules, has been immunolocalized to the amorphous component (Kagan et al 1986), and emilin, or gp115, has been identified at the interface between amorphous elastin and the microfibrils (Bressan et al 1993).

The important structural/mechanical function performed by fibrillin has been confirmed by the identification of the gene for fibrillin (*FBN1*) as the defective gene in Marfan's syndrome, the first heritable disorder of connective tissue to be defined as such (McKusick 1955). The cardinal phenotypic features of Marfan's syndrome include skeletal, cardiovascular and ocular manifestations. In addition, there are also manifestations in the skin, lungs and CNS (Pyeritz 1993). Fibrillin-containing microfibrils are ubiquitously distributed throughout the body and are particularly abundant in the cardinal systems affected in Marfan's syndrome.

We have been interested in how fibrillin is assembled into microfibrils, which molecular interactions occur between fibrillin and other molecules associated with microfibrils and elastic fibres, and why mutations in *FBN1* result in the Marfan phenotype. In order to answer these questions, we first cloned and sequenced *FBN1* cDNA (Maslen et al 1991, Corson et al 1993) to determine the primary structure of fibrillin. We have used our cDNA to express recombinant peptides of fibrillin 1 in human cells in order to define functional domains in fibrillin and to provide the basis for understanding the effects of mutations on its function and structural integrity.

## The primary structure of fibrillin

The primary structure of fibrillin has been deduced from the cloning and sequencing of approximately 10 kb of *FBN1* cDNA (Maslen et al 1991, Lee et al 1991, Corson et al 1993, Pereira et al 1993). A second gene for fibrillin, *FBN2*, was discovered through cloning (Lee et al 1991) and is highly homologous in amino acid sequence and secondary structure to fibrillin 1, except for one notable domain (Zhang et al 1994). Hence, the following description of the structure of fibrillin 1 also applies to fibrillin 2.

The structure of fibrillin can be visualized schematically as shown in Fig. 1. Fibrillin contains a long central region composed of many cysteine-rich repetitive motifs. There are regions of unique sequence at the N- and C-terminal ends. The unique sequence comprising the N-terminal end of fibrillin

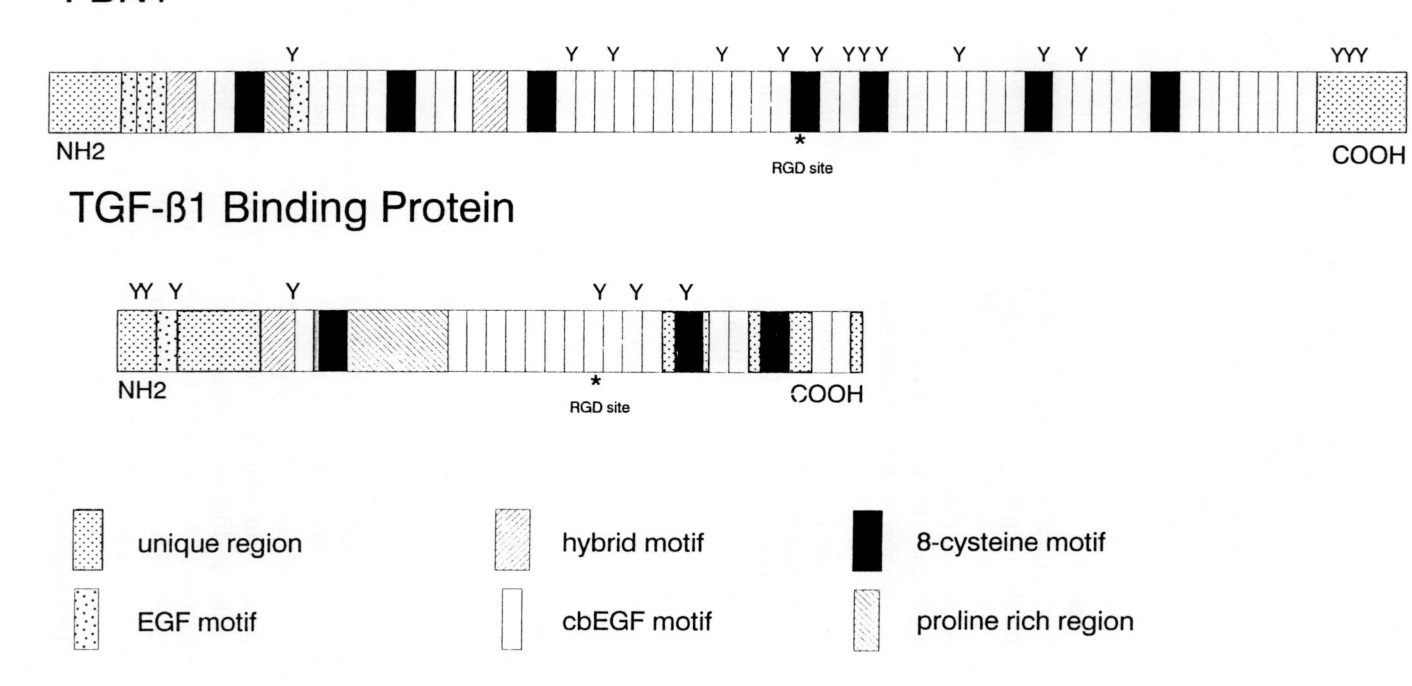


FIG. 1. The structure of fibrillin 1 and TGF-$\beta$1-binding protein, as deduced from cDNA cloning. Y indicates potential site for *N*-linked glycosylation.

is 84 residues long (including the signal sequence) and contains four cysteine residues. The unique sequence comprising the C-terminal end is 184 residues long and contains only two cysteines. There is one other region of unique sequence present in fibrillin. In fibrillin 1, this region is proline rich and contains no cysteines; the homologous region in fibrillin 2 is glycine rich.

In between the two terminal regions, fibrillin contains 47 epidermal growth factor (EGF)-like modules, which are characterized by the presence of six cysteine residues. Four of the 47 EGF-like motifs are of the following generic form:

CXXX CXXXXX CXXXXX CXC XXXXXXXX CXXXXX.

Three of these generic EGF-like motifs are adjacent to the N-terminal region, and the fourth is next to the proline-rich region. The remaining 43 EGF-like modules fit the following consensus sequence:

DID/NE CXXXXXX CXXGX CXNTXGSY/FX CXC
XXGXXXXXXXXX CX.

This sequence has been identified in many other proteins, including extracellular matrix proteins like fibulin 1 and 2, nidogen and versican. Modules of this type have been demonstrated to bind $Ca^{2+}$ (Dahlbäck et al 1990, Handford et al 1991, Selander-Sunnerhagen et al 1992).

Another cysteine-rich motif is found seven times in fibrillin. This motif contains eight cysteine residues in the following consensus sequence:

DXRXXX CYXXXXXXXCXXXXXXXXX KXX CCCXXXGXAWGXP
CEXCPXXXTXEFXXL CPXGPGFXXXXXXXXX.

This unusual sequence motif is present in only one other molecule, transforming growth factor $\beta$1 (TGF-$\beta$1)-binding protein (Kanzaki et al 1990).

A third motif, which we have described as a 'hybrid' motif (Corson et al 1993), is repeated twice in fibrillin. Through most of the module, this motif resembles the eight-cysteine motif; at the C-terminal end of the motif, it more closely resembles the EGF-like sequence. This unusual sequence motif is also present in only one other molecule; again, the other molecule is TGF-$\beta$1-binding protein (Kanzaki et al 1990).

Although TGF-$\beta$1-binding protein is smaller than fibrillin and does not have exactly matching homologous domains like fibrillin 1 and fibrillin 2, TGF-$\beta$1-binding protein contains similar repeating modules. A proline-rich region can also be identified in TGF-$\beta$1-binding protein. A schematic comparison between fibrillin and TGF-$\beta$1-binding protein is shown in Fig. 1. Other 'fibrillin-like' molecules, which may be more like TGF-$\beta$1-binding protein than fibrillin, have been identified recently (Gibson et al 1993). These proteins, and perhaps also TGF-$\beta$1-binding protein, may be associated with microfibrils.

## The 5′ end of *FBN1*

Three alternate exons, termed A, B and C, were identified upstream of the exon containing the first ATG (exon M) (Corson et al 1993). The sequence upstream of the first methionine remains in frame through exon A, the most abundant of the three. The sequence through exon B also remains in frame, whereas a stop codon is present in exon C. The genomic region containing exons B, A, C and M is very GC rich. Further analysis revealed a 1.8 kb CpG island spanning these exons, suggesting the proximity of the start site(s) for transcription.

In order to determine whether the first ATG in exon M is the start site of translation, or whether additional upstream sequences are required, we have tested specific regions of the gene for promoter activity in chloramphenicol acetyltransferase (CAT) assays. The region of *FBN1* between the *Kpn*I (K) and *Not*I (N) restriction sites, containing exons B, A, C and a portion of exon M, can drive the expression of CAT. When regions upstream of exons B, A and M were tested in this assay, each of these regions was also able to promote CAT activity. These results are summarized in Fig. 2. A segment of the 5′ *FBN1* genomic clone, from the *Kpn*I site shown in Fig. 2 upstream to the *Hin*dIII site (this genomic clone, obtained from a *Hin*dIII library, is described in detail in Corson et al 1993), failed to promote CAT activity in this assay. These results suggest that the majority, if not all, of the *FBN1* transcripts utilize the ATG codon in exon M to initiate translation.

## Recombinant peptides of fibrillin 1

In order to determine the function of fibrillin, we have prepared recombinant peptides of fibrillin 1. Utilizing the expression vector, pCis (Gorman et al 1990), and the signal peptide from BM40 (also known as SPARC), the recombinant peptides shown in Fig. 3 have been expressed in human cells (HT-1080 fibrosarcoma or 293 embryonal kidney cells from American Type Culture Collection) according to previously developed methods (Fox et al 1991, Reinhardt et al 1993) and purified from the medium. Recombinant peptides have been confirmed by N-terminal amino acid sequencing, analysed by rotary shadowing, and tested for $Ca^{2+}$ binding and reactivity with monoclonal antibodies. By all methods tested so far, the recombinant peptides appear to be properly folded and functional.

N-terminal amino acid sequence analysis of rF11 and rF8 did not result in the expected sequences. The constructs are shown in Fig. 3. However, the N-terminal sequence of rF8, which should have begun with the last $Ca^{2+}$-binding EGF-like motif, was instead ST(N)ETDASNIED. The four amino acids directly preceding this sequence are RKRR, which fits the consensus sequence

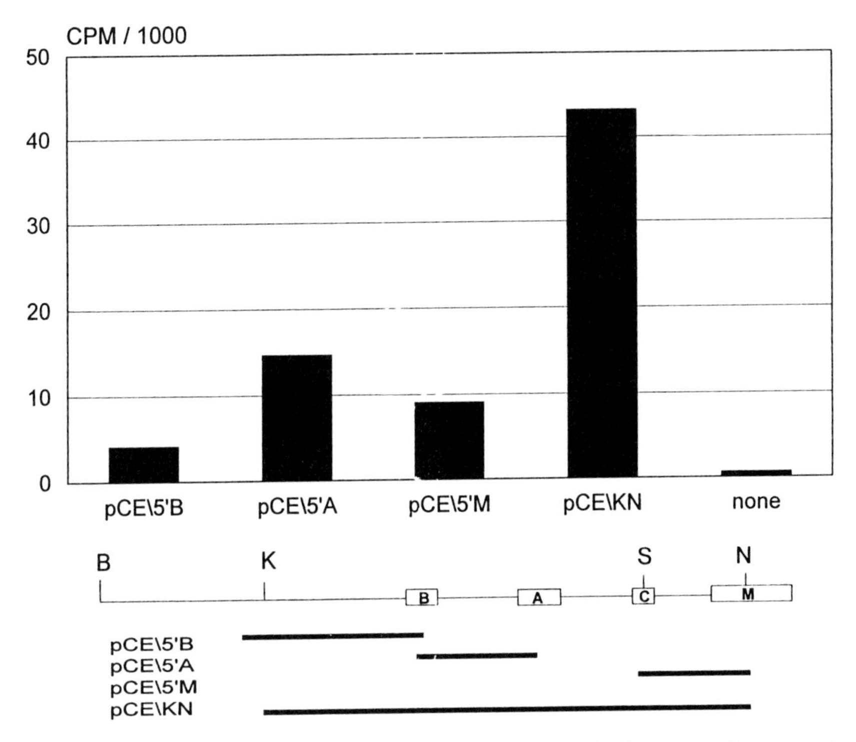


FIG. 2. Results of CAT assays testing four *FBN1* genomic fragments for promoter activity. PCR products from various regions within the *Bam*H I/*Not*I fragment of the *FBN1* gene were cloned into the CAT expression vector pCAT-Enhancer (Promega) and used to transfect human normal skin fibroblasts. The various expression vectors (*bottom*) were incubated with cells in the presence of serum-free medium and modified bovine serum (Stratagene) for 4–20 hours. Cells were incubated an additional 20–24 hours in complete Dulbecco's modified Eagle's medium. At the completion of the transfection protocol, cells were lysed and incubated with [$^{14}$C]chloramphenicol and n-butyryl coenzyme A for approximately 20 hours. After three xylene extractions, the amount of butyrylated [$^{14}$C]chloramphenicol generated by the cell lysates was determined by liquid scintillation counting. CAT activity in this figure is shown as average counts per minute (CPM) of butyrylated [$^{14}$C]chloramphenicol, based upon results of two separate experiments in which cells were transfected in triplicate with the various vector constructs. B, K, S and N represent restriction sites for *Bam*HI, *Kpn*I, *Sac*I and *Not*I, respectively. B, A, C and M (in boxes) are exons described in Corson et al (1993).

RXK/RR that signals cleavage of precursor molecules within the secretory pathway (Hosaka et al 1991). In addition, the N-terminal sequence of rF11 was RGGGGHDALKGP, and the preceding amino acids are RAKR, which also fits the consensus sequence RXK/RR signalling a cleavage site. These data indicate that fibrillin 1 may normally be processed at these sites.

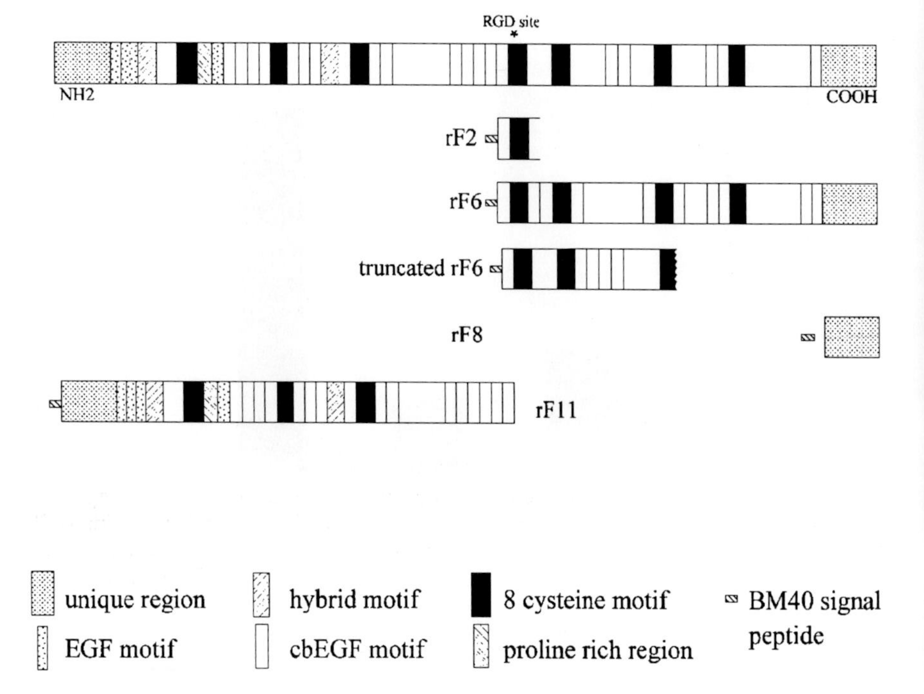


FIG. 3. Recombinant fibrillin peptides. Shown here are the expected peptides from five different *FBN1* constructs, preceded by the BM40 signal peptide.

**$Ca^{2+}$-binding**

Since fibrillin contains 43 EGF-like domains of the '$Ca^{2+}$-binding' type (cb EGF-like), we tested whether fibrillin monomers obtained from cell culture medium (Sakai et al 1991) bind $Ca^{2+}$. We found that, in ligand-binding blots, radioactive $Ca^{2+}$ bound specifically to fibrillin. $Ca^{2+}$ did not bind to fibrillin which had been treated with a disulphide bond-reducing agent (Corson et al 1993). These results are consistent with the notion that the EGF-like domains, which are presumably stabilized by three intrachain disulphide bonds, are responsible for binding $Ca^{2+}$.

Recombinant fibrillin peptide, rF2, which is composed of the fourth eight-cysteine domain (which contains the single RGD site present in fibrillin) and the flanking cb EGF-like domains on each side (see Fig. 3), was tested by equilibrium dialysis for $Ca^{2+}$ binding. This recombinant fragment contained two $Ca^{2+}$ binding sites, by Scatchard analysis, and had a dissociation constant of around 400 $\mu$M (Fig. 4). These results confirm that cb EGF-like domains in fibrillin bind $Ca^{2+}$ and suggest that each cb EGF-like domain binds $Ca^{2+}$ with equal affinity. In addition, these results suggest that the eight-cysteine modules

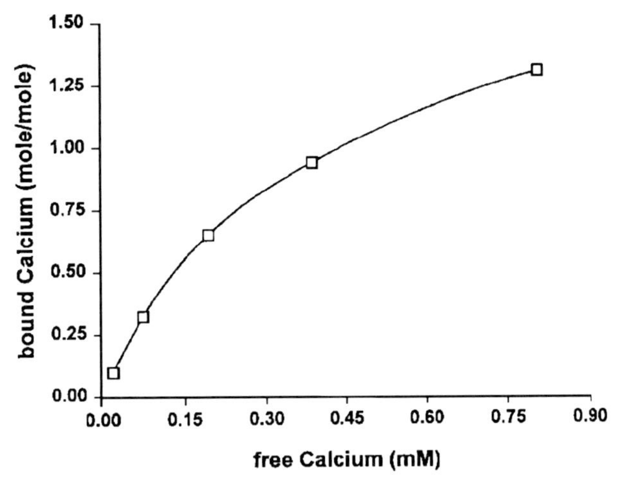


FIG. 4. $Ca^{2+}$ binding of recombinant fibrillin peptide rF2. Purified rF2 fragments were treated with EDTA and dialysed extensively against $Ca^{2+}$-free 20 mM Tris-HCl, pH 7.5, containing 0.15 M NaCl. Equilibrium dialysis was then performed in a Spectrum equilibrium dialyser (0.15 ml per half cell) with known $^{45}CaCl_2$ concentrations for 16 hours at 25 °C. Bound and free $Ca^{2+}$ were determined by scintillation counting of triplicates (40 $\mu$l) of each half cell.

do not bind $Ca^{2+}$. In order to substantiate these suggestions, additional $Ca^{2+}$ binding experiments are being performed with the larger recombinant fragments. Preliminary data obtained with the truncated rF6 have also suggested that each cb EGF-like domain has one $Ca^{2+}$ binding site of equal affinity (again the dissociation constant was determined to be around 400 $\mu$M). The dissociation constant determined in these experiments is similar to those determined for other proteins containing cb EGF-like domains. With these values for dissociation constants, $Ca^{2+}$ binding sites should be fully saturated *in vivo*.

$Ca^{2+}$ binding may be important in stabilizing the conformation of fibrillin molecules. It is thought that $Ca^{2+}$ stabilizes the $\beta$-sheet conformation between two strands in each EGF-like module. In fibrillin, since several cb EGF-like domains are present one after another in blocks of tandem repeats (the longest stretch of tandem cb EGF-like repeats has 12 in a row), $Ca^{2+}$ binding may be particularly important in stabilizing tertiary structure of these blocks of repeats.

In addition, $Ca^{2+}$ binding may be important in mediating specific protein–protein interactions. Blood coagulation proteins, which contain one or a few more cb EGF-like domains, particpate in $Ca^{2+}$-mediated interactions (Dahlbäck et al 1990), and *Drosophila* Notch, which contains many tandemly repeated cb EGF-like domains, utilizes specific ones for certain $Ca^{2+}$-mediated interactions, suggesting the possibility that Notch can be a receptor for multiple different ligands during development (Fehon et al 1990). By analogy to Notch, it may be possible that fibrillin has the potential for many different protein–protein interactions, which may be determined developmentally or in a tissue-specific manner. We are currently investigating binding of possible ligands to fibrillin, using recombinant peptides of fibrillin 1.

### Assembly of fibrillin into microfibrils

Initially, fibrillin was proposed to be a major structural component of microfibrils because (1) fibrillin is immunolocalized to all tissues where microfibrils can be visualized ultrastructurally and (2) microfibrils contain fibrillin molecules which are periodically spaced along the lengths of microfibrils (Sakai et al 1986). Previously, Ross & Bornstein (1969) demonstrated that microfibrils are stabilized by disulfide bonds. The finding that fibrillin is a cysteine-rich protein which rapidly forms intermolecular disulfide bonds (Sakai 1990, Sakai et al 1991) was consistent with the proposal of fibrillin as a major structural component of microfibrils.

How are fibrillin molecules assembled into microfibrils? We know that fibrillin monomers are linear flexible molecules about 148 nm long (Sakai et al 1991), and we know that fibrillin antibody labelling of microfibrils yields a periodicity of about 55 nm (Sakai et al 1991). Rotary-shadowed and negative-stained images of structures called 'beaded fibrils' (reviewed in Sakai & Keene 1995) have been shown to contain fibrillin and to represent alternative visualizations of microfibrils (Keene et al 1991). Fragments of microfibrils have also been characterized by rotary shadowing, amino acid sequencing and reactivity with fibrillin antibodies (Maddox et al 1989). These results have suggested that the beaded fibril is composed of linear arms between periodically spaced globular domains, and that similarly periodically spaced fibrillin molecules contribute at least to the arms, but may also contribute to the globular domains.

Immunolocalization studies with monoclonal antibodies have also been useful in suggesting the arrangement and orientation of fibrillin monomers within the structure of the microfibril. Monoclonal antibody 69 (mAb 69) binds asymmetrically just to one side of each globular domain in the beaded fibril (Keene et al 1991) (Fig. 5). MAb 69 displays the same antibody labelling periodicity on microfibrils (about 55 nm) as a different monoclonal antibody (mAb 201), and the two antibodies incubated with tissues at the same time display a double labelling with a 55 nm periodicity (Sakai et al 1991). These data suggested that linear molecules of fibrillin are assembled into microfibrils in a head-to-tail orientation (Sakai et al 1991).

MAb 201 was used previously to identify a pepsin-resistant fragment of fibrillin 1, PF1 (Maddox et al 1989) (Figs 5, 6). The N-terminal sequence of

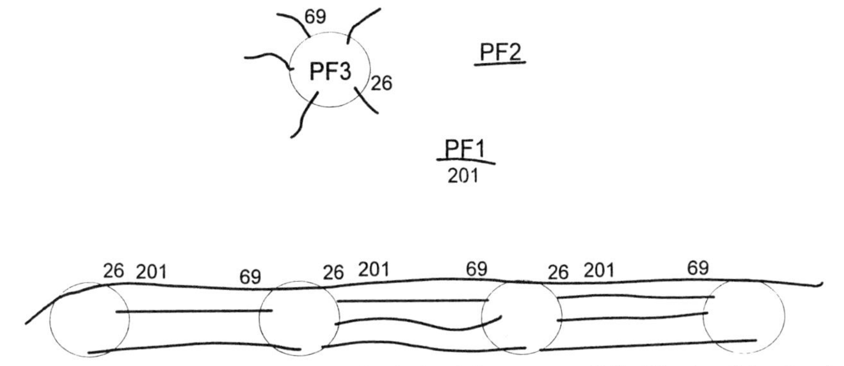


FIG. 5. Schematic drawing of pepsin-derived fragments of fibrillin 1, of the beaded fibril (microfibril), and of epitopes of monoclonal antibodies.

PF1 begins in the proline-rich region, and amino acid sequences from PF1 have been found in the deduced sequence from *FBN1* to extend through the 16th EGF-like motif (Corson et al 1993) (Fig. 6).

MAb 69 identified a pepsin-resistant aggregate called PF3, which resembles the globular domain of the beaded fibril, and did not recognize PF1 or a seond pepsin-resistant fragment of fibrillin, PF2 (Maddox et al 1989) (Figs 5, 6). PF2 has been positioned using sequence information to the region shown in Fig. 6 (Maslen et al 1991). The region in fibrillin containing the epitope for mAb 69 has been recently mapped by the use of recombinant fibrillin 1 peptides. MAb 69 reacts with rF6, the C-terminal half of fibrillin 1, but not with rF11, the N-terminal half. Since mAb 69 does not react with the truncated rF6 peptide, nor with PF2, the region containing this epitope lies between the C-terminal region and the sixth eight-cysteine domain. Also, since the epitope recognized by mAb 69 is stabilized by disulfide bonds and since the C-terminal domain may be processed, the region containing the mAb 69 epitope does not include the extreme C-terminal region.

Another monoclonal antibody, mAb 26, which recognizes PF3 but not PF1 or PF2, has been mapped to the N-terminal region of fibrillin. MAb 26 reacts with rF11, and not with rF6. These results suggest that the epitope recognized by mAb 26 lies between the N-terminal region and the proline-rich region (Fig. 6).

All together, these data, which are summarized in Figs 5 and 6, suggest that the globular domain of the beaded fibril contains overlapping N- and C-terminal regions of fibrillin 1. These overlapping regions may extend from the N-terminus to the proline-rich region (the region containing the epitope for mAb 26) and from the C-terminal region to the sixth eight-cysteine domain (the region containing the epitope for mAb 69).

In order to confirm these suggestions from our immunochemical investigations, we are currently trying to identify the domains required for

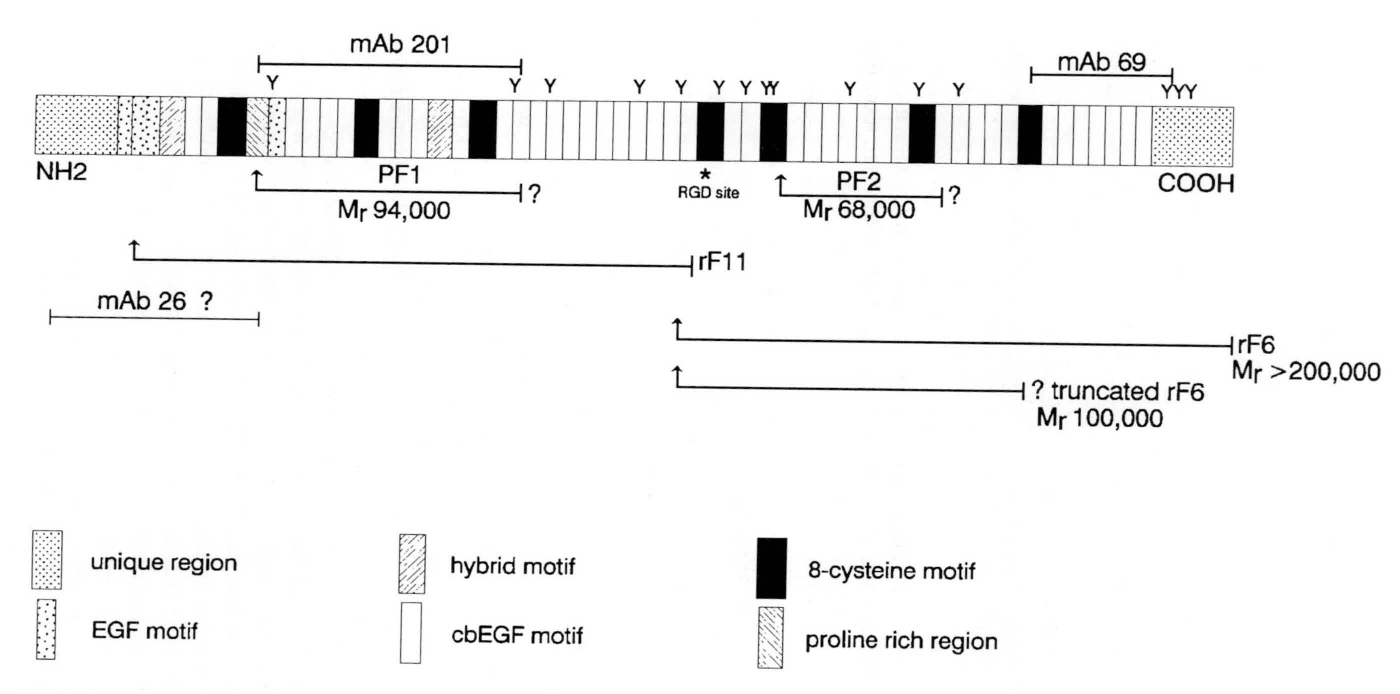


FIG. 6. Schematic drawing mapping the positions of pepsin fragments, recombinant peptides and epitopes of monoclonal antibodies. Y indicates potential site for *N*-linked glycosylation.

intermolecular disulfide bond formation and for microfibril assembly, using recombinant fibrillin peptides.

## Marfan's Syndrome

More than 30 different mutations in *FBN1* have been identified in individuals with Marfan's syndrome. These mutations include deletions, insertions and premature stop codons, but the majority of the mutations described to date are single base pair mutations which result in the substitution of a single amino acid residue in a cb EGF-like domain. The type of mutation and position of the mutation are depicted in Fig. 7 (reviewed in Dietz et al 1994).

Because Marfan's syndrome is a dominant disorder, each affected individual will have one mutant and one wild-type allele. The defective protein product of the mutant allele may exert a dominant-negative effect on the normal protein product, because multiple fibrillin molecules are assembled into an aggregate structure, the microfibril, which is dependent for its proper function upon the adequacy of its parts. Presumably, if even a few of its subunits are defective, the function of the entire microfibril can be compromised. Therefore, it is currently thought that a mutation in one fibrillin allele may cause Marfan's syndrome if the mutant fibrillin adversely affects the secretion of normal fibrillin, if it perturbs the proper assembly of the microfibril, if it destabilizes the structure of the microfibril or if it disturbs a critical functional domain (e.g. a domain utilized for binding an important ligand) such that microfibrils do not perform appropriately.

An alternative possibility to the dominant negative explanation is that Marfan's syndrome may be caused by the simple reduction in the amount of normal functional fibrillin. In this case, either a null allele, or a mutant allele which results in a mutant fibrillin that does not associate with normal fibrillin and therefore does not adversely affect microfibrils, would be sufficient to cause Marfan's syndrome. In osteogenesis imperfecta, functional null alleles result in mild disease (Byers 1993). It has been suggested that functional null alleles of fibrillin will not result in Marfan's syndrome, but in a much milder phenotype. A patient with the MASS phenotype (mitral valve involvment, mild aortic dilatation, skin and skeletal involvement) has been identified with a 4 bp insertion in one of his *FBN1* alleles, but this mutant allele is expressed at an extremely low level (6% of normal). In this case, a preponderance of normal fibrillin molecules, even if these are reduced in amount by one half, leads to mild disease and not to Marfan's syndrome. On the other hand, individuals with mutations expressed at somewhat higher levels, but significantly below those of normal (16% and 25%), had severe disease, suggesting that some critical threshold of abnormal molecules may be required to produce the Marfan phenotype (Dietz et al 1993).

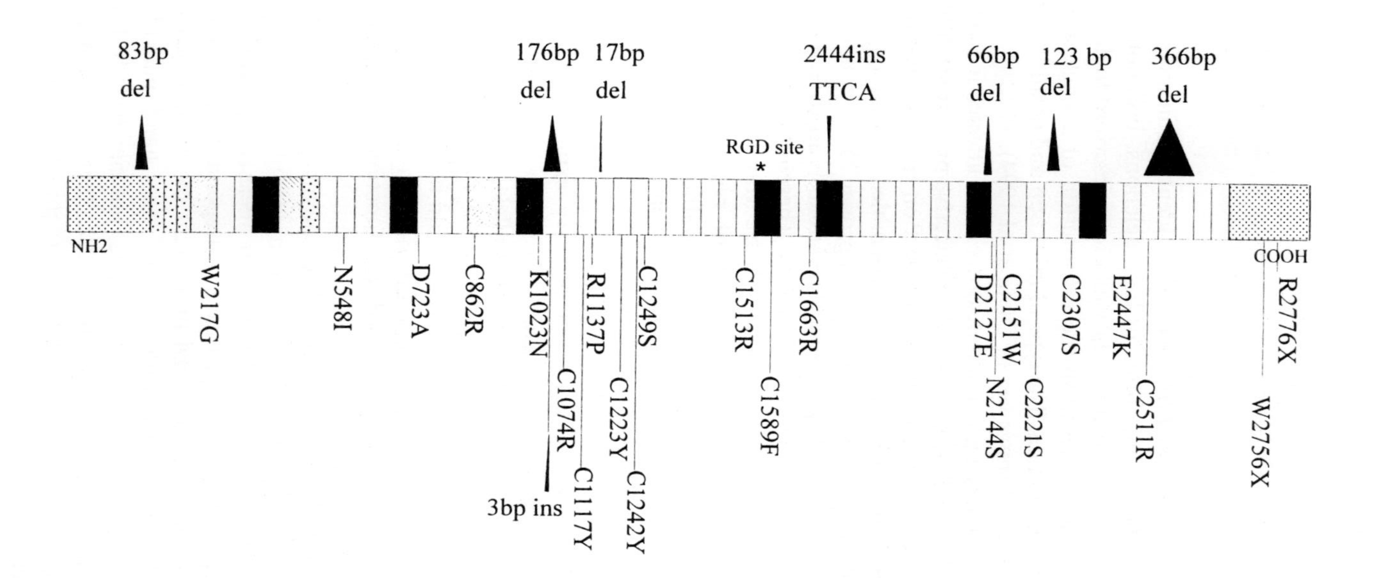


FIG. 7. Mutations in *FBN1* found in individuals with Marfan's syndrome. del, deletion; ins, insertion. For key to shaded areas, see Fig. 6.

Much more information is required before genotype–phenotype correlations can be made. The production of recombinant peptides and development of assays to test for function will provide the basic knowledge necessary for us to begin to make these correlations. Functional domains can then be tested with peptides which have engineered mutations similar to those found in Marfan's syndrome.

Since microfibrils may be complex structures containing proteins other than fibrillin 1, it is interesting to note that only *FBN1* has been linked with Marfan's syndrome. *FBN2* is not linked to Marfan's syndrome, which suggests that fibrillin 2 may play a role in the structure and function of microfibrils somewhat different to that of fibrillin 1. *FBN2* is linked to congenital contractural arachnodactyly (Lee et al 1991), raising the possibility that microfibril-related genes may underlie diseases which share some, but not all, of the phenotypic features of Marfan's syndrome. These diseases may one day be classified as diseases of microfibrils. As these disease genes are identified, additional clues to the structure and function of microfibrils will be gained.

*Acknowledgements*

We would like to thank Dr Hans Peter Bachinger for help with the $Ca^{2+}$ binding assays, Noe Charbonneau for computer graphics and technical assistance, and the Deutsche Forschungsgemeinschaft (to DPR), and the Shriners Hospital for Crippled Children for funding.

## References

Breathnach SM, Melrose SM, Bhogal B et al 1981 Amyloid P component is located on elastic fibre microfibrils in normal human tissue. Nature 293:652–654

Bressan GM, Daga-Gordini D, Colombatti A et al 1993 Emilin, a component of elastic fibers preferentially located at the elastin–microfibrils interface. J Cell Biol 121:201–212

Byers P 1993 Osteogenesis imperfecta. In: Royce P, Steinmann B (eds) Connective tissue and its heritable disorders. Wiley-Liss, New York, p 317–350

Corson GM, Chalberg SC, Dietz HC, Charbonneau NL, Sakai LY 1993 Fibrillin binds calcium and is coded by cDNAs that reveal a multidomain structure and alternatively spliced exons at the 5′ end. Genomics 17:476–484

Dahlbäck B, Hildebrand B, Linse S 1990 Novel type of very high affinity calcium-binding sites in β-hydroxy-asparagine-containing epidermal growth factor-like domains in vitamin K-dependent protein S. J Biol Chem 265:18481–18489

Dahlbäck K, Löfberg H, Alumets J, Dahlbäck B 1989 Immunohistochemical demonstration of age-related deposition of vitronectin (S-protein of complement) and terminal complement complex on dermal elastic fibers. J Invest Dermatol 93:727–733

Dietz HC, McIntosh I, Sakai LY et al 1993 Four novel FBN1 mutations: significance for mutant transcript level and EGF-domain calcium binding in the pathogenesis of Marfan syndrome. Genomics 17:468–475

Dietz HC, Ramirez F, Sakai LY 1994 Marfan's syndrome and other microfibrillar diseases. Adv Hum Genet 22:153–186

Fehon RG, Kooh PJ, Rebay I et al 1990 Molecular interactions between the protein products of the neurogenic loci *notch* and *delta*, two EGF-homologous genes in Drosophila. Cell 61:523–534

Fox JW, Mayer U, Nischt R et al 1991 Recombinant nidogen consists of three globular domains and mediates binding of laminin to collagen type IV. EMBO J 10:3137–3146
Gibson MA, Kumaratilake JS, Cleary EG 1989 The protein components of the 12-nanometer microfibrils of elastic and nonelastic tissues. J Biol Chem 264:4590–4598
Gibson MA, Sandberg LB, Grosso LE, Cleary EG 1991 Complementary DNA cloning establishes microfibril-associated glycoprotein (MAGP) to be a discrete component of the elastin-associated microfibrils. J Biol Chem 266:7596–7601
Gibson MA, Davis E, Filiaggi M, Mecham RP 1993 Identification and partial characterization of a new fibrillin-like protein (FLP). Am J Med Genet 47:148(abstr)
Gorman CM, Gies DR, McCray G 1990 Transient production of proteins using an adenovirus transformed cell line. DNA Protein Eng Tech 2:3–10
Handford PA, Mayhew M, Baron M et al 1991 Key residues involved in calcium-binding motifs in EGF-like domains. Nature 351:164–167
Horrigan SK, Rich CB, Streeten BW et al 1992 Characterization of an associated microfibril protein through recombinant DNA techniques. J Biol Chem 267:10087–10095
Hosaka M, Nagahama M, Kim WS et al 1991 Arg-X-Lys/Arg-Arg motif as a signal for precursor cleavage catalyzed by furin within the constitutive secretory pathway. J Biol Chem 266:12127–12130
Kagan HM, Vaccaro CA, Bronson RE et al 1986 Ultrastructural immunolocalization of lysyl oxidase in vascual connective tissue. J Cell Biol 103:1121–1128
Kanzaki T, Olofsson A, Morén A et al 1990 TGF-$\beta$1 binding protein: a component of the large latent complex of TGF-$\beta$1 with multiple repeat sequences. Cell 61:1051–1061
Keene DR, Maddox BK, Kuo HJ, Sakai LY, Glanville RW 1991 Extraction of extendable beaded structures and their identification as fibrillin-containing extracellular matrix microfibrils. J Histochem Cytochem 39:441–449
Kobayashi R, Tashima Y, Masuda H et al 1989 Isolation and characterization of a new 36-kDa microfibril-associated glycoprotein from porcine aorta. J Biol Chem 264:17437–17444
Lee B, Godfrey M, Vitale E et al 1991 Linkage of Marfan syndrome and a phenotypically related disorder to two different fibrillin genes. Nature 352:330–334
Maddox BK, Sakai LY, Keene DR, Glanville RW 1989 Connective tissue microfibrils. J Biol Chem 264:21381–21385
Maslen CL, Corson GM, Maddox BK, Glanville RW, Sakai LY 1991 Partial sequence of a candidate gene for the Marfan syndrome. Nature 352:334–337
McKusick VA 1955 The cardiovascular aspects of Marfan's syndrome: a heritable disorder of connective tissue. Circulation 11:321–342
Pereira L, D'Alessio M, Ramirez F et al 1993 Genomic organization of the sequence coding for fibrillin, the defective gene product in Marfan syndrome. Hum Mol Genet 2:961–968
Pyeritz RE 1993 The Marfan syndrome. In: Royce P, Steinmann B (eds) Connective tissue and its heritable disorders. Wiley-Liss, New York, p 437–468
Reinhardt DP, Mann K, Nischt R et al 1993 Mapping of nidogen binding sites for collagen type IV, heparan sulfate proteoglycan, and zinc. J Biol Chem 268:10881–10887
Ross R, Bornstein P 1969 The elastic fiber. J Cell Biol 40:366–381
Sakai LY 1990 Disulfide bonds crosslink molecules of fibrillin in the connective tissue space. In: Tamburro A, Davidson JM (eds) Elastin: chemical and biological aspects. Congedo Editore, Galatina, Italy, p 213–227
Sakai LY, Keene DR 1995 Fibrillin: monomers and microfibrils. Methods Enzymol 245:29–52
Sakai LY, Keene DR, Engvall E 1986 Fibrillin, a new 350-kD glycoprotein, is a component of extracellular microfibrils. J Cell Biol 103:2499–2509

Sakai LY, Keene DR, Glanville RW, Bächinger HP 1991 Purification and partial characterization of fibrillin, a cystein-rich structural component of connective tissue microfibrils. J Biol Chem 266:14763–14770
Selander-Sunnerhagen M, Ullner M, Persson E et al 1992 How an epidermal growth factor (EGF)-like domain binds calcium. J Biol Chem 267:19642–19649
Zhang H, Apfelroth SD, Hu W et al 1994 Structure and expression of fibrillin-2, a novel microfibrillar component preferentially located in elastic matrices. J Cell Biol 124:855–863
Zimmermann DR, Dours-Zimmerman MT, Schubert M, Bruckner-Tuderman L 1994 Versican is expressed in the proliferating zone in the epidermis and in association with the elastic network of the dermis. J Cell Biol 124:817–825

## DISCUSSION

*Rosenbloom:* Your work with the recombinant fibrillin peptides is going to answer a lot of questions.

With these peptides, or with the native fibrillin, have you been able to define where the cleavage occurs at the N-terminus? How long is the signal peptide? This is important because there are some people with Marfan's syndrome in whom very short fibrillin peptides are produced which cause a very serious phenotype.

*Sakai:* There is one mutation which involves deletion of the second exon. When we reported this finding (Dietz et al 1993a), we pointed out that for the individual with this mutation to have Marfan's syndrome and for this to be a dominant negative effect would require something very special about the first 54 amino acids (which includes the signal peptide). This is because the second exon is deleted and then there is a frame shift and a stop codon. If cleavage of the N-terminus actually occurs *in vivo*, the 54 amino acids are reduced to 10 residues of fibrillin. I find it difficult to imagine that this is a dominant negative effect, unless this mutation somehow prevents processing and this negatively impacts wild-type fibrillin.

*Uitto:* As I understand, at least from the work of Hal Dietz (McIntosh et al 1993), the major effect of nonsense mutations in Marfan's syndrome and other heritable diseases is a reduction in the mRNA steady-state levels rather than truncation of the affected protein. How does this fit with the dominant negative theory?

*Sakai:* The support for the dominant negative theory comes primarily from the work of Hal Dietz, where he has measured the level of expression of the mutant allele (Dietz et al 1993a, 1993b). In one example, there is a 4 bp insertion in the middle of the molecule which leads to a frame shift and a premature stop codon (Dietz et al 1993a). The level of expression of this mutant allele is very small (6%) relative to normal. In that particular individual, the phenotype is not Marfan's syndrome, but instead is what Reed

Pyeritz has called the 'MASS' phenotype (Glesby & Pyeritz 1993). It is characterized by involvement of mitral valve (M), aortic dilation (A), and skin (S) and skeletal (S) involvement. Although these features are present, they are not extreme, and this individual is not clinically classified as having Marfan's syndrome.

Other mutations that Hal Dietz has looked at have levels of the mutant mRNA of 16% (Dietz et al 1993a) and 25% (Dietz et al 1993b), and these two individuals have severe Marfan's syndrome. It was on the basis of this evidence that we originally suggested that the dominant negative model applies to Marfan's syndrome. But we need to examine this closely, and determine whether reduction of total wild-type fibrillin can also result in severe disease.

*Uitto:* You mentioned that fibrillins 1 and 2 are structurally very closely related, yet as far as I know there is no Marfan mutation in *FBN2*. There are linkage data between *FBN2* and congenital contractural arachnodactyly (CCA) (Tsipouras et al 1992). What's known of the differences in spatial developmental expression or function between fibrillin 1 and fibrillin 2?

*Sakai: FBN2* is clearly not linked with any families with Marfan's syndrome. As you said, there is linkage of *FBN2* to CCA. Before I answer your question I would like to make it clear that the suggestion that there may be an array of microfibrillar proteins responsible for Marfan-like syndromes really comes from this observation that *FBN2* is linked to CCA. *FBN2*, as I understand from Checco Ramirez (personal communication) is not expressed except very early on in development; this may be a reason why defects in *FBN2* do not result in something like Marfan's syndrome.

*Mecham:* We have been looking at fibrillin 2 expression with Checco Ramirez in various animal models we study in our laboratory. It is expressed early in development, and there is some indication that it might be the fibrillin that is used to initiate fibre assembly, although these data are pretty soft at the moment. Fibrillin 2 expression persists throughout development, especially in blood vessels. There are clearly tissue-specific differences in fibrillin 1 and 2 expression.

Lynn Sakai, have you tested your RGD-containing fibrillin fragment for cell adhesion activity?

*Sakai:* Yes, this was the reason we made that small peptide. It contains the eight-cysteine motif that has the RGD site and two flanking EGF-like motifs. The flanking EGF-like motifs were necessary because the eight-cysteine motif by itself didn't appear to be secreted: it was probably not folded properly and was retained in the cell. This peptide has been used in cell attachment assays where it promotes cell adhesion rather weakly. The larger C-terminal recombinant peptide that contains within it the eight-cysteine module with the RGD site does promote the attachment of smooth muscle cells and fibroblasts. Dieter Reinhardt has also looked with his recombinant peptides for interactions with integrins, and the $\alpha_v\beta_3$ integrin (the vitronectin receptor) binds to the RGD-containing sequence in fibrillin 1. We are not sure about the significance of this result.

*Mecham:* In your studies where you blocked the assembly of fibrillin in cell culture with the C-terminal fragment, which detecting antibody did you use?

*Sakai:* It was mAb 201.

*Mecham:* And that epitope is not contained in the solubilized fragment you were adding?

*Sakai:* No.

*Boyd:* Is there fibrillin 3 on chromosome 17? There was some initial data from Checco Ramirez which suggested that there was a fibrillin-like gene on chromosome 17 (Lee et al 1991). However, I haven't seen any evidence suggesting that it is expressed.

*Sakai:* Checco Ramirez has stated that this was an artefact.

*Robert:* Are there any estimates as to how much of the phenotype of Marfan's syndrome can be linked to fibrillin mutations? Can fibrillin anomalies be linked to problems other than Marfan's syndrome, such as isolated aortic aneurysms?

*Sakai:* The genotype–phenotype correlations are not there yet. The linkage data and the discovery of mutations suggest that the mutations that have been identified are responsible for the phenotype of Marfan's syndrome. Mutations in *FBN1* are also related to a disease called ectopia lentis, and Leena Peltonen has described a mutation in a family with ectopia lentis that is a missense mutation in one of the cb EGF-like motifs (Kainulainen et al 1994). But that particular case is a little controversial, because ectopia lentis is a difficult diagnosis to make. Many people would disagree that a particular family has isolated ectopia lentis and not Marfan's syndrome.

*Robert:* I'm absolutely sure that there are many more aneurysms than we suspected. We have a famous case because General De Gaulle had a sudden rupture of such an aneurysm. Little is known about them.

*Sakai:* A current idea is that Marfan's syndrome displays a broad range of phenotypes and somehow these mutations within the fibrillin molecule may be able to explain all of them. It's possible that some mutations, for example the 4 bp insertion, will lead to a very mild phenotype (MASS)—other mutations will lead to more severe phenotypes, and perhaps others will lead to something like isolated ectopia lentis. A number of people have already started to look at familial aortic aneurysm, but so far the results are not in.

*Boyd:* Marfan's patients develop aortic dissections, not aortic aneurysms; there's an enormous difference between the two phenotypes. Hal Dietz has looked at 30 or 40 of our aortic aneurysm patients and found no convincing evidence for *FBN1* mutations in any of them (personal communication). There is some concern as to whether the variants that Uta Francke has found (and I know that she has expressed this concern herself) are possibly rare polymorphisms, because she has found them in asymptomatic individuals. It is difficult to tell whether a defect is a mutation or a polymorphism because the age of onset in aneurysm patients is often very late.

*Robert:* At least in one case, we got a piece of aorta with a dissecting aneurysm, and we put it in culture to study elastin biosynthesis by incorporating the lysine. The next day everything was dissolved. There was a fantastic lytic activity in the vessel wall (Derouette et al 1981).

*Rosenbloom:* There are two other proteins which we think are in the microfibrils. The first of these, microfibril associated protein 1 (MFAP1), was cloned by Judy Foster (Horrigan et al 1992) and was originally called AMP (for associated microfibril glycoprotein). The human gene maps to the same locus as fibrillin 1 (Yeh et al 1994). Another protein, which we have just named MFAP3, has no resemblance to any of these other microfibrillar proteins (Abrams et al 1995). It maps near the locus for *FBN2* on chromosome 5, but it can be separated by chromosome translocation and so forth. We now have loci which are very close to the fibrillins. Genetically, it would be very difficult to dissociate *MFAP1* from *FBN1*.

*Pierce:* With your monoclonal antibodies, have you been able to determine if processing of fibrillin actually occurs in microfibrils?

*Sakai:* We're trying to answer that question. We have made several synthetic peptides and raised antisera to them. These peptides are from the regions that we presume are cleaved. To date, most of our data suggest that those regions are missing in fibrillin.

## References

Abrams WR, Ma R-I, Kucich U et al 1995 Molecular cloning of the microfibrillar protein MFAP3 and assignment of the gene to human chromosome 5q32–q33.2. Genomics, in press

Dietz HC, McIntosh I, Sakai LY et al 1993a Four novel FBN1 mutations: significance for mutant transcript level and EGF-like domain calcium binding in the pathogenesis of Marfan syndrome. Genomics 17:468–475

Dietz HC, Calle D, Francomano C et al 1993b The skipping of constitutive exons in vivo induced by nonsense mutations. Science 259:680–683

Derouette S, Hornbeck W, Loisance D, Godeau G, Cachera JP, Robert L 1981 Studies on elastic tissue of aorta in aortic dissections and Marfan syndrome. Path Biol 29:537–547

Glesby MJ, Pyeritz RE 1993 Association of mitral valve prolapse and systemic abnormalities of connective tissue: a phenotypic continuum. J Am Med Assoc 262:523–528

Horrigan SK, Rich CB, Streeten BW, Li Z-Y, Foster JA 1992 Characterization of an associated microfibrillar protein through recombinant DNA techniques. J Biol Chem 267:10087–10095

Kainulainen K, Karttunen L, Puhakka L et al 1994 Ten novel mutations of FBN1 resulting in a wide variety of clinical phenotypes. Nature Genetics 6:64–69

Lee B, Godfrey M, Vitale E et al 1991 Linkage of Marfan syndrome and a phenotypically related disorder to two different fibrillin genes. Nature 352:330–334
McIntosh I, Hamosh A. Dietz HC 1993 Nonsense mutations and diminished mRNA levels. Nature Genet 4:219
Tsipouras P, Del Mastro R, Sarfarazi M et al 1992 Genetic linkage of the Marfan syndrome, ectopia lentis and congenital contractural arachnodactyly to the fibrillin genes on chromosome 15 and 5. N Engl J Med 326:905–909
Yeh H, Chow M, Abrams WR et al 1994 Structure of the human gene encoding the associated microfibrillar protein (MFAP1) and localization to chromosome 15q15–q21. Genomics 23:443–449

# Elastin gene mutations in transgenic mice

Jan L. Sechler§, Lawrence B. Sandberg‡, Philip J. Roos‡, Ida Snyder, Peter S. Amenta*, David J. Riley† and Charles D. Boyd

*Departments of Surgery, *Pathology and †Medicine, UMDNJ-Robert Wood Johnson Medical School, New Brunswick, NJ 08903 and ‡Department of Pathology, Jerry L Pettis Memorial Veterans Hospital, Loma Linda University, Loma Linda, CA 92357, USA*

*Abstract.* We have constructed several rat tropoelastin minigene recombinants encoding the complete sequence of rat tropoelastin, two isoforms of rat tropoelastin and a truncated tropoelastin lacking the domains encoded by exons 19–31 of the rat gene. Coding and non-coding domains in all these recombinants were placed under the transcriptional control of 3 kb of the promoter domain of the rat tropoelastin gene. These minigenes were used to prepare a total of 28 separate founder lines of transgenic mice. A species-specific reverse-transcriptase polymerase chain reaction (RT-PCR) assay was established to demonstrate the synthesis of rat and mouse tropoelastin mRNA in several tissues obtained from both neonatal and adult transgenic mice. Thermolytic digestion of insoluble elastin isolated from several neonatal mouse tissues revealed the presence of rat tropoelastin peptides in progeny from all those founder mice in which detectable levels of rat tropoelastin mRNA were noted. Phenotypic and histopathological assessment of transgenic and non-transgenic animals revealed the development of two diverse elastic tissue disorders. The progeny of two separate founder lines overexpressing the rat tropoelastin isoform lacking exon 33, developed an emphysematous phenotype in early adulthood. In contrast, transgenic mice, in which expression of the truncated rat tropoelastin minigene lacking exons 19–31 had been observed, died of a ruptured ascending aortic aneurysm. Tropoelastin gene mutations, therefore, will result in heritable disorders of elastic tissue. Moreover, different mutations in the tropoelastin gene will be responsible for very different abnormalities in elastic tissue function.

*1995 The molecular biology and pathology of elastic tissues. Wiley, Chichester (Ciba Foundation Symposium 192) p 148–171*

Elastic fibres are extracellular matrix structures responsible for the properties of resilience and elastic recoil in all elastic tissues (Parks & Deak 1990, Rosenbloom 1984). There are two morphological components to elastic fibres,

§Present address: Department of Molecular Biology, Princeton University, Pinceton, NJ 08544, USA.

the microfibrillar component and the amorphous component (Mecham & Heuser 1991). The microfibrillar component is made up of 10–12 nm microfibrils that are composed of at least seven different glycoproteins, the best characterized of which are two genetically distinct 340 kDa glycoproteins called fibrillins (Gibson et al 1991, Lee et al 1991). Elastin constitutes the amorphous component of elastic fibres and is assembled on the microfibrillar scaffold from a soluble precursor, tropoelastin (Indik et al 1990). Over the last five years, extensive sequence analysis of cDNA and genomic DNA recombinants encoding tropoelastin from several phylogenetically diverse species has revealed that tropoelastin is indeed a family of hydrophobic proteins, all approximately 70 kDa and synthesized from a single-copy, multiexon gene by extensive alternate usage of several exons (Indik et al 1990, Boyd et al 1991, Pierce et al 1990). A typical isoform of tropoelastin is characterized by alternating hydrophobic domains (rich in glycine, valine and proline) and alanine-rich, lysine-containing sequences. These lysine-containing regions serve as a substrate for the formation of lysine-derived cross-links (desmosines and isodesmosines) that are formed during the assembly of elastin (Rosenbloom 1984). The formation of desmosine and isodesmosine cross-links is catalysed by an elastic fibre-associated enzyme, lysyl oxidase (Kagan 1986). Both the hydrophobic domains and the desmosine cross-links in elastin are necessary for the property of elastic recoil.

There are a variety of disorders characterized by abnormal elastin synthesis and a concomitant deposition of aberrant elastic fibres. Common, multifactorial diseases such as hypertension (Mecham et al 1987, Deyl et al 1985, Iredale et al 1989), atherosclerosis (Kramsch & Hollander 1973), actinic elastosis (Uitto et al 1989) and even some forms of breast cancer (Glaubitz et al 1984) involve increased accumulation of elastin and an associated deposition of morphologically atypical elastic fibres. Several, less common, heritable diseases are also characterized by an aberrant deposition of elastic fibres associated with either increased or decreased accumulation of elastin. For example, abnormal deposition of elastic fibres is characteristic of Marfan's syndrome (Hollister et al 1990) and supravalvular aortic stenosis (SVAS) (O'Connor et al 1985). Cutis laxa is a heritable cutaneous disorder characterized by reduced elastin synthesis (Olsen et al 1988). In contrast, pseudoxanthoma elasticum and the Buschke–Ollendorff syndromes are examples of heritable skin diseases in which an increased deposition of elastin has been demonstrated (Neldner 1988, Uitto & Shamban 1987).

In all of the rare and more common diseases of elastic tissue, the presence of mutations in genes responsible for the synthesis of elastin fibre proteins has been causally implicated. However, only in Marfan's syndrome have mutations in a fibrillin gene been identified that are clearly responsible for the pathogenesis of this disease (Dietz et al 1991). Although several recent reports of large mutations involving the tropoelastin gene have been described in patients with SVAS (Curran et al 1993, Ewart et al 1994), there is no evidence to date that unambiguously shows that tropoelastin gene mutations will result in any elastic tissue phenotype.

Transgenic mice have been used extensively in the past to show that gene mutations will result in the pathogenesis of a wide range of heritable diseases (Jaenisch 1988, Palmiter & Brinster 1986). Recently, for example, mutations in the gene encoding type X collagen have been introduced into transgenic mice and have been shown to be responsible for the development of skeletal deformities in transgenic progeny (Jacenko et al 1993). Using a similar approach, in this paper we describe the construction of four rat tropoelastin minigenes, the introduction of these recombinants into transgenic mice and a biosynthetic, phenotypic and histopathological examination of mouse elastic tissue synthesizing normal and aberrant rat tropoelastin. The results demonstrate, for the first time, that mutations in the tropoelastin gene will cause an elastic tissue disorder. The diverse phenotypes that develop as a consequence of different tropoelastin gene mutations, moreover, establish an important precedent for a role for mutations in the tropoelastin gene in analogous human disorders of elastic tissue, including SVAS.

## Results and discussion

Elastin is a polymer assembled from one or more isoforms of a monomeric subunit (Mecham & Heuser 1991). Our reasoning behind the construction of several founder strains of transgenic mice was that the introduction of an exogenous tropoelastin gene containing a mutation would result in the synthesis of abnormal tropoelastin monomers that would be incorporated into elastin together with the normal, endogenous mouse tropoelastin. If the exogenous tropoelastin is expressed at high enough levels and if the domain of tropoelastin affected by the mutation is important to elastin synthesis, then disruption of elastic fibre assembly and subsequent elastic tissue function should be apparent. In other words, the synthesis of the endogenous mouse tropoelastin should not mask a phenotype resulting from a mutation introduced into a tropoelastin gene. Indeed, the synthesis of endogenous, normal tropoelastin may be an important prerequisite to maintaining a non-lethal phenotype.

In preparation for the development of several founder transgenic animals, we prepared four constructs using our previously characterized cDNA and genomic DNA recombinants (Pierce et al 1990, 1992, Alatawi 1994) encoding rat tropoelastin (Fig. 1). All four recombinants contained approximately 3 kb of genomic DNA upstream of exon 1 of the rat tropoelastin gene. By DNA sequencing and expression studies (using the chloramphenicol acetyltransferase [CAT] reporter gene), we have shown that this 3 kb fragment contained many of the promoter elements necessary for the initiation and possible regulation of transcription of the rat tropoelastin gene (Sechler et al 1995a). This promoter domain was followed by rat tropoelastin cDNA sequence corresponding to coding domains within exons 1–35. A single intron (intron 35) from the rat tropoelastin gene was introduced downstream of the cDNA sequence, followed by exon 36 which contained the termination codon and the complete 3′ untranslated domain.

The first of these minigene constructs (Tropoprom-1) contained the complete coding sequence for rat tropoelastin. Tropoprom-2 was identical to Tropoprom-1 except it lacked the cDNA sequence encoded by exon 33. Tropoprom-3 lacked the coding domain complementary to exons 13–15. The final construct, Tropoprom-4, lacked the coding sequence contained within exons 19–31 (Fig. 1).

Tropoprom-1 is essentially a control construct that was developed to establish that a normal elastic fibre will develop as a chimaera of mouse tropoelastin and the predominant, constitutively spliced isoform of rat tropoelastin. Tropoprom-2 and Tropoprom-3 should result in the synthesis of alternatively spliced isoforms of rat tropoelastin that lack exon 33 and exons 13–15, respectively. If the developmentally regulated and tissue-specific pattern of alternate usage of exon 33 (Heim et al 1991) is important to elastic fibre assembly, then the appearance of abnormal amounts of the tropoelastin isoform lacking the domain encoded by exon 33 should influence elastin assembly at least in some tissues. Similarly, Tropoprom-3 (lacking exons 13–15) should, if expressed in transgenic mice, also test the functional significance of splice variants of tropoelastin that lack the amino acid sequence encoded by these three exons (Pierce et al 1990). Tropoprom-4 contains a mutation lacking 13 exons at the 3′ end of the rat tropoelastin gene. If a disruption of coding sequence within the tropoelastin gene (not an aberration of synthesis of the pattern of tropoelastin isoforms) will result in an elastic tissue disorder, expression of this large mutation in the Tropoprom-4 construct should result in an elastic tissue pathology.

These four rat tropoelastin minigene constructs were separately injected into the male pronuclei of approximately 600 fertilized mouse embryos. The injected embryos were then implanted into pseudopregnant mice. A total of 113 live births were recorded. Using mouse tail genomic DNA and a PCR assay specific for the rat minigene recombinants, we demonstrated that 28 of these 113 founder mice contained the rat constructs integrated into genomic DNA. Four founder mice contained Tropoprom-1, 11 animals contained Tropoprom-2, 10 transgenic mice had integrated Tropoprom-3 into genomic DNA and three founder mice contained Tropoprom-4.

*Expression of rat tropoelastin minigenes in transgenic mice*

Founder transgenic mice ($F_0$) were bred with normal mice and tail clip DNA from the resulting progeny ($F_1$) was analysed for the incorporation of the rat

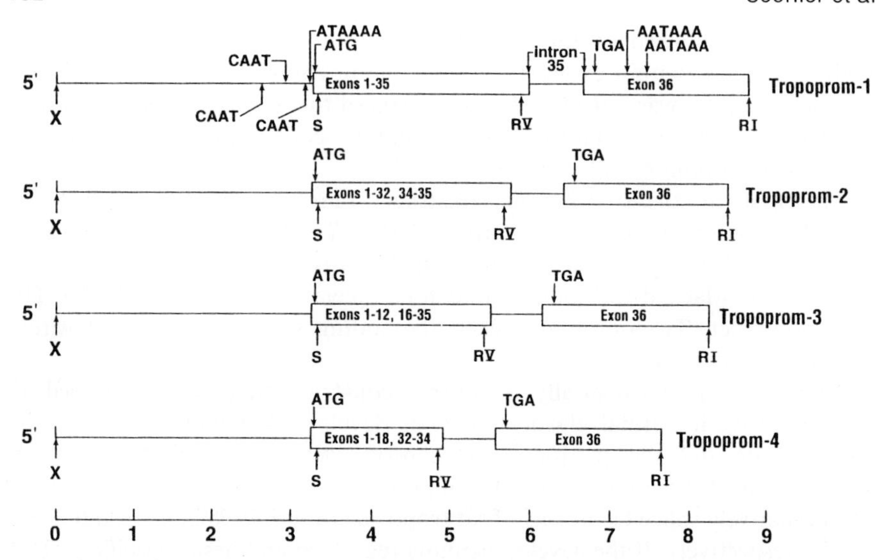


FIG. 1. The rat tropoelastin minigene contructs. Four minigene constructs (referred to as Tropoprom 1–4) were assembled from a combination of previously characterized rat tropoelastin cDNA and genomic DNA recombinants (Alatawi 1994, Pierce et al 1990, 1992). Each minigene contained 3.2 kb of the promoter domain 5′ of exon 1. This region is indicated as an *Xho*I (X) and *Sau*3A1 (S) restriction fragment which also included the ATG initiation codon. The positions of a TATA box motif (ATAAA) and three CAAT boxes located within this restriction fragment are also indicated. These *cis*-acting consensus sequences are present in all the Tropoprom constructs but are indicated only in the diagram of the Tropoprom-1 recombinant. Each Tropoprom construct also contains a single intron (intron 35) and the entire exon 36. Exon 36 encodes the cysteine-containing C-terminus of tropoelastin, a termination codon and a 3′ untranslated domain that includes two polyadenylation signals (AATAAA). This region of each Tropoprom construct was isolated as an *Eco*RV (RV)–*Eco*RI (RI) restriction fragment from a previously characterized genomic DNA recombinant. Each Tropoprom construct contained different exon domains represented in several rat tropoelastin cDNA recombinants. Tropoprom-1 contains a continuous cDNA sequence derived from exons 1–35. Tropoprom-2 contains the identical cDNA sequence but lacks the DNA sequence encoded by exon 33. Coding sequence derived from exons 13–15 is missing in Tropoprom-3. Exons 19–31 are absent in the Tropoprom-4 recombinant.

transgene into genomic DNA. Transgenic $F_1$ animals, hemizygous for the integrated copy or copies of the transgene were then used in a series of brother–sister matings to generate an $F_2$ generation, 25% of which will be homozygous for the integrated transgenes. Confirmed homozygous $F_2$ transgenic animals and, where necessary, hemizygous $F_1$ and $F_2$ transgenic progeny were then used to determine both trangsene copy number and levels of expression of the integrated rat tropoelastin minigene constructs.

Southern blot analysis of transgenic mouse tail genomic DNA, together with the appropriate copy number calibration standards derived from the rat minigene constructs, confirmed that a range of copy numbers for the various constructs existed. As few as two copies per diploid genome of Tropoprom-3 were integrated in one founder line and 52 copies of Tropoprom-2 were integrated in another transgenic founder (Table 1).

To assess the transcriptional expression of the integrated transgenes through the appearance of mature rat tropoelastin mRNA in transgenic mouse RNA preparations, we initially determined the complete derived amino acid sequence of mouse tropoelastin from several overlapping RT-PCR products obtained from mouse tropoelastin mRNA (Wydner et al 1994). A comparison of the coding sequence from mouse and rat tropoelastin message allowed us to synthesize mouse- and rat-specific oligomers. Species-specific primer pairs were then used to confirm, by RT-PCR analysis, the presence of rat tropoelastin mRNA in several tissues from $F_1$ progeny of the original founder transgenic mice. Screening progeny from 25 of the founder lines (three founder transgenic mice containing Tropoprom-2 constructs did not produce any progeny) revealed that 32% of these original founder lines did not synthesize detectable levels of rat tropoelastin mRNA. All the other founder lines expressed varying

**TABLE 1 Relative levels of expression of rat and mouse tropoelastin mRNA in transgenic mouse tissue**

| *Construct* | *Founder line number* | *Relative levels of rat tropoelastin mRNA* | | | | | *Transgene copy number* |
|---|---|---|---|---|---|---|---|
| | | *Skin* | *Lung* | *Kidney* | *Liver* | *Aorta* | |
| Tropoprom-1 | 508 | 809 | 26 | 189 | 812 | ND | ND |
| | 507 | 325 | 28 | 66 | 508 | ND | 25 |
| Tropoprom-2 | 510 | 32 | 11 | 269 | 1080 | 5 | 2 |
| | 510* | ND | 84 | 1111 | 1250 | ND | 2 |
| | 513 | 400 | 13 | 433 | 1101 | 6 | 52 |
| | 516 | 66 | 10 | 88 | 633 | ND | 19 |
| Tropoprom-3 | 525 | 95 | 23 | 633 | 1320 | 7 | ND |
| Tropoprom-4 | 539 | 46 | 5 | 22 | 177 | 4 | 3 |
| | 539* | 115 | 35 | 304 | 1036 | ND | 3 |
| | 540 | 196 | 20 | 768 | 767 | 20 | ND |

Total RNA was isolated from several tissues from progeny of eight founder lines of transgenic mice. Rat and mouse tropoelastin mRNA was quantitated by laser densitometric scanning of autoradiograms of radiolabelled PCR products synthesized as described in the legend to Fig. 2. Levels of endogenous mouse tropoelastin mRNA were set in each tissue at 100%. Values presented reflect steady-state levels of rat tropoelastin mRNA. We obtained all the results by using RNA isolated from neonatal animals (2–3 days old) except for the determinations using progeny from the founder lines marked with an asterisk: tissue was obtained from these animals at 5 weeks of age. ND, not done.

levels of a mature, correctly processed rat tropoelastin mRNA that, appropriately, lacked intron 35.

Levels of expression of transcriptionally active transgenes were assessed by the use of a quantitative RT-PCR assay in which relative steady-state levels of both mouse and rat tropoelastin mRNA were measured (Fig. 2). A set of oligonucleotide primers were designed that were complementary to a region within exon 9 and exon 17 that was identical in both mouse and rat tropoelastin mRNA. RT-PCR amplification using these primers resulted in the synthesis of a 510 bp product from mouse total RNA and an almost identically sized 516 bp product from rat total RNA. To identify these DNA fragments, we took advantage of some unique restriction sites. The 516 bp rat amplimer, for example contained a single *Apa*I site. Digestion with *Apa*I produced 156 bp and 360 bp fragments. The mouse amplimer lacked this restriction site. *Apa*I digestion of co-amplified RT-PCR products from both mouse and rat tropoelastin mRNA, therefore, would produce restriction fragments readily separable by polyacrylamide gel electrophoresis (Fig. 2). Radiolabelling these restriction fragments allowed us to quantitate the amount of each species-specific amplimer by densitometric scanning of autoradiograms.

Using this assay, we screened total RNA preparations from newborn progeny obtained from several founder transgenes and the results are summarized in Table 1.

It is clear from these integration and expression studies that the 3.2 kb promoter domain used in the construction of the rat tropoelastin minigene recombinants is, in many tissues, a strong promoter responsible for the synthesis of abundant levels of rat tropoelastin mRNA readily detectable by our RT-PCR assay. The regulation of transcription by this promoter domain is, however, complex; evaluation of tissue-specific expression of the rat minigenes indicated that appropriate expression varied according to the type of tissue. For example, levels of expression of the rat transgenes in skin was usually comparable to or exceeded that of the endogenous expression of the mouse gene. In lung and aorta, however, transgene expression levels were significantly lower than the endogenous gene. In mouse kidney and liver, rat tropoelastin mRNA levels far exceeded the expression of the endogenous gene. This varied expression of the rat minigene constructs in different tissues suggests that different *cis*-acting elements within or flanking the tropoelastin gene are responsible for tissue-specific regulation of transcription. The *cis*-acting elements responsible for expression of the tropoelastin gene in skin seem to be represented within the 3.2 kb promoter domain present in the rat transgenic constructs. In contrast, the *cis*-acting DNA sequences necessary for appropriate expression in lung and aortic tissue are not present in the transgenic construct. In particular, the *cis*-acting elements required for inhibition of expression in non-elastogenic tissues such as kidney and liver are absent from

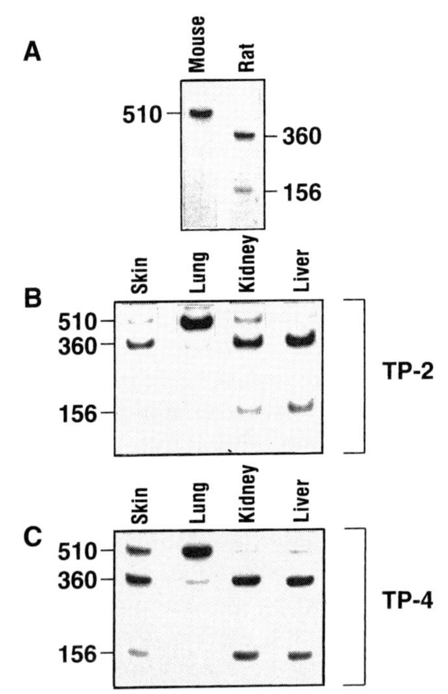


FIG. 2. A RT-PCR assay for the relative quantitation of rat and mouse tropoelastin mRNA in elastic tissue from transgenic mice. Oligonucleotide primers were designed that were complementary to a cDNA sequence encoded by exon 9 (MR9) and exon 17 (MR17) in the rat tropoelastin gene that was identical to coding sequence within the analogous domains in mouse tropoelastin mRNA. (A) Following reverse transcription of tropoelastin mRNA using Moloney murine leukaemia virus reverse transcriptase and random hexamer primers, polymerase chain amplification using MR9 and MR17 primers resulted in a 516 bp fragment from rat tropoelastin mRNA and a 510 bp fragment from mouse tropoelastin mRNA. *Apa*I digestion of PCR products generated 360 bp and 156 bp fragments from amplimers derived from rat tropoelastin mRNA. In contrast, the mouse-derived PCR product did not contain an *Apa*I site. PCR products were radiolabelled by including [$^{32}$Pd]CTP in the amplification reaction. Radiolabelled DNA fragments were size-separated by electrophoresis through 8% polyacrylamide. Polyacrylamide gels were dried and exposed to X-ray film. The sizes of restriction fragments (in bp) were calculated from molecular weight DNA markers (1 kbp DNA ladder) run in parallel. Panel (A) illustrates the recovery of PCR products, following *Apa*I digestion, obtained from RNA isolated from non-transgenic neonatal mouse and rat skin. Panels (B) and (C) illustrate the recovery of *Apa*I restriction fragments obtained from PCR amplimers prepared from RNA isolated from tissue samples of neonatal Tropoprom-2 (TP-2) and Tropoprom-4 (TP-4) transgenic mice.

the transgenic recombinants. The low levels of endogenous expression in these tissues are most likely to originate from the vasculature.

Another striking observation, clearly evident from these expression studies, is that the regions of the tropoelastin gene necessary for the developmental regulation of expression are completely absent from the rat minigene constructs (Table 1). While endogenous levels of mouse tropoelastin mRNA were shown to decline in total RNA isolated from older transgenic progeny (five-week-old mice) rat tropoelastin mRNA continued to be expressed at levels comparable to those detected in neonatal tissue.

### *Incorporation of rat tropoelastin into elastic fibres in transgenic mice*

It is clear from the analysis of rat tropoelastin mRNA levels described earlier that all four minigene constructs are transcriptionally active in the majority of the founder transgenic lines created. As an additional prerequisite to an evaluation of the effect of expression of a rat tropoelastin gene mutation on elastic fibre formation in transgenic mice, we also analysed the incorporation of rat tropoelastin into insoluble elastin isolated from several transgenic mouse elastic tissues.

We could not distinguish between mouse and rat elastin or tropoelastin using an immunohistochemical approach because antibodies specific to either rat or mouse tropoelastin are not available. To distinguish therefore between rat and mouse tropoelastin within a preparation of insoluble elastin, we took advantage of the observation by Sandberg and co-workers that high-performance liquid chromatography (HPLC) profiles of thermolytic digests of isolated insoluble elastin can be species specific (Sandberg et al 1990). These investigators had previously shown clear differences in HPLC profiles of thermolytic digests of insoluble elastin isolated from sheep and rat elastic tissue (Sandberg et al 1990). More recently, these investigators also demonstrated profile differences between preparations of insoluble elastin isolated from mouse and rat elastic tissue (L. B. Sandberg, P. J. Roos, unpublished observations). These observations immediately suggested a means to identify peptides from rat and mouse tropoelastin isolated from the same tissue.

In collaboration with Dr Sandberg, we demonstrated that differences in HPLC profiles between thermolytic digests obtained from rat and mouse insoluble elastin arose principally through differences in the frequency of hydrophobic peptide repeats within elastin from the two species (Fig. 3). For example, from a comparison of the complete amino acid sequence of mouse and rat tropoelastin, it is apparent that the hexapeptide VGGVPG is repeated six times in mouse tropoelastin and only four times in rat tropoelastin (Wydner et al 1994). The largest peak in the HPLC profile prepared from mouse elastin is obtained after 37 minutes of elution (we have referred to this peak as P37). Protein sequence analysis of P37 revealed that it contained six copies of VGGVPG. Sequence analysis of a thermolytic peptide recovered from rat elastin after 37 minutes of elution from the HPLC column revealed, predictably, only three copies of the same hexapeptide repeat. The size of the P37 peak obtained from rat elastin is, as expected, proportionately smaller than the P37 peak in the HPLC profile obtained from mouse elastin.

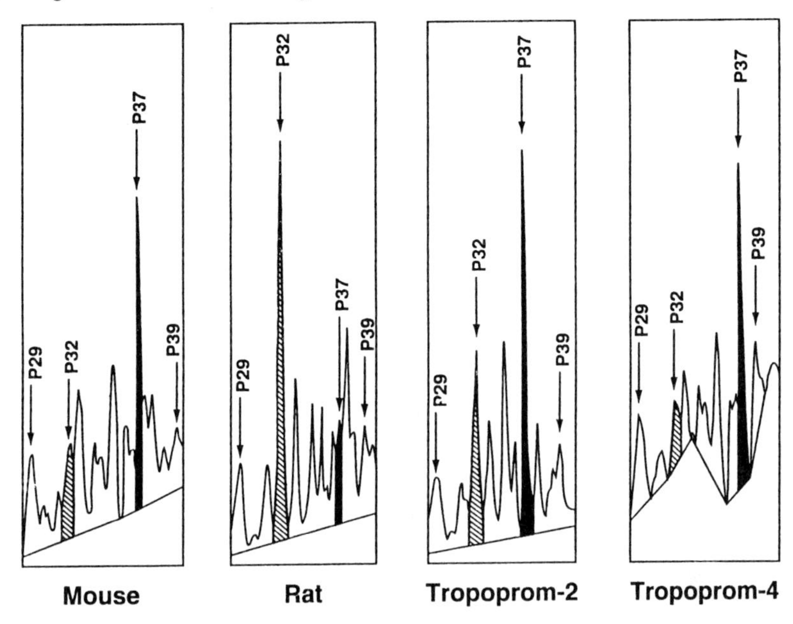


FIG. 3. HPLC profiles of thermolytic peptides obtained from insoluble elastin isolated from rat, mouse and transgenic mouse skin. HPLC of peptides isolated from insoluble elastin following digestion with thermolysin was carried out as previously described (Sandberg et al 1990). The largest peak obtained from rat elastin was evident after 32 min elution off the HPLC column (P32). The largest peak obtained from mouse elastin was evident after 37 min elution (P37). These peaks are indicated on the chromatograms together with the elutions times of several other peaks (in min). The various chromatograms were obtained using neonatal skin tissue from various sources, as indicated.

Similarly, the largest peak in an HPLC profile of a thermolytic digest of rat elastin was recovered after 32 min of elution (a peak we have called P32). Sequence analysis of this peptide revealed the presence of the tetrapeptides ALPG and AVPG. ALPG is repeated four times and AVPG is present at three copies in rat tropoelastin. A P32 peak is present in the HPLC profile obtained from mouse elastin; the size of the P32 peak correlates exactly with the representation of ALPG and AVPG as single copy peptides in mouse elastin.

The size of both the P37 and P32 peaks are clearly an indication of the phylogenetic origins of the peptides represented by these peaks (Fig. 3). Therefore, we used the P32:P37 ratio to define the difference between rat and mouse elastin. It is apparent from Table 2 that a P32:P37 ratio of 1.414 is typical for elastin isolated from rat skin. In contrast, a P32:P37 ratio of 0.598 was obtained from mouse skin elastin.

**TABLE 2 P32:P37 ratios obtained from HPLC profiles of thermolytic digests of rat, mouse and transgenic mouse skin**

| *Tissue source* | *P32 : P37 ratio* |
|---|---|
| Rat | 1.085 |
| Rat | 1.285 |
| Rat | 1.872 |
| Normal mouse | 0.635 |
| Normal mouse | 0.560 |
| Tropoprom-1 (#507) | 1.567 |
| Tropoprom-2 (#513) | 1.082 |
| Tropoprom-2 (#518) | 1.067 |
| Tropoprom-3 (#537) | 2.151 |
| Tropoprom-4 (#539) | 0.292 |

P32 and P37 peaks were isolated as fractions from the HPLC column. An amino acid composition was then determined for each peak. Total peptide concentration in each peak was determined from this compositional analysis (in nanomoles). The amount of peptide recovered in each peak was then normalized to the total amount of peptide applied to the HPLC column. These normalized values (nanomoles/mg) were then compared to provide the ratios reported in this table. Determinations from skin obtained from three different neonatal rats are presented. The average of these ratios is 1.414. Two determinations are presented from non-transgenic neonatal mouse skin. The average of these ratios is 0.598.

P32:P37 ratios were then calculated from HPLC profiles obtained from thermolytic digests of insoluble elastin obtained from the skin of age-matched progeny of several founder transgenic mice. The calculated ratios ranged from 1.067 to 2.151. This peak ratio was significantly different to the ratio recovered from non-transgenic mouse skin and clearly demonstrated the contribution of rat peptides to the increased size of the P32 peak and concomitant increase in the P32:P37 ratio. Ratios recovered from some of the transgenic mice were intermediate between ratios obtained from normal rats and mice, and reflected the presence of chimeric elastin, composed of a mixture of rat and mouse tropoelastin. In skin from other transgenic mice, the P32:P37 ratio is approximately the same as the peak ratio recovered from normal rat skin. These skin samples were obtained from progeny in which we had previously shown high levels of expression of the transgenic constructs. The tetrapeptide repeats present in the P32 peak are three and four times more abundant in rat elastin than mouse elastin. Overexpression of a rat transgene, therefore, will readily mask any contribution of these peptides synthesized from the mouse tropoelastin gene.

The P32 : P37 ratio determined from skin isolated from transgenic mice containing transcriptionally active forms of the minigene constructs Tropoprom 1–3 were all significantly different from non-transgenic mouse ratios. Clearly, the rat tropoelastin mRNA we had observed earlier in transgenic mouse tissue was responsible for the synthesis of tropoelastin that was not only secreted but also incorporated into elastic fibre.

The P32 : P37 ratio recovered from skin obtained from transgenic mice expressing the Tropoprom-4 construct is strikingly different to any ratio previously seen in skin samples from either transgenic or non-transgenic mice (Fig. 3 and Table 2). The Tropoprom-4 construct lacks exons 19–31. The repeating tetrapeptides ALPG and AVPG are encoded by exon 22 in the rat tropoelastin gene (Pierce et al 1990). Consequently, expression of the truncated rat tropoelastin gene present in the Tropoprom-4 recombinant should not contribute thermolytic peptides to the P32 peak. Similarly, a single copy of the hexapeptide VGGVPG, encoded within exon 24 of the rat tropoelastin gene, should also not be represented in the truncated tropoelastin synthesized by the Tropoprom-4 construct. However, exon 7 of the rat gene encodes two copies of VGGVPG and these hexapeptides will be represented in the aberrant protein synthesized from Tropoprom-4. The presence of a truncated rat tropoelastin containing two copies of VGGVPG should contribute to an increased size of the P37 peak which already represents six copies of the hexapeptide contributed from the endogenous expression of the normal mouse tropoelastin gene. No change in the size of the P32 peak but an increased P37 peak as a consequence of Tropoprom-4 expression would result in a P32 : P37 ratio that would be less than the ratio recovered from normal mouse skin elastin. This is precisely the result we obtained (Table 2). The P32 : P37 ratio from transgenic mice containing the Tropoprom-4 construct demonstrated the presence of a truncated tropoelastin incorporated into elastin. The unusual ratio also confirmed the validity of our assay.

### *The phenotypic consequences of expression of the Tropoprom constructs in transgenic mice*

The first indications of an abnormal phenotype were evident in the founder transgenic mice and progeny containing the Tropoprom construct (Tropoprom-2) that lacked exon 33. Of the 11 founder mice established with integrated Tropoprom-2 recombinants, no transgenic progeny could be established from four as a consequence of an unusually high incidence of maternal cannibalism, a well-known indicator of possible defects in newborn animals. In addition, from those progeny that were successfully propagated from founder mice containing Tropoprom-2 constructs, no homozygous animals were ever obtained from hemizygous brother–sister matings, once again as a result of maternal cannibalism.

While difficulties in breeding are very suggestive of defects in newborn mice, the first real evidence demonstrating that such defects exist was found in the progeny from two separate founder lines expressing Tropoprom-2. Several of the progeny from these two founder transgenics died at 5–7 weeks of age as a result of respiratory distress. Autopsy of multiple progeny revealed abnormally small and non-compliant lungs. Histological sections prepared from pulmonary tissue from affected transgenic mice and non-transgenic litter mates revealed marked enlargement of alveoli and extensive septal damage (Fig. 4). These pulmonary changes were typical of pulmonary alterations associated with the development of emphysema. These emphysematous changes occurred in the progeny from separate founder transgenic mice actively transcribing the Tropoprom-2 construct. No such lung defects were present in transgenic progeny containing transcriptionally active Tropoprom-1 recombinants, constructs containing the complete sequence for rat tropoelastin. Emphysema did not develop, therefore, in the Tropoprom-2-containing progeny as a consequence of the incorporation of rat tropoelastin into lung elastin nor did a pulmonary defect arise as a result of the disruption of a mouse gene at a site of integration of a Tropoprom-2 recombinant into mouse genomic DNA. Many of our transgenic ($F_1$) mice, therefore, hemizygous for the Tropoprom-2 recombinant, develop an emphysematous phenotype as a result of the aberrant overexpression of a tropoelastin isoform lacking the hydrophobic peptide encoded by exon 33.

We observed a second, completely different, phenotype in transgenic mice expressing high levels of the Tropoprom-4 construct. This recombinant lacks exons 19–31 of the rat tropoelastin gene. We successfully established three founder lines in which Tropoprom-4 had been integrated into the mouse genome. One of these founder lines did not express the transgene but two separate founder mice expressed this truncated minigene construct. Progeny from founder mice expressing Tropoprom-4 were abnormally small. Hemizygous $F_2$ progeny from $F_1$ females continued to express the minigene. Litter sizes were small in number, transgenic offspring were small in size and the mice died within 2–3 days of birth. The $F_1$ progeny and founder mice died between 7–9 months of age. Autopsy revealed massive internal haemorrhage. The lungs, however, appeared normal. Histopathological evaluation revealed a very abnormal ascending aorta (Fig. 4), characterized by hypoplastic and disorganized medial elastic laminae. In some histopathological sections in particular, a complete loss of medial elastic fibres was noticed together with a disrupted adventitia and even evidence for complete rupture of the aortic wall. This histological evidence is entirely consistent with the development of an ascending aortic aneurysm.

In contrast to the dramatic and contrasting phenotypes observed in some of the transgenic mice containing Tropoprom-2 and Tropoprom-4 recombinants, founder transgenes containing Tropoprom-1 and Tropoprom-3 constructs did

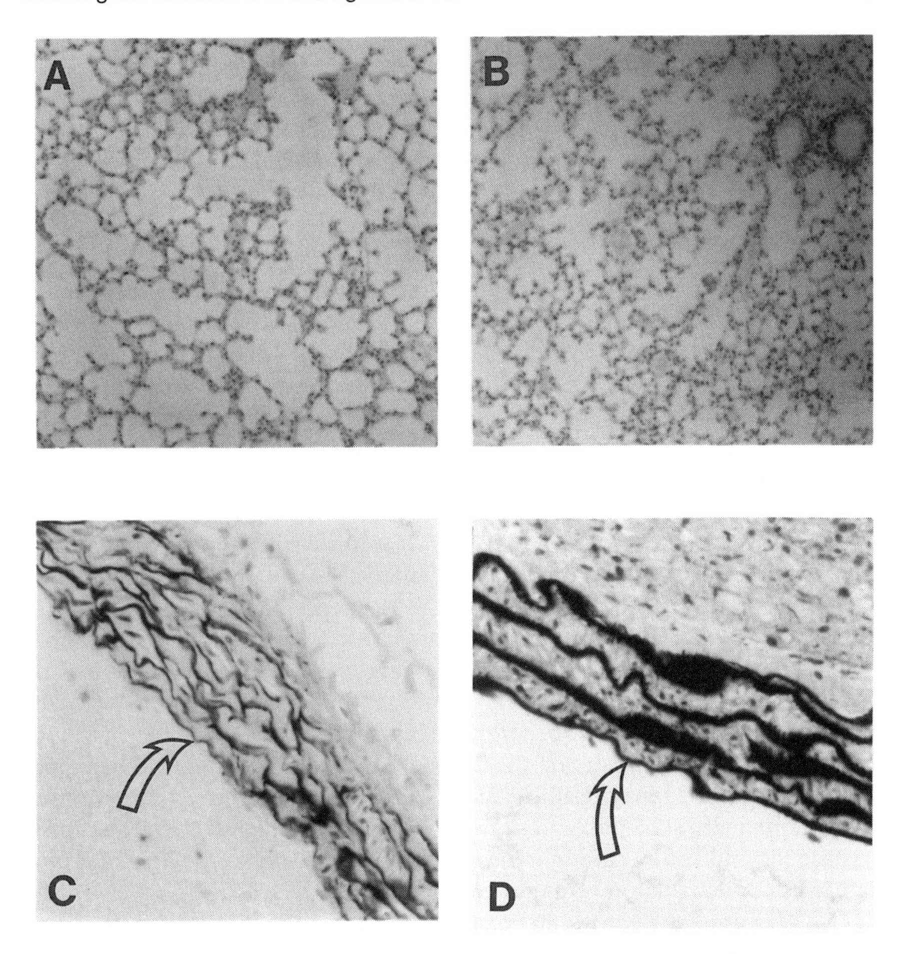


FIG. 4. Histopathologic sections of lung and aortic tissue obtained from transgenic mice containing Tropoprom-2 and Tropoprom-4 recombinants. Tropoprom-2 (−exon 33). (A) Haematoxylin and eosin-stained lung sections from a transgenic mouse expressing Tropoprom-2 and (B) a non-transgenic litter mate. Original magnification in both sections is × 100. Tropoprom-4 (−exons 19–31). (C) Verhoeff van Giesen elastin fibre stain of a section of the ascending aorta from a transgenic mouse expressing high levels of Tropoprom-4 and (D) an age-matched transgenic mouse expressing Tropoprom-1 (+exons 1–36). The arrows indicate the position of the internal elastic lamina. Original magnification in both sections is ×133.

not reveal any elastic tissue phenotype or any abnormalities in breeding patterns or behaviour. Transcriptionally active Tropoprom-1 (encoding the complete sequence of rat tropoelastin) and Tropoprom-3 (a construct lacking exon 12–15) minigenes were successfully established in a total of eight separate founder lines. Several of these transgenes were shown to express some of the highest levels of rat tropoelastin mRNA of any of the transgenes analysed. Large litters, however, were common and maternal cannibalism was not encountered. Moreover, histopathological evaluation of lung and aortic tissue from both newborn and neonatal transgenes revealed normal elastic fibre morphology.

## Conclusions

The synthesis of the complete sequence of rat tropoelastin in the progeny of multiple founder transgenic mice and the assembly of a mouse–rat elastin chimaera does not seem to influence elastic fibre assembly or elastic tissue function. Similarly, overexpression of the rat tropoelastin isoform lacking exons 13–15 does not result in any obvious aberrant phenotype. It would seem therefore that, within the elastic tissues analysed (aorta, lung and skin), this particular isoform of tropoelastin is unnecessary, during mouse development, for normal elastic fibre assembly and function.

In contrast, overexpression of the rat tropoelastin isoform lacking the domain encoded by exon 33 results in a relatively early-onset emphysematous phenotype. Emphysema is thought to be a disease caused by the elastase-catalysed degradation of pulmonary elastic fibres, leading to the destruction of alveolar septa, consequent alveolar enlargement and eventual loss of pulmonary function (Snider et al 1986, Janoff 1985). The results we have obtained with our Tropoprom-2 transgenic mice are not inconsistent with this hypothesis. Overexpression of the tropoelastin isoform lacking the domain encoded by exon 33 may result in the assembly of a structurally compromised elastic fibre, more susceptible to degradation either through the direct action of elastase or indirectly through mechanical stress, leading to alveolar wall damage and degradation of damaged elastic fibres by elastases.

The Tropoprom-4 construct resulted in the incorporation of a truncated tropoelastin into insoluble elastin in several elastic tissues in transgenic mice. The presence of this aberrant elastin in the ascending aorta produced a severe aneurysmal phenotype.

Recently, studies by Keating and co-workers (Curran et al 1993, Ewart et al 1994) have shown that large deletions on human chromosome 7q, involving the loss of part of the tropoelastin gene, are associated with the vascular disorder SVAS. Specifically, in two separate and unrelated patients, these investigators have shown the presence of a large deletion (Ewart et al 1994) and a balanced translocation (Curran et al 1993), both of which involved a breakpoint located at the 3′ end of the tropoelastin gene. There is a clear association between the appearance of these mutations and the SVAS phenotype. Linkage studies moreover have established that a mutation causing SVAS resides in the locus on the long arm of chromosome 7 that contains the tropoelastin gene (Ewart et al 1993a). It is not clear, however, whether the disruption of the tropoelastin gene *per se* actually causes the SVAS phenotype. The large deletions observed in SVAS patients and in patients with a related disorder, Williams syndrome, also involve DNA flanking the tropoelastin gene (Ewart et al 1993b). Disruption of a gene in close proximity to the tropoelastin gene could actually be responsible therefore for SVAS and/or Williams syndrome.

The deletions affecting the 3′ end of the tropoelastin gene in some SVAS patients may result in the creation of a null allele. The mutation could also result in the synthesis of a truncated tropoelastin that, if incorporated into insoluble elastin, may result in the assembly of aberrant elastic fibres. Support for this latter view may be forthcoming from the curious observation that deletion of exons 19–31 in our Tropoprom-4 mice results in an ascending aortic abnormality with no other obvious elastic tissue dysfunction. It was not possible to determine whether these transgenic mice developed SVAS; the histopathology of elastic fibres in the ascending aorta is remarkably similar, however, to the disorganized elastic fibre morphology typical of SVAS (O'Connor et al 1985). Aneurysms, moreover, have been reported as a clinical complication of SVAS (Beitzke et al 1986). An exciting speculation therefore would be that the aberrant transcripts from our in-frame, intragenic deletion of exons 19–31 in the rat tropoelastin gene, that will not disrupt any other gene in the vicinity of the endogenous mouse tropoelastin gene, result in a severe form of SVAS that leads to aortic rupture. Should SVAS indeed be directly due to a disruption of elastic fibre assembly, through the synthesis of aberrant protein rather than a dosage effect as a consequence of a null allele, then it would seem very likely that the non-vascular components of Williams syndrome, particularly the neurodegenerative features of this disorder, may be due to disruption of gene(s) proximal to the tropoelastin gene on the long arm of human chromosome 7.

## *Methods*

The details of the methods summarized in this manuscript have been fully described in several recent publications (Wydner et al 1994, Sechler et al 1995b, Sechler 1994, Alatawi 1994).

## *Acknowledgements*

This work was supported by NIH grants HL 37438, HL 42798 and HL 39869. We gratefully acknowledge the assistance of DNX (Princeton, NJ) in preparing the founder transgenic mice and Gary Benson for typing the manuscript.

## References

Alatawi A 1994 Characterization of the complete structure of the rat tropoelastin gene. PhD thesis, Rutgers University, NJ, USA

Beitzke A, Becker H, Rigler B, Stein JI, Suppan C 1986 Development of aortic aneurysms in familial supravalvular aortic stenosis. Pediatr Cardiol 6:227–229

Boyd CD, Christiano AM, Pierce RA, Stolle CA, Deak SB 1991 Mammalian tropoelastin: multiple domains of the protein define an evolutionarily divergent amino acid sequence. Matrix 11:235–241

Curran ME, Atkinson DL, Ewart AK, Morris CA, Leppert MF, Keating MT 1993 The elastin gene is disrupted by a translocation associated with supravalvular aortic stenosis. Cell 73:159–168

Deyl Z, Horakova M, Macek K 1985 Changes in elastin composition in aorta of spontaneously hypertensive (SHR) rats. Biochem Biophys Res Commun 129:179–186

Dietz HC, Cutting GR, Pyeritz RE et al 1991 Marfan syndrome caused by a recurrent *de novo* missense mutation in the fibrillin gene. Nature 352:337–339

Ewart AK, Morris CA, Ensing GJ et al 1993a A human vascular disorder, supravalvular aortic stenosis, maps to chromosome 7. Proc Natl Acad Sci USA 90:3226–3230

Ewart AK, Morris CA, Atkinson D et al 1993b Hemizygosity at the elastin locus in a developmental disorder, Williams syndrome. Nature Genet 5:11–16

Ewart AK, Jin W, Atkinson D, Morris CA, Keating MT 1994 Supravalvular aortic stenosis associated with a deletion disrupting the elastin gene. J Clin Invest 93: 1071–1077

Gibson MA, Sandberg LB, Grosso LE, Cleary EG 1991 Complementary DNA cloning establishes microfibril-associated glycoprotein (MAGP) to be a discrete component of the elastin-associated microfibrils. J Biol Chem 266:7596–7601

Glaubitz LC, Bowen JH, Cox EB, McCarty KS 1984 Elastosis in human breast cancer. Arch Pathol Lab Med 108:27–30

Heim RA, Pierce RA, Deak SB, Riley DJ, Boyd CD, Stolle CA 1991 Alternative splicing of rat tropoelastin pre-mRNA is developmentally regulated. Matrix 11: 359–366

Hollister DN, Godfrey M, Sakai LY, Pyeritz RE 1990 Immunohistologic abnormalities of the microfibrillar-fiber system in the Marfan syndrome. N Engl J Med 323:152–159

Jacenko O, LuValle PA, Olsen BR 1993 Spondylometaphyseal dysplasia in mice carrying a dominant negative mutation in a matrix protein specific for cartilage-to-bone transition. Nature 365:56–61

Jaenisch R 1988 Transgenic animals. Science 240:1468–1474

Janoff A 1985 Elastases and emphysema: current assessment of the protease–antiprotease hypothesis. Am Rev Respir Dis 132:417–433

Indik Z, Yeh H, Ornstein-Goldstein N, Rosenbloom J 1990 Structure of the elastin gene and alternative splicing of elastin mRNA. In: Sandell LJ, Boyd CD (eds) Extracellular matrix genes. Academic Press, San Diego, CA, p 221–250

Iredale RB, Eccleston-Joyner CA, Rucker RB, Gray SD 1989 Ontogenic development of the elastic component of the aortic wall in spontaneously hypertensive rats. Clin Exp Hypertens Part A Theory Pract 11:173–187

Kagan HM 1986 Characterization and regulation of lysyl oxidase. In: Mecham RP (ed) Regulation of matrix accumulation. Academic Press, San Diego, CA, p 321–398

Kramsch DM, Hollander W 1973 The interaction of serum and arterial lipoproteins with elastin of the arterial intima and its role in the lipid accumulation in atherosclerotic plaques. J Clin Invest 52:236–247

Lee G, Godfrey M, Vitale E et al 1991 Linkage of Marfan syndrome and a phenotypically related disorder to two different fibrillin genes. Nature 352:330–334

Mecham RP, Heuser JE 1991 The elastic fiber. In: Hay ED (ed) The cell biology of the extracellular matrix, 2nd edn. Plenum Press, New York, p 79–110

Mecham RP, Whitehouse LA, Wrenn DS et al 1987 Smooth muscle-mediated connective tissue remodeling in pulmonary hypertension. Science 237:423–426

Nelder K 1988 Pseudoxanthoma elasticum. Clin Dermatol 6:1–159

O'Connor W, Davis J, Geissler R, Cottrill C, Noonan J, Todd E 1985 Supravalvular aortic stenosis: clinical and pathologic observations in six patients. Arch Pathol Lab Med 109:179–185

Olsen DR, Fazio MJ, Shamban AT, Rosenbloom J, Uitto J 1988 Cutis laxa: reduced elastin gene expression in skin fibroblast cultures as determined by hybridizations with a homologous cDNA and an exon 1-specific oligonucleotide. J Biol Chem 263:6465–6467

Palmiter RD, Brinster RL 1986 Germ-line transformation of mice. Annu Rev Genet 20:465–499

Parks WC, Deak SB 1990 Tropoelastin heterogeneity: implications for protein function and disease. Am J Respir Cell Mol Biol 2:399–406

Pierce RA, Alatawi A, Deak SB, Boyd CD 1992 Elements of the rat tropoelastin gene associated with alternative splicing. Genomics 12:651–658

Pierce RA, Deak SB, Stolle CA, Boyd CD 1990 Heterogeneity of rat tropoelastin mRNA revealed by cDNA cloning. Biochemistry 29:9677–9683

Rosenbloom J 1984 Elastin: relation of protein and gene structure to disease. Lab Invest 51:605–623

Sandberg LB, Roos PJ, Pollman MJ, Hodgkin DD, Blankenship JW 1990 Quantitation of elastin in tissues and culture: problems related to the accurate measurement of small amounts of elastin with special emphasis on the rat. Connect Tissue Res 25:139–148

Sechler JL 1994 Elastin gene mutations in transgenic mice. PhD thesis, Rutgers University, NJ, USA

Sechler JL, Cardone JC, Friedmann J et al 1995a Altered tropoelastin gene expression is associated with the development of hypertension in the spontaneously hypertensive (SHR) rat. Connect Tissue Res, in press

Sechler JL, Sandberg LB, Roos PJ, Amenta PS, Riley DJ, Boyd CD 1995b Overexpression of an alternative splice variant of tropoelastin in transgenic mice results in lung abnormalities resembling pulmonary emphysema. J Biol Chem, submitted

Snider GL, Lucey EC, Stone PJ 1986 Animal models of emphysema. Am Rev Respir Dis 133:149–169

Uitto J, Shamban A 1987 Heritable skin diseases with molecular defects in collagen or elastin. Clin Dermatol 5:63–84

Uitto J, Fazio MJ, Olsen DR 1989 Molecular mechanisms of cutaneous aging. J Am Acad Dermatol 21:614–622

Wydner KS, Sechler JL, Boyd CD, Passmore HC 1994 Use of an intron length polymorphism to localize the tropoelastin gene to mouse chromosome 5 in a region of linkage conservation with human chromosome 7. Genomics 23:125–131

## DISCUSSION

*Parks:* You showed that transgenic mice expressing Tropoprom-2 had altered lung morphology (Fig. 4), yet your PCR data showed very little expression of Tropoprom-2 in the lung (Table 1). However, it was expressed at high levels in the skin. Do you know if the morphology of the elastic fibres in the skin of these transgenic animals is altered?

*Boyd:* We haven't looked yet.

*Parks:* How do you suppose that such a low level of Tropoprom-2 expression would alter lung morphology in these mice?

*Boyd:* We have speculated that there are two essential effects. One is that even low levels of expression of this particular isoform at an inappropriate time in lung development will result in the assembly of a fibre that can't cope with the mechanical forces that are required of it in the alveolar septum. Consequently, the elastic fibres rupture, the alveolar walls are disrupted and this leads to the larger spaces that are seen in the histopathology. This degenerative process is a characteristic of the emphysematous phenotype. Our other speculation is that this has nothing to do with mechanical stress, but instead fibres are assembled that are much more susceptible to proteolysis. As a result, there is rapid degradation of elastic fibres by elastase in the pulmonary tissue. This is the scenario we favour, because apart from the lung damage we see no other phenotype in these animals. It is possible that pulmonary tissue is more severely affected than any other tissue, independent of the very low level of Tropoprom-2 expression in pulmonary tissue, because that's where proteolytic susceptibility may be much more critical.

*Parks:* But in the micrographs you showed there was no overt inflammation, which makes me wonder where these potential proteases would be coming from. I noticed that there was a high level of expression of Tropoprom-2 mRNA in the kidney and liver, but was there any evidence of abnormal deposition of elastic fibres in these organs?

*Boyd:* If your *in situ* hybridizations had worked, we would know!

*Parks:* No, *in situ* hybridization is an assay of mRNA expression, not protein deposition. Your PCR results showed that there is significant expression of the transgene in kidney and the liver, so I was wondering if these tissues have the capacity to deposit elastic fibres outside of their vasculature?

*Boyd:* We haven't started a systematic analysis of the various tissues in which we have elicited expression—this obviously needs to be done.

*Starcher:* I was also surprised that even though there was very little expression of Tropoprom-2, you still found an effect in the lung. But I caution you about calling the lung damage you saw 'emphysema'. You really have to show that there's actual degradation of the alveolar walls to be able to do this, and I didn't notice this in your micrographs. In emphysematous lungs we usually see balls of retracted elastin where the alveoli are broken. The degree of abnormal lung development in your micrographs was really very slight, when you look at the changes in the mean linear intercepts. We and other people have done experiments where we pour elastases into lungs and create emphysema, and even though there is not much of the lung structure left in these animals, they still manage to breathe and they survive. I've seen lungs where you can hardly find a piece of lung without a hole in it, and the animals don't suffocate. I would have also anticipated that you would have seen more of a fibrotic effect than emphysema, because I did not see any inflammation. We've done experiments where we put lipopolysaccharide into the lungs and you get a continuous progression of neutrophils coming in the lungs. Over a long period of time we see some eroding of the lung, but these animals live and you eventually get a fibrotic condition.

*Boyd:* We've been told by our pulmonary expert physician that the mean linear intercept differences we've seen are consistent with the differences that are often seen in human emphysematous conditions, particularly in early-onset emphysema (David Riley, personal communication). Remember, in order to do the mean linear intercept calculations we use animals a couple of weeks before they start dying of a breathing disorder (these are three-week-old animals; they start dying between five and seven weeks of age). At this stage, the development of the disease is still fairly mild.

*Starcher:* This is another problem. When you look at very young animals, the alveoli are still being formed. The alveolar walls are being formed at around Day 10 to 12, so you may not have reached maturity in the lung by three weeks. You may be seeing a problem in lung maturation rather than degradation.

*Boyd:* The Verhoeff van Giesen stains of the pulmonary tissue show lots of evidence for elastic fibres that have been disrupted in the alveolar wall. There's no doubt about that.

*Sandberg:* I just want to comment on the problem of the mean linear intercept that Barry Starcher raised. The definition of emphysema is really based on mean linear intercepts and it matters less what the histology looks like. Bad-looking lungs can still have normal mean linear intercepts and not have emphysema. What is important is that Charles Boyd was able to show in these very young animals that there was a statistically significant difference in mean linear intercepts. Enzymically induced lung damage involves an entirely different mechanism.

*Uitto:* There was obviously quite a lot of variation in the level of Tropoprom expression. Do you know whether this correlates with the transgene copy number and perhaps the site of integration? In fact, do you know where these transgenes are integrated?

*Boyd:* We haven't looked at the sites of integration, although we do know that the sites are different in the different founders. In certain tissues, such as skin, there seems to be a correlation between expression and copy number, but in other tissues there isn't. We have taken this to mean that there isn't any obvious correlation between copy number and expression.

*Uitto:* In the mouse line 513, I noted that there were 52 copies of the Tropoprom-2 transgene. Do you know if they all went into a single integration site?

*Boyd:* From the gene dosage results of Jan Sechler, it seems that there was indeed a single integration site for these 52 copies.

*Davidson:* Is the elastin content of the lungs in the Tropoprom-2 transgenic mice different from control litter mates?

*Boyd:* We don't know.

*Davidson:* How many of the different Tropoprom-2 lines express this phenotype?

*Boyd:* All those lines expressed high levels of the Tropoprom-2 construct.

*Davidson:* You have got a total of 28 lines using presumably the same *cis*-acting elements and the same 3′ UTR domain. Is there a consistent pattern of tissue-specific or developmentally related expression that relates to these regulatory portions of the molecule?

*Boyd:* There are a number of guidelines: it's hard to say whether there's an overall expression pattern that can be generalized. One certainly sees inappropriate expression of the transgenes in kidney and liver; this is the one thing that I think is consistent. The other is that the expression of all the founders continues into adult tissue. It is at times very hard to quantitate individual expression between various tissues, because there are enormous differences.

*Davidson:* Why hasn't anyone looked at whether or not the amount of elastin gene expression in SVAS patients is altered? Why hasn't this been more fully explored?

*Boyd:* This is because of the relatively small number of patients we've got. All of them are children, and parents are not necessarily that keen on providing skin fibroblast cultures. So at the present time, ourselves, Steve Thibodeau and Mark Keating have only been able to obtain informative genomic DNA samples.

*Robert:* I think your results with the transgenic mice showing different lung structures are very interesting and it is not necessary to look for a perfect analogy to any of the human situations. If a similar situation occurred in humans they would develop, for instance, more severe asthma or emphysema. This is a very good basis for medical speculations. Some of the diseases you mentioned in your introduction concerning anomalies in vascular development could also be connected. Why would the aorta get narrower if you have less elastin expression? It could instead get much larger and rupture. Modifications of elastin gene expression may lead to explanations concerning the mechanism of genetic regulation of the structure and morphology of blood vessels during development.

*Mungai:* Have you carried out any studies on the actual characteristics of the elastic fibres in those animals where the aorta ruptures?

*Boyd:* No; obviously that's a major priority. Presumably you are suggesting electron microscope and biochemical studies.

*Mungai:* Yes, and also biophysical studies: for instance, what is the elastic limit in relation to the normal physiological requirements?

*Boyd:* Those are terribly important experiments that ought to be done.

*Ooyama:* Most of the aneurysms we see are atherosclerotic. Do you think your tropoelastin knockout mice could be a model for atherosclerotic aneurysm?

*Boyd:* We were surprised to find aortic rupture in the ascending aorta, because most aneurysms develop in the infrarenal aorta. We looked at the histopathology at seven months of age in the Tropoprom-4 construct and found no evidence for any disrupted elastic fibre morphology in that region of the aorta, which suggests that this may not be a model. But, of course, the seven-month-old mouse is not equivalent to a 60-year-old human—it may well be that we will have to leave these animals for longer. We've had a great deal of difficulty breeding these animals, because they die at seven months. The trick is to get a Tropoprom-4 mouse that expresses the construct at relatively low levels, see whether it will live beyond the seven-month limit we have experienced in the few constructs we've had so far, and then ask if histopathological abnormalities in the infrarenal aorta will develop.

*Mariani:* Could you describe your histochemical staining procedure?

*Boyd:* We used perfusion-fixed sections from three-week-old mice, and did both Verhoeff van Giesen and haemotoxylin and eosin staining. You can't really do quantitative morphology on a non-perfusion-fixed section. We were able to reproduce what we saw in the non-perfusion-fixed sections taken from an animal that died at five weeks of age. With the Verhoeff van Giesen stain we were able to see clear evidence for elastic fibre breakage and disruption, although at that very crude level we couldn't see any obvious morphological difference between the elastic fibres in the alveolar septum. We are going to have to do electron microscopy to be able to see those morphological differences.

*Keeley:* If you look at the bronchi in these animals, could their suffocation be due not to the lung changes but to bronchial changes?

*Boyd:* We haven't seen any elastic fibre differences in the bronchus.

*Keeley:* I'm interested in the Tropoprom-1 mouse, which is really just expressing the normal tropoelastin gene. This tropoelastin continues to be expressed in these animals throughout development, not turning off as it normally does, and, as well, it is produced in tissues not normally expressing elastin. This might result in some phenotypic changes in these animals, at least later in life. Did you see anything like that?

*Boyd:* We haven't looked, but what we are doing is to breed the transgenic Tropoprom-1 mouse with a hairless mouse, because the levels of expression in skin are very high. The progeny will be studied in a collaboration with Unilever, involving quantitative confocal microscopy on skin elastic fibres in order to address that very question.

*Rosenbloom:* I wonder if Bill Parks would like to comment on the continued expression of this kind of transgene, when his or Richard Pierce's data suggest that even if it were to be continually transcribed, the mRNA would probably be degraded.

*Parks:* Because he has shown continued expression of the transgene into the adult, Charles Boyd's results agree with ours. Our data show that the endogenous tropoelastin gene continues to be transcribed well into adulthood.

*Boyd:* Of course, our assay is for message, not pre-message.

*Parks:* That's fine: thus the elements in the pre-mRNA or the processed mRNA that control the post-transcriptional regulation of tropoelastin are not present in your minigenes. Our favourite hypothesis is that this mechanism is mediated by something in the 3′ UTR. Since you're seeing continued expression of minigene mRNA, which does include the 3′ UTR, I think we might have to modify our idea a bit and consider other parts of the gene, such as intron 1, which is a great candidate for a regulatory element and is not present in your minigenes.

*Boyd:* We've come to the same conclusion; that some intron sequence absent in the minigenes was obviously mediating a post-transcriptional control in pre-message availability which would then be consistent with the results you've got.

*Parks:* We think of elastin assembly as a process that requires numerous products and that there must be a precise, co-ordinate regulation of these products and their stoichometry. In your model, you're supplying more of the principal substrate, namely, tropoelastin. Do you think that more elastin will be made in these animals and how does that affect our thinking about the mechanisms of fibre assembly?

*Boyd:* Good quantitative biochemistry is going to be important to answer this. At a very gross level, there's no obvious thickening of the elastic fibres (as I showed you in that elastic fibre stain of the ascending aorta from the Tropoprom-1 construct; Fig. 4D). There's a huge increase in the total amount of tropoelastin available for elastic fibre assembly in those animals, but we don't see an obvious increase in the thickness of the elastic lamina. We haven't done quantitative morphometry on that, nor the biochemistry; until we have it's hard to say. It is, however, possible that, even though these cells are secreting a lot more tropoelastin, it's not incorporated, because you're saturating a fibre assembly process which cannot cope with any more, and the rest is getting degraded.

*Parks:* Have you considered looking for tropoelastin peptides circulating in blood or urine?

*Boyd:* Yes, we are starting such a study.

*Keeley:* We know from our stretch experiments in aortic tissue that the rate of synthesis of elastin can be almost doubled, without any increases in steady-state levels of tropoelastin in the tissue. In other words, the assembly mechanisms are capable of handling twice the rates of synthesis without tropoelastin backing up and accumulating in the tissue, even in day-old chicks which are already making elastin at a very rapid rate.

*Pierce:* Is there a theoretical P32 : P37 ratio that you could deduce for elastin isolated from your Tropoprom-4 transgenes?

*Boyd:* Limitations in the assay prevent us from deducing an actual number. We predicted, however, a ratio less than the P32 : P37 ratio recovered from mouse elastin. A you will remember from the results I presented in Table 2, that's exactly what we found.

*Pierce:* That would indicate that there was overwhelming production primarily of the transgene as opposed to the endogenous gene.

*Boyd:* I would be cautious about making a quantitative judgement on those HPLC assays. One thing we've learnt from Larry Sandberg is that it's hard to say exactly how much tropoelastin from any one of the transgenes will contribute to the P32 : P37 ratio. All we're saying is that the ratio is different enough from the endogenous mouse to illustrate clearly that there is rat tropoelastin present in transgenic mouse elastin.

*Davidson:* Do you have any information to indicate whether it's pre-transcriptional, transcriptional or post-transcriptional control that regulates these mRNA levels?

*Boyd:* No; that's obviously a priority.

# Elastic fibre assembly: macromolecular interactions

Robert P. Mecham*, Thomas Broekelmann*, Elaine C. Davis*, Mark A. Gibson† and Patricia Brown-Augsburger*

**Department of Cell Biology and Physiology, Washington University School of Medicine, Box 8228, 660 South Euclid, St. Louis, MO 63110, USA and †Department of Pathology, University of Adelaide, Adelaide, South Australia 5001, Australia*

*Abstract.* To investigate the mechanisms behind elastic fibre assembly, we studied the molecular interactions between elastin and microfibrillar components using solid-phase binding assays. Fibrillin 1, purified from tissue using reductive-saline extraction, showed no binding to microfibril-associated glycoprotein (MAGP) or tropoelastin. MAGP, however, was found to bind specifically to tropoelastin in a divalent-cation independent manner. Antibody inhibition studies indicated that the C-terminus of tropoelastin defined the interactive site with MAGP. MAGP and fibrillin were also substrates for transglutaminase, which may provide an important mechanism for stabilizing microfibrillar structure. In other studies we found that a major cross-linking region in elastin is formed through the association of domains encoded by exons 10, 19 and 25 of tropoelastin and that the three chains are joined together by one desmosine and two lysinonorleucine cross-links.

*1995 The molecular biology and pathology of elastic tissues. Wiley, Chichester (Ciba Foundation Symposium 192) p 172–184*

Elastic fibres are among the most complex structures of the extracellular matrix. Unlike the fibril-forming collagens, elastic fibres do not spontaneously assemble. Their organization requires the coordinate expression of at least five different proteins that direct the alignment and cross-linking of the elastin monomer. The inherent complexity of the assembly process combined with the unique physical properties of the component proteins have made this one of the most intractable problems in matrix biology.

Shortly after the discovery and isolation of collagen, many laboratories began to investigate its synthesis and assembly in the extracellular matrix. Over the last five decades, much has been learned about this process (Yurchenco et al 1994). These studies have been aided by the ability to extract soluble collagen monomers that self assemble *in vitro*. For elastin, in contrast, similar approaches have not been possible. The highly cross-linked nature of elastic fibres has precluded the extraction of monomeric units that reassemble. Other than understanding the chemical mechanism of cross-link formation (Partridge et al 1964), nothing is known about how tropoelastin monomers interact one with another to form the functional polymer.

Structural studies during the mid 1970s found that two cross-links unique to elastin, desmosine and isodesmosine, were clustered in helical, alanine-rich sequences (Foster et al 1976, Gerber & Anwar 1975). At that time, however, it was impossible to determine which of the 16 possible cross-linking domains aligned one with another in the fully cross-linked molecule. This problem arose because mature, insoluble elastin could only be solubilized using relatively non-specific proteases, such as thermolysin or elastase. As a result, the solubilized peptides that were isolated and studied were extremely small—in most cases containing the cross-linking amino acid and three to 10 other residues (Mecham & Foster 1978). A second problem was that the complete amino acid sequence of elastin was not known so it was impossible to assign a cross-link-containing sequence to a precise location in the molecule. As a result, for almost 20 years there has been no new information about assembly of elastin!

A significant advance in our knowledge of elastic fibre assembly was the identification and characterization of cDNAs that encode elastin and microfibrillar proteins. The complete amino acid sequence derived from these clones has given us domain maps that now permit the unambiguous assignment of functional sequences to specific regions of these proteins. We have used this sequence information to identify the domains that are involved in the formation of a cross-linked peptide that we feel defines the nucleation site of tropoelastin polymerization. We have also studied potential interactions between tropoelastin and the microfibrillar proteins MAGP and fibrillin. Our findings indicate that MAGP may be the component of the microfibril responsible for the binding and alignment of tropoelastin during assembly.

## Tropoelastin binds to native MAGP

The central dogma of elastic fibre assembly is that microfibrils direct the alignment and polymerization of tropoelastin. It is only recently, however, that reagents have been available to investigate molecular associations between tropoelastin and specific microfibrillar proteins. Using a solid-phase binding assay, we found that MAGP bound specifically to tropoelastin but not fibrillin 1. Binding to tropoelastin was divalent-cation independent and saturable, but required that both proteins be in their native state; reduction of either tropoelastin or MAGP abolished binding (Fig. 1). Native MAGP used in this study was generated using a baculovirus expression system (Brown-Augsburger et al 1994).

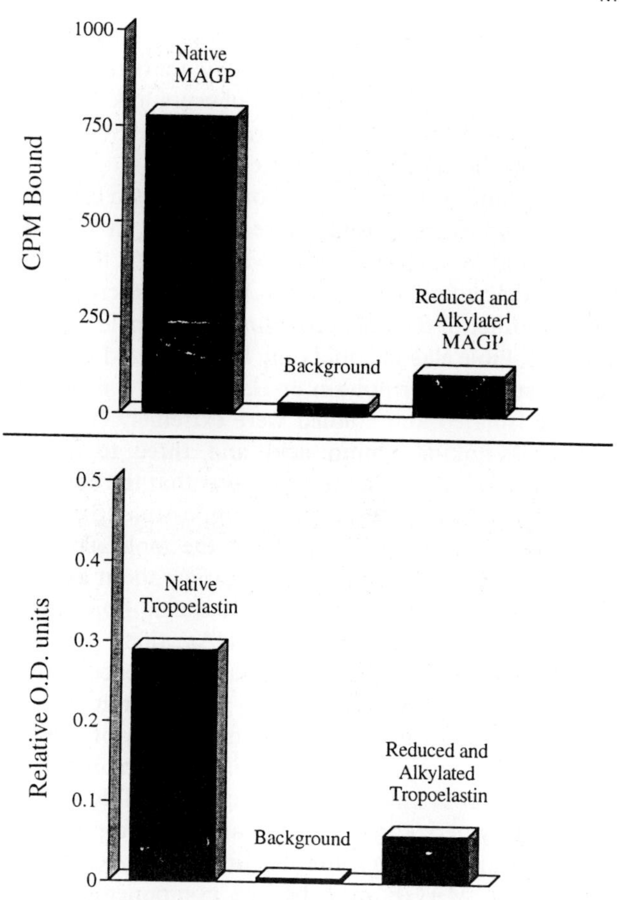


FIG. 1. (*Top*) binding of reduced and alkylated microfibril-associated glycoprotein (MAGP) and native MAGP to tropoelastin. Recombinant MAGP purified from Sf9 cells (ovary, *Spodoptera frugiperda*) was reduced, alkylated and added to wells of microtitre plates coated with tropoelastin or casein. Native MAGP was included as a positive control. Bound MAGP was detected with a $^{125}$I-labelled MAGP antibody. CPM, counts per minute. (*Bottom*) MAGP binding to native versus reduced and alkylated tropoelastin. Microtitre plates were coated with tropoelastin that had been reduced and alkylated, or tropoelastin subjected to mock reduction and alkylation. Equal coating of wells by tropoelastin was verified by immunoassay. The extent of MAGP binding was detected with a MAGP-specific antibody. O.D., optical density.

The region of tropoelastin that promotes stable binding to MAGP appears to be the C-terminus. This was suggested from the observation that reduction destroys the ability of tropoelastin to bind MAGP. Since the only two cysteine residues in tropoelastin are in the C-terminus, inactivation of binding by reduction implicates this site. We provided further evidence that the

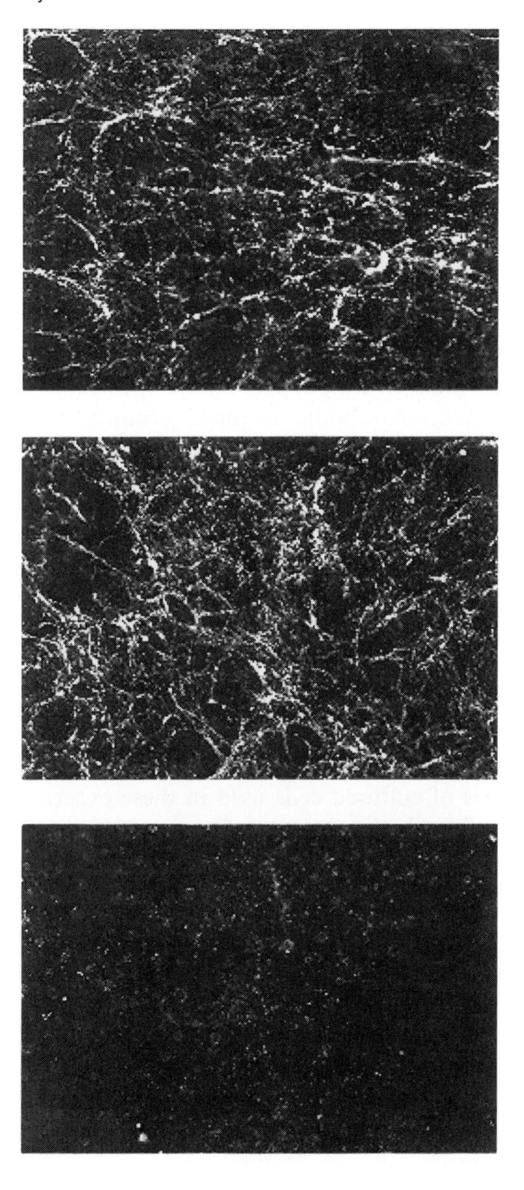

FIG. 2. Antibody to the C-terminus of tropoelastin blocks elastin deposition into fibres. First-passage FBC-180 cells were plated and grown in the presence of Fab fragments of antibodies raised against the C-terminus (*bottom panel*) or exon 4/5 (*middle panel*) of tropoelastin. Fresh antibody was added after 3 days, and at 6 days cultures were stained for elastin using an anti-elastin monoclonal antibody (BA4) visualized with a goat anti-mouse FITC-conjugated secondary antibody. The top panel shows control elastin staining of cells grown in the absence of blocking Fabs.

C-terminus mediates this interaction with MAGP by culturing fetal bovine auricular chondroblasts from a 180 d fetus (FBC-180 cells) in the presence and absence of Fab fragments from domain-specific antibodies for tropoelastin. Elastin deposition in the matrix was then analysed by immunofluorescence using a mouse monoclonal antibody for elastin. Early passage FBC cells produce abundant amounts of elastin in culture and process tropoelastin into elastic fibres. Figure 2 shows the typical 'beads on a string' pattern for elastin in FBC cultures grown in the absence of blocking antibodies. Labelled MAGP and fibrillin exhibit the same arrangement without the beaded pattern. Double-labelling of tropoelastin and MAGP showed that tropoelastin co-localizes with MAGP and fibrillin in these cultures (Brown-Augsburger et al 1994). Figure 2 also shows the effect of Fab fragments reactive with exon 4/5 and with the C-terminus of tropoelastin. Antibodies to the C-terminus completely blocked the accumulation of elastin into the extracellular matrix whereas antibody to exon 4/5, which is near the N-terminus of the molecule, had no effect. Importantly, the C-terminal antibody did not disrupt the formation of microfibrils as assessed by MAGP and fibrillin 1 staining.

To determine whether antibodies to microfibrillar proteins could alter elastic fibre assembly, we plated freshly harvested cells onto dishes in the presence of Fab fragments from antibodies to MAGP and fibrillin 1. Cultures treated with fibrillin antibody showed moderate disorganization and a slight decrease in elastin accumulation. The inclusion of an antibody to the N-terminal domain of MAGP, however, completely blocked elastin deposition into fibres without disrupting the microfibrillar network (data not shown).

Northern analysis of cultured cells used in these experiments demonstrated that the addition of antibodies had no effect on the level of elastin synthesis. When we analysed the incorporation of tropoelastin into the extracellular matrix by quantifying the level of desmosine in the various cultures, we found that the antibody to the C-terminus of tropoelastin reduced desmosine levels by 85% relative to control cultures. Similarly, desmosine levels in cultures treated with MAGP antibody were 50% of control. Together these results confirm the biochemical observation of binding between tropoelastin and MAGP. Further, they strengthen the argument that MAGP is the anchoring protein for tropoelastin in the microfibril.

## Domain alignment in cross-linked elastin

To study how domains of tropoelastin align to form the cross-linked polymer, we exposed insoluble elastin, purified from copper-deficient pigs and reduced with sodium borotritide to radiolabel reducible cross-links, to successive treatments with trypsin and chymotrypsin. The solubilized peptides were fractionated on a PRP3 reverse-phase column and fractions containing the most radioactivity were again fractionated using size exclusion and C18

| | Exon |
| --- | --- |
| FPGVGVLPGV | 10 |
| GAAGGLVPGAPGFGPGVG | 19 |
| GLGPGIGVAPGVGVAPGVGV | 25 |

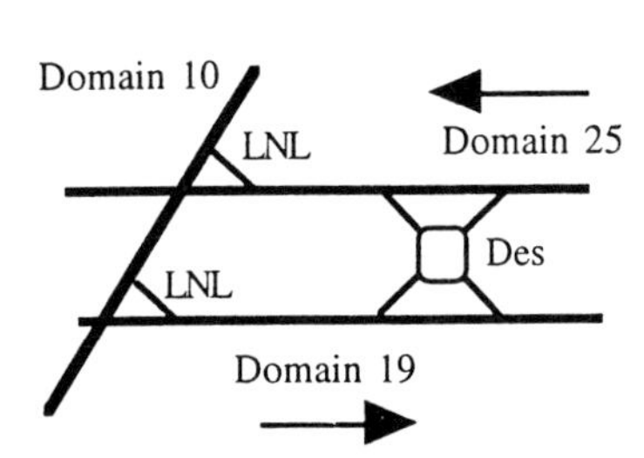


FIG. 3. (*Top*) multiple sequences for the cross-linked peptide chy62. (*Bottom*) model showing the three peptide chains joined by one desmosine (Des) and two lysinonorleucine (LNL) cross-links. Arrows indicate that domains 19 and 25 are in an antiparallel orientation.

reverse-phase chromatography. Amino acid analysis was used to verify the presence of cross-linking amino acids and to determine the type of cross-link present. Several peptides were purified to homogeneity and were found to contain either a single oxidized lysine (allysine) that had not yet condensed to form a cross-link, or the cross-links desmosine, isodesmosine and lysinonorleucine. One peptide (chy62) was characterized completely and was found to contain sequences from three exons: 10, 19 and 25 (Fig. 3). That these chains were actually cross-linked as opposed to three separate peptide fragments that co-purified was indicated by an identical repetitive yield for all three chains during the sequencing reaction.

The cross-links in peptide chy62 consisted of one desmosine (derived from four lysine residues) and two lysinonorleucines (two lysines each). Interestingly, exons 19 and 25 are the only two cross-linking exons in tropoelastin that contain three lysines, thus it is logical to predict that these two domains must be associated during cross-linking. However, because of the orientation of the lysine side chains on the face of the helix, molecular modelling predicts that only two of the three lysines are in a position to form an intermolecular cross-link between the two chains (Mecham & Davis 1994). We know from previous studies that two lysines on each chain condense to form a desmosine cross-link (Foster et al 1976). The remaining lysines, however, cannot react with each other but are perfectly positioned to interact with lysines in exon 10 contributed by a third chain, forming two lysinonorleucine cross-bridges. A simplified model of these data is shown in the bottom of Fig. 3. Three-dimensional

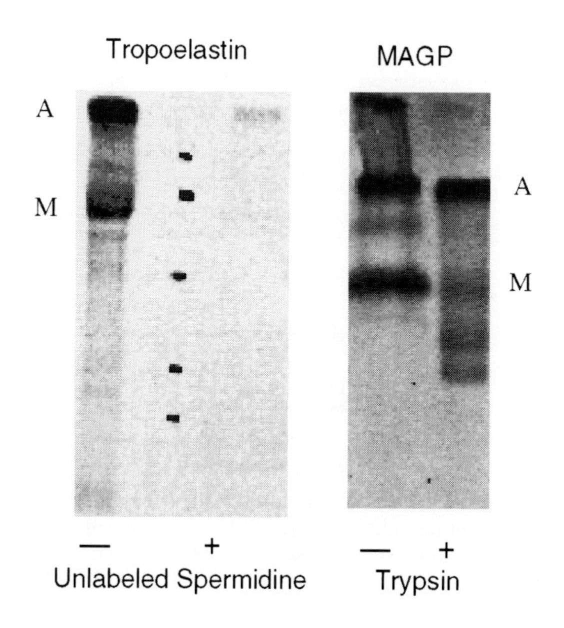


FIG. 4. Autoradiogram showing transglutaminase-catalysed incorporation of [$^{14}$C]spermidine into tropoelastin and MAGP. For both proteins, incorporation could be inhibited with excess cold spermidine (shown for tropoelastin). Radioactive bands at the top of the gel represent aggregates (A) formed from the monomer (M) as a result of transglutaminase cross-linking. Only the monomeric form of MAGP is degraded by trypsin.

modelling using the Sybyl program confirms that the lysine residues in exons 19, 25 and 10 are appropriately positioned to form the proposed structure.

## Modification of tropoelastin and microfibrillar components by transglutaminase

Several recent studies have documented that cross-linking between extracellular matrix proteins can occur through transglutaminase-mediated pathways. Fibronectin, vitronectin, collagen type III and nidogen have all been shown to be specific substrates for transglutaminase (reviewed in Aeschlimann & Paulsson 1991). To examine whether transglutaminase might play a role in stabilizing elastic fibre interactions, we incubated MAGP, fibrillin 1 and tropoelastin with transglutaminase in the presence of [$^{14}$C]spermidine. Figure 4 shows the incorporation of [$^{14}$C]spermidine into tropoelastin and MAGP. For all proteins, additional radioactive bands appeared at the top of the gel after incubation with transglutaminase, indicating that covalent complexes are formed. We conclude from these experiments that MAGP, fibrillin 1, and tropoelastin contain glutamine residues which can act as amine acceptor sites in tissue transglutaminase-catalysed cross-linking. The propensity of these molecules to aggregate following modification with transglutaminase also suggests that lysine residues within the protein serve as amine donors. It will be interesting to determine whether cross-linking of MAGP to tropoelastin or fibrillin by transglutamination is a common and necessary feature of elastic fibre assembly.

The ability of MAGP and fibrillin 1 to serve as substrates for transglutaminase may answer several perplexing questions about other molecules that have been suggested to be components of microfibrils. Both fibronectin and vitronectin have been found to co-localize with microfibrils in some tissues (Dahlbäck et al 1989, Schwartz et al 1985, Tomasini-Johansson et al 1993). Although they do not seem to be integral components of microfibrils (Gibson et al 1989), a covalent association with microfibrillar proteins could occur through transglutaminase-mediated cross-linking.

**Specificity of cross-linking sites to enzymic modification by lysyl oxidase**

An important question raised by our sequencing data is whether certain cross-linking domains in tropoelastin are preferred sites for oxidation by lysyl oxidase during fibre assembly. While 33 of the 38 total lysines in tropoelastin eventually serve as cross-links, not all need be modified by lysyl oxidase since all of the cross-links except allysine aldol require the participation of an unoxidized lysine side chain in their formation. To begin to address the question of which lysines are modified by lysyl oxidase during fibre assembly, we have raised antibodies to short synthetic peptides that model the Lys-Pro-containing cross-linking sequences of bovine tropoelastin. The lysine residues provide the major epitope recognized by several of these antibodies since succinylation of ε-amino groups greatly reduces the antigenicity of the peptide. In addition, each of the antibodies has been tested by Western blot analysis for reactivity with bovine tropoelastin (Wrenn et al 1987). When it was used for electron microscopic immunolocalization of elastin in developing ligamentum nuchae, we found that the antibody to exon 10 reacted with tropoelastin inside the cell but not with elastin associated with the growing elastic fibre in the extracellular matrix. Antibodies to the cross-linking site in exon 35 and to a non-cross-linking site in exon 26, however, reacted with both intracellular and extracellular elastin. We interpret this result to suggest that exon 10 is rapidly oxidized by lysyl oxidase upon secretion of tropoelastin. The lysines in exon 35, in contrast, either remain unoxidized or are oxidized at a much slower rate.

Additional evidence that exon 10 is a preferred site of oxidation by lysyl oxidase was provided by immunoblot analysis of recombinant tropoelastin purified from a bacterial expression system (Grosso et al 1991) and tropoelastin isolated from minces of bovine ligamentum nuchae incubated overnight in the presence of the lysyl oxidase inhibitor β-aminopropionitrile (BAPN). An antibody to the cross-linking sequence in exon 4/5 reacted with both tropoelastin preparations whereas an antibody to exon 10 reacted only with recombinant tropoelastin. Upon further analysis, we found that tropoelastin from the tissue mince contained reducible aldehydes, indicating that some of the lysine residues had been oxidized by lysyl oxidase, even in the presence of BAPN. The failure of the exon 10 antibody to react suggests that the lysines in this domain are strong candidates for oxidation. It is unlikely that exon 10 is absent since this exon has not been shown to be alternatively spliced in bovine ligamentum nuchae (Parks & Deak 1990).

## Conclusions

Cross-linking of tropoelastin requires the precise alignment of cross-linking sites within each monomer. Unlike collagen, where the information for assembly is inherent in the primary structure, elastin does not self-assemble. Instead, tropoelastin alignment requires helper proteins associated with elastic fibre microfibrils. We have shown that MAGP binds specifically to tropoelastin, making this microfibrillar protein an ideal candidate for the tropoelastin alignment protein *in vivo*. With the elucidation of the primary sequence of elastin and several microfibrillar components, it will now be possible to identify macromolecular domains that mediate interactions between these elastic fibre components.

## References

Aeschlimann D, Paulsson M 1991 Cross-linking of laminin–nidogen complexes by tissue transglutaminase: a novel mechanism for basement membrane stabilization. J Biol Chem 266:15308–15317

Brown-Augsburger P, Broekelmann T, Mecham L et al 1994 Microfibril associated glycoprotein (MAGP) binds to tropoelastin and is a substrate for transglutaminase. J Biol Chem 269:28443–28449

Dahlbäck K, Löfberg H, Alumets J, Dahlbäck B 1989 Immunochemical demonstration of age-related deposition of vitronectin (S-protein of complement) and terminal complement complex on dermal elastic fibers. J Invest Dermatol 92:727–733

Foster JA, Bruenger E, Rubin L et al 1976 Circular dichroism studies of an elastin cross-linked peptide. Biopolymers 15:833–841

Gerber GE, Anwar RA 1975 Comparative studies of the cross-linked regions of elastin from bovine ligamentum nuchae and bovine, porcine, and human aorta. Biochem J 149:685–695

Gibson MA, Kumaratilake JS, Cleary EG 1989 The protein components of the 12-nanometer microfibrils of elastic and nonelastic tissues. J Biol Chem 264:4590–4598

Grosso LE, Parks WC, Wu L, Mecham RP 1991 Fibroblast adhesion to recombinant tropoelastin expressed as a protein A fusion protein. Biochem J 273:517–522

Mecham RP, Davis EC 1994 Elastic fiber structure and assembly. In: Yurchenko PD, Birk DE, Mecham RP (eds) Extracellular matrix assembly and structure. Academic Press, San Diego, CA, p 281–314

Mecham RP, Foster JA 1978 A structural model for desmosine cross-linked peptides. Biochem J 173:617–625

Parks WC, Deak SB 1990 Tropoelastin heterogeneity: implications for protein function and disease. Am J Respir Cell Mol Biol 2:399–406

Partridge SM, Elsden DF, Thomas J 1964 Biosynthesis of the desmosine and isodesmosine crossbridges in elastin. Biochem J 93:30c–33c

Schwartz E, Godlfischer S, Coltoff-Schiller B, Blumenfeld OO 1985 Extracellular matrix microfibrils are composed of core proteins coated with fibronectin. J Histochem Cytochem 33:268–274

Tomasini-Johansson BR, Ruoslahti E, Pierschbacher MD 1993 A 30 kD sulfated extracellular matrix protein immunologically crossreactive with vitronectin. Matrix 13:203–214

Wrenn DS, Parks WC, Whitehouse LA et al 1987 Identification of multiple tropoelastins secreted by bovine cells. J Biol Chem 262:2244–2249

Yurchenco PD, Birk DE, Mecham RP 1994 Extracellular matrix assembly and structure. Academic Press, San Diego, CA.

## DISCUSSION

*Robert:* The transglutaminase results you showed reminded me of the action of cysteamine which is one of the very few physiologically occurring amine structures to interfere with normal elastin development in the aorta. Its presence in high concentrations in the blood, or its infusion, induce atherosclerotic-type lesions as seen in homocystinuria (Boers et al 1985). Do you think there could be interference with the transglutaminase cross-linking you described?

*Mecham:* I think this is possible, although I'm not certain how much of a role transglutaminase plays in fibrillogenesis. The extractability of elastic fibre components from elastic tissues using reductive saline techniques suggests that they are not covalently cross-linked.

*Kagan:* You results bring to mind some unpublished experiments we performed in our laboratory several years ago. We assessed the time dependency of [$^{14}$C]valine-pulsed elastin being incorporated into the alkali-insoluble fraction of chick embryo aortas in organ culture. What intrigued me was that there was a biphasic kinetic response; that is, there was a relatively slow incorporation within the first hour of pulsing, followed by a rapid steady-state incorporation of [$^{14}$C]valine-labelled material into the alkali-insoluble fraction. Following up on that result, we observed that the first slow phase of incorporation was resistant to inhibition by BAPN and other lysyl oxidase inhibitors. This result might reflect a lysyl oxidase-independent mechanism of cross-linking of tropoelastin at the very earliest stage of fibre formation, such as might be catalysed by transglutaminase. Alternatively, the BAPN-resistant phase of elastin insolubilization may reflect a phase of cross-link formation catalysed by lysyl oxidase in which the enzyme is not accessible to solvent molecules. In that regard, was the transglutaminase activity you identified derived from an extracellular or intracellular enzyme?

*Mecham:* Tissue transglutaminase has been found in a number of tissues, and in many cases is a component of the extracellular matrix.

*Keeley:* Is formation of the transglutaminase peptide bond a reversible process? Are there specific proteases which break this kind of isopeptide bond? If so, these lysines may not be permanently removed from possible involvement in cross-linking, since they could be regenerated later.

*Mecham:* It is a covalent interaction that's not reversible. In fact, it's probably more stable than lysyl oxidase-derived cross-links, because it does not go through a reversible aldehyde intermediate. An important point to remember is that transglutaminase-derived cross-links will permanently tie up lysines that might normally be oxidized by lysyl oxidase.

*Robert:* I'm not sure that the $\gamma$-glutamine linkage could be split, but plasmin could split very close to the cross-linked fibrin clot which is easily dissolved.

*Bailey:* Bob Mecham is right; the bond is not reversible. The way to isolate the isopeptide bond is by using a mixture of enzymes and digesting the protein right down to the enzyme-resistant isopeptide cross-link. If you take transglutaminase, you can cross-link collagen with it, but we know *in vivo* that collagen is not cross-linked by transglutaminase. This sort of *in vitro* treatment of proteins can't always be extrapolated to the *in vivo* situation. Have you got any further evidence that transglutaminase is involved? Have you actually located the enzyme there by antibody?

*Mecham:* We have shown that there is transglutaminase activity in an extract of bovine aorta, but, as you say, there is no evidence that elastic fibre components are natural substrates *in vivo*.

*Kagan:* Is there a specific inhibitor of transglutaminase?

*Mecham:* At high concentrations, the diamines, such as spermidine or cadaverine, can inhibit cross-link formation between two proteins. I'm not certain how you would use them to inhibit the enzyme *in situ*.

*Keeley:* Just to clarify my point: I didn't mean that the bond was non-covalent. However, specific peptidases breaking these isopeptide bonds might release these lysines again, making them available for cross-linking.

*Hinek:* There is more than one transglutaminase enzyme which can do the job, and one obvious candidate is cell-surface-associated transglutaminase, which can be blocked by dansylcadaverine. The obvious experiment to do would be to check if dansylcadaverine or other inhibitors of transglutaminase affect the assembly of elastic fibrils. Electron microscopy reveals that the formation of the microfibrils precedes the deposition of elastin (Hinek & Thyberg 1977). The first microfibrils are always tightly attached to the cell surface. This suggests that the cell-surface transglutaminase plays an important role in initiation of microfibril formation.

*Mecham:* I would like to emphasize that we have no evidence whatsoever that transglutaminase is playing a role, only that these proteins are substrates in *in vitro* assays.

*Sakai:* I was interested in your transglutaminase experiments as well, and I agree with Allen Bailey that the important thing is to try and find these cross-links *in vivo*. Rob Glanville has evidence for those cross-links in microfibrils isolated from tissue. The cross-link is labile to acid hydrolysis, so in the extraction procedures and the sequencing procedures it's somewhat more difficult to detect these cross-links. At the 3rd International Symposium on the Marfan Syndrome (Berlin, September 6–9, 1994) Rob Glanville reported that he had found in the beaded fibrils that are extracted from tissues (fetal membranes) evidence for six lysine residues that were probably involved in transglutaminase-mediated cross-links (R.W. Glanville, R.-Q. Qian, unpublished results). Two of these lysines came from fibrillin; these two lysines are present in specific $Ca^{2+}$-binding EGF-like domains in the interbend domain. But that still leaves four other lysine residues that may not be accounted for by fibrillin–fibrillin interactions.

*Urry:* In your cartoon of elastogenesis, you show one MAGP molecule associated with one tropoelastin molecule. Don't you have to have subsequent processing? Otherwise the stoichiometry in the elastic fibre would be one MAGP to every tropoelastin molecule.

*Mecham:* MAGP may bind tropoelastin via a reversible ionic interaction, yet be covalently linked to fibrillin via transglutaminase.

*Urry:* In terms of the assembly of the structure, you showed that the tropoelastin molecule is attached, one to each MAGP molecule.

*Mecham:* We feel that MAGP is interacting non-covalently with the C-terminus of tropoelastin.

*Urry:* Could you explain your preparative method for looking at tropoelastin by electron microscopy?

*Mecham:* Tropoelastin is absorbed onto mica flakes, quick frozen, and then etched and shadowed.

*Urry:* What concerns me is that the natural elastic fibres are 50% water. All structure formation depends on water and interactions with water. If you remove water, I don't see how you can retain a structure that's relevant.

*Mecham:* It may be an artefact of the system, but at this point it is the best we can do.

*Grant:* What's the current thinking on the interactions of the fibrillin-like protein (FLP) with tropoelastin or any of these microfibrillar components?

*Mecham:* FLP was first identified by Mark Gibson, who used an antibody to fibrillin 1 to screen a bovine ligamentum nuchae expression library. One of the selected clones was not fibrillin 1, but encoded a protein that was fibrillin-like in that it contained a similar domain structure to fibrillin 1 and fibrillin 2. Cloning studies suggested that FLP was also similar to TGF-$\beta$1-binding protein and that the FLP mRNA undergoes alternative splicing. Furthermore, antibodies to FLP localized to microfibrils, indicating that FLP is closely associated with these structures.

Like the fibrillins, FLP cannot be extracted from tissues without reduction and denaturation. Thus, it is not possible to do binding studies with native protein and I don't think that the denatured proteins are going to tell us much about protein–protein interactions.

*Grant:* What about the possibility of heteropolymers of fibrillin in these microfibrils? Are fibrillin 1 and 2 found together?

*Mecham:* We've done immunogold labelling with antibodies to fibrillin 1 and 2. It's quite clear that you can find fibrillin 1 and 2 in the same microfibrillar bundles. What we haven't been able to determine is whether microfibrils contain both fibrillin 1 and 2, or if they are homopolymers of only fibrillin 1 or 2.

*Mariani:* Given the presence of lysine residues in many microfibrillar proteins, what is the possible contribution of these proteins to desmosine cross-links?

*Mecham:* It's going to be very interesting to express other microfibrillar proteins and see if they also interact with elastin in the way I showed for MAGP. Obviously, with MAGP we have the reagents to do a thorough investigation. We have to do the same thing with the other proteins. It might be that there are synergistic or competitive effects between these proteins that might mediate the kinds of morphological differences we see in various tissues.

*Rosenbloom:* A comment about the co-appearance of these various proteins. Essentially all these proteins which have been clearly identified can all be identified in ciliary zonules which are essentially pure microfibrils. These include, in addition to the fibrillins, microfibril-associated protein 1 (MFAP1), MFAP3 and MAGP.

## References

Boers GHJ, Smalls AGH, Trijbels FJ et al 1985 Heterozygosity for homocystinuria in premature peripheral and cerebral occlusive arterial disease. N Engl J Med 313:709–715

Hinek A, Thyberg J 1977 Electron microscopic observations on the formation of elastic fibers in primary cultures of aortic smooth muscle cells. J Ultrastruct Res 60:12–20

# The 67 kDa spliced variant of β-galactosidase serves as a reusable protective chaperone for tropoelastin

Aleksander Hinek

*Division of Cardiovascular Research, The Hospital for Sick Children, 555 University Avenue, Toronto, Ontario M5G 1X8, Canada*

*Abstract.* Numerous cell types express the 67 kDa galactolectin related to the alternatively spliced variant of β-galactosidase. This 67 kDa protein, while present on cell surfaces, mediates cell contacts with elastin, laminin and collagen type IV. In elastin-producing tissues, the 67 kDa protein also co-localizes with intracellular tropoelastin and mature elastic fibres. We have established that this elastin binding protein (EBP) serves as a molecular chaperone for tropoelastin. The EBP binds this highly hydrophobic and unglycosylated ligand intracellularly, protecting it from intracellular self aggregation and premature proteolytic degradation, and mediates its orderly assembly upon the microfibrillar scaffold. While some of this protein is incorporated as a permanent component of elastic fibres, most of the EBP, after extracellular dissociation from its ligand, recycles back to the intracellular endosomal compartment and re-associates with the newly synthesized tropoelastin. We suggest that recycling of this reusable shuttle protein is imperative for the effective extracellular deposition of insoluble elastin.

*1995 The molecular biology and pathology of elastic tissues. Wiley, Chichester (Ciba Foundation Symposium 192) p 185–196*

The specific chemotactic response of human leukocytes and fibroblasts to tropoelastin and proteolytic fragments of insoluble elastin described by Senior et al (1980, 1984) suggested the existence of a true elastin receptor. A classical receptor–ligand binding assay, showing that iodinated tropoelastin bound to ligamentum nuchae fibroblasts and their isolated plasma membranes in a time-dependent, saturable and reversible manner, further confirmed this hypothesis (Wrenn et al 1988). Scatchard analysis indicated that there are approximately $2 \times 10^6$ binding sites per cell with the binding affinity ($K_d$) of 8 nM. Using affinity chromatography techniques, we have demonstrated that numerous cell types, including fibroblasts, vascular smooth muscle cells, chondroblasts, monocytes, neutrophils and several tumour cells, express a cell-surface protein capable of interacting with tropoelastin and with its cross-linked polymer. This

67 kDa peripheral membrane protein is associated with 61 kDa and 55 kDa integral membrane proteins which are linked to intracellular actin (Hinek et al 1988, Mecham et al 1989a,b). One of the most striking features of this 67 kDa elastin binding protein (EBP) was its galactolectin character which enabled purification on glycoconjugate affinity columns containing β-galactosugars. Moreover, β-galactosugars, such as galactose and lactose, also eluted this protein from elastin affinity columns, while glucose or fucose had no effect. This indicated that the association between elastin and the 67 kDa EBP is a tight protein–protein interaction which can be allosterically abolished by galactosugars binding to the lectin site of EBP (Hinek et al 1988, 1992, Mecham et al 1991). Consistent with this, we have shown that lactose abolished the chemotactic response of numerous cell types to elastin and to VGVAPG peptides, but did not alter directed migration to platelet-derived growth factor (PDGF), a chemoattractant that binds to a different cell surface receptor (Mecham et al 1989b, Hinek 1994). Moreover, we have also shown that either a deficiency in EBP or shedding of the EBP from cell surfaces by exogenous galactosugars disrupts assembly of elastic fibres in cultures of auricular chondrocytes and vascular smooth muscle cells (Hinek et al 1988, 1991, Hinek & Rabinovitch 1993). This led us to speculate that the EBP presents tropoelastin molecules on the cell surface and then utilizes its lectin site to dock to a galactosylated microfibrillar protein(s) and subsequently releases tropoelastin onto the microfibrillar scaffold.

In the course of our study leading to complete characterization of the EBP, we discovered that the EBP isolated from sheep aorta smooth muscle cells (SMCs) is related to an alternatively spliced form of human β-galactosidase (S-GAL), described by Morreau et al (1989). This product of the β-galactosidase gene has lost its catalytic activity and lysosomal targeting, but has retained its ability to bind galactosugars and has acquired a unique sequence containing the elastin and laminin binding domain (Hinek et al 1993). Since an antibody (anti-S-GAL) raised to the synthetic peptide representing the elastin/laminin binding domain of human S-GAL reacts with the 67 kDa sheep EBP, we obtained a powerful tool for further studies of this protein. Using this antibody (which does not cross-react with the enzymically active form of β-galactosidase) we also localized the EBP to intracellular compartments of sheep aortic SMCs and human skin fibroblasts. Double immunolabelling indicated that EBP and tropoelastin coexist in the same intracellular compartments of those cells. Further biochemical data verified that EBP and tropoelastin form a stable molecular complex. Moreover, we have shown that binding of the EBP to tropoelastin not only prevents intracellular self-aggregation of highly hydrophobic tropoelastin molecules but also protects them from association with serine proteinases. Therefore, we have proposed that the EBP serves as a protective companion protein (molecular chaperone) for tropoelastin which binds this highly hydrophobic and unglycosylated

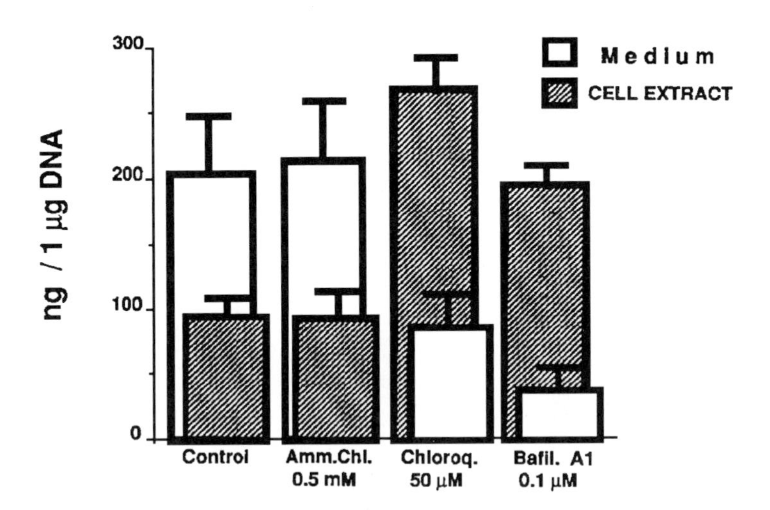


FIG. 1. Chloroquine and bafilomycin $A_1$, blockers of endosomal acidification that arrest the EBP recycling, cause an intracellular accumulation of tropoelastin in cultured aortic myocytes (AoSMC). Ammonium chloride, which alkalinizes the most acidic endosomal compartments and lysosomes, and does not disrupt EBP recycling, does not block secretion of soluble elastin (tropoelastin). Tropoelastin was assessed separately by ELISA in the media and cell layers of cultures maintained in the presence or absence of ammonium chloride (0.5 mM), chloroquine (50 $\mu$m) and bafilomycin $A_1$ (0.1 $\mu$M). The amount of tropoelastin present in each sample was calculated from a standard curve prepared with known concentrations of tropoelastin, and the final results (ng of tropoelastin/1 $\mu$g DNA) were expressed as the mean $\pm$ SD calculated from triplicates of two separate experiments.

protein and escorts it through the secretory pathways (Hinek & Rabinovitch 1994). This is in accord with the general theory of intracellular chaperoning of unfolded proteins by the heat shock proteins (Hsps) (Agard 1993). The blockade of exposed hydrophobic patches on numerous secreted proteins requires a factor that can reversibly bind to them and protect against premature aggregation (Langer et al 1992, Landry et al 1992). The 67 kDa EBP, like Hsp70 and its relatives, meets this requirement. In the case of Hsp70, the release of bound protein requires stimulation of endogenous ATPase, while the EBP can dissociate from tropoelastin under the influence of galactosugars which bind to the lectin site of this 67 kDa chaperone.

Because the production of tropoelastin by arterial SMCs exceeds production of EBP, we speculated that some of this 67 kDa chaperone might recycle and bind again to newly synthesized tropoelastin. To test this hypothesis, we labelled cultured arterial SMCs externally with fluorescein-conjugated $F(ab')_2$

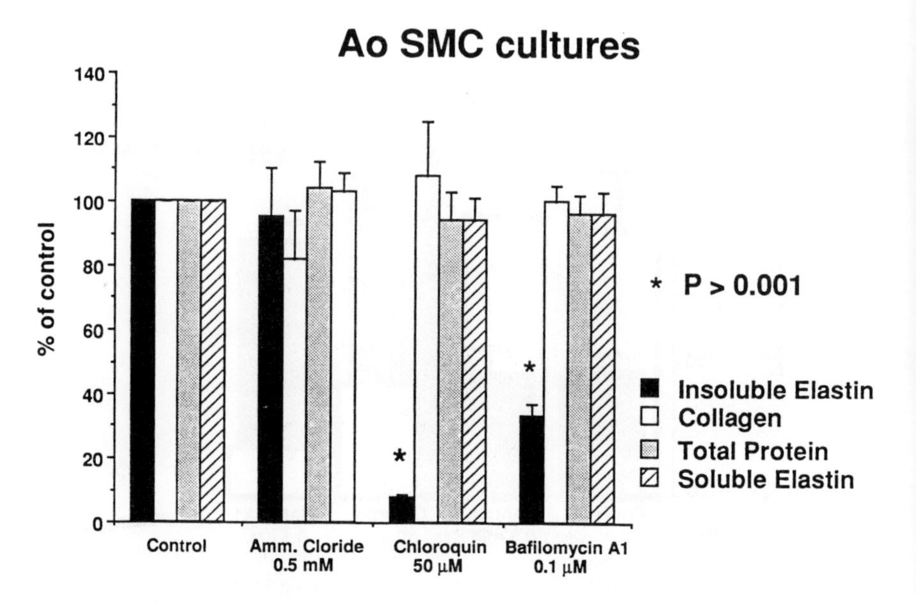


FIG. 2. Chloroquine (50 $\mu$M) and bafilomycin $A_1$ (0.1 $\mu$M), but not ammonium chloride (0.5 mM), significantly inhibit deposition of insoluble elastin in 24 h cultures of aortic myocytes. Neither of these reagents affect the level of total protein synthesis or the synthesis of soluble elastin and collagen. Deposition of insoluble elastin was measured by incorporation of [$^{14}$C]valine into the cyanogen bromide-insoluble precipitate. Total protein synthesis was assessed by incorporation of [$^3$H]leucine to TCA precipitate from medium and cell layers. Soluble elastin present in the medium and cell layers extracts of [$^{14}$C]valine-labelled cultures was immunoprecipitated with the mono-specific polyclonal antibody to tropoelastin. Collagen synthesis was reflected by levels of radioactive hydroxyproline produced by cultures labelled with [$^3$H]proline. All results (CPM/$\mu$g DNA) were calculated from triplicate cultures of two separate experiments and finally expressed as the % of control (mean $\pm$ SD).

fragments of anti-S-GAL immunoglobulin to follow trafficking of their 67 kDa receptor. Our results indicate that the majority of this protein residing on the cell surface and traceable by external labelling can be internalized and recycled back to the plasma membranes within 60–90 min. Moreover, we have shown that some blockers of normal endosomal acidification arrest the EBP recycling in cultured aortic myocytes (Fig. 1). For example, chloroquine inhibits the initial steps of EBP endocytosis, while bafilomycin $A_1$, a specific inhibitor of vacuolar $H^+$-ATPase, which blocks acidification of the distal endosomal compartment, arrests the exit of the elastin receptor from the endocytic pathway and inhibits its return to the plasma membrane. Both of these reagents also cause an intracellular accumulation of tropoelastin and selectively inhibit deposition of insoluble elastin in cultures of ovine aortic

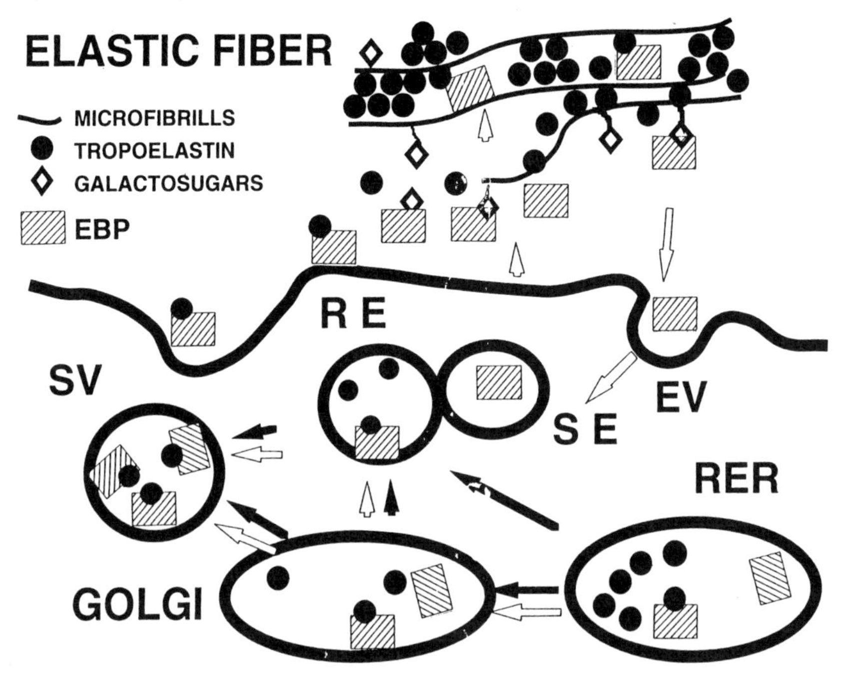


FIG. 3. In elastin-producing cells the 67 kDa protein binds tropoelastin in rough endoplasmic reticulum (RER) and serves as its protective molecular chaperone on its secretory pathway through the Golgi and secretory vacuoles (SV). When secreted, the EBP–tropoelastin complex is placed on the cell surface. Binding of galactosugar residues protruding from microfibrillar glycoprotein(s) to the lectin site of the EBP serves to lower its affinity for tropoelastin and facilitate transfer of tropoelastin monomer to an assembly site on the microfibrillar scaffold of the growing elastic fibre. Some of this protein is incorporated as a permanent component of elastic fibres, but the majority of the EBP, after dissociation from its ligand, recycles back to the cell endosomal compartments. It is taken up by endocytic vesicles (EV), passes the sorting endosome (SE) and enters the recycling endosomes (RE), where it re-associates with newly synthesized tropoelastin.

SMCs and in organ cultures of chicken aorta (Fig. 2). Neither chloroquine nor bafilomycin $A_1$ (50 $\mu$M and 0.1 $\mu$M, respectively) inhibit total protein synthesis, or synthesis of tropoelastin or collagen. Interestingly, ammonium chloride (5 mM), a weak base which accumulates in the most acidic endosomal compartments and lysosomes, did not disrupt EBP recycling or affect deposition of insoluble elastin. This suggests that the majority of the EBP does not enter lysosomes during its recycling. In fact, our electronmicrographs showing immunolocalization of the EBP over the extracellular elastic fibres suggest that extracellular deposition but not intracellular degradation may be responsible for final elimination of this protein from the recycling pathways.

In summary, our recent results suggest that the 67 kDa EBP serves as a molecular chaperone escorting tropoelastin through the secretory pathways and mediating its orderly extracellular assembly onto the microfibrillar scaffold. Some of this protein gets incorporated as a permanent component of elastic fibres, but the majority of the EBP, after dissociation from its ligand, recycles back to the endosomal compartment of the cell to re-associate with newly-synthesized tropoelastin (Fig. 3). We also suggest, that undisturbed recycling of this reusable shuttle protein is essential for the extracellular deposition of insoluble elastin.

*Acknowledgements*

This work was supported by a grant from the Heart and Stroke Foundation of Ontario (B 2040) and a Program grant from the Medical Research Council of Canada (PG 12351). The author is a Career Investigator of the Heart and Stroke Foundation of Ontario.

## References

Agard DA 1993 To fold or not to fold. Science 260:1903–1904

Hinek A 1994 Nature and multiple functions of the 67-kD elastin/laminin binding protein. Cell Adhes Commun 2:185–193

Hinek A, Rabinovitch M 1993 The ductus arteriosus migratory smooth muscle cell phenotype processes tropoelastin to a 52-kDa product associated with impaired assembly of elastic laminae. J Biol Chem 268:1405–1413

Hinek A, Rabinovitch M 1994 67-kD elastin-binding protein is a protective companion of extracellular insoluble elastin and intracellular tropoelastin. J Cell Biol 126:563–574

Hinek A, Wrenn DS, Mecham RP, Barondes SH 1988 The elastin receptor: a galactoside-binding protein. Science 239:1539–1541

Hinek A, Mecham RP, Keeley F, Rabinovitch M 1991 Impaired elastin fiber assembly related to reduced 67-kD elastin-binding protein in fetal lamb ductus arteriosus and in cultured aortic smooth muscle cells treated with chondroitin sulfate. J Clin Invest 88:2083–2094

Hinek A, Boyle J, Rabinovitch M 1992 Vascular smooth muscle cell detachment from elastin and migration through elastic laminae is promoted by chondroitin sulfate-induced shedding of the 67-kDa cell-surface elastin binding protein. Exp Cell Res 203:344–353

Hinek A, Rabinovitch M, Keeley F, Okamuroaho Y, Callahan J 1993 The 67-kD elastin/laminin-binding protein is related to an enzymatically inactive, alternatively spliced β-galactosidase. J Clin Invest 91:1198–1205

Landry SJ, Jordan R, McMacken R, Gierasch LM 1992 Different conformations for the same polypeptide bound to chaperones DnaK and GroEL. Nature 355:455–457

Langer T, Lu C, Echols H, Flanagan J, Hayer MK, Ulrich Hartl F 1992 Successive action of DnaK, DnaJ and GroEL along the pathway of chaperone-mediated protein folding. Nature 356:683–689

Mecham RP, Hinek A, Entwistle R, Wrenn DS, Griffin GL, Senior RM 1989a Elastin binds to a mutlifunctional 67-kilodalton peripheral membrane protein. Biochemistry 28:3716–3722

Mecham RP, Hinek A, Griffin GL, Senior RM, Liotta LA 1989b The elastin receptor shows structural and functional similarities to the 67-kDa tumor cell laminin receptor. J Biol Chem 264:16652–16657

Mecham RP, Whitehouse L, Hay M, Hinek A, Sheetz MP 1991 Ligand affinity of the 67-kD elastin/laminin binding protein is modulated by the protein's lectin domain—visualization of elastin/laminin receptor complexes with gold-tagged ligands. J Cell Biol 113:187–194

Morreau H, Galjart NJ, Gillemans N, Willemsen R, Vanderhorst GTJ, Dazzo A 1989 Alternative splicing of $\beta$-galactosidase messenger RNA generates the classic lysosomal enzyme and a $\beta$-galactosidase-related protein. J Biol Chem 264:20655–20663

Senior RM, Griffin GL, Mecham RP 1980 Chemotactic activity of elastin-derived peptides. J Clin Invest 66:859–862

Senior RM, Griffin GL, Mecham RP, Wrenn DS, Prasad KU, Urry DW 1984 Val-Gly-Val-Ala-Pro-Gly, a repeating peptide in elastin, is chemotactic for fibroblasts and monocytes. J Cell Biol 99:870–874

Wrenn DS, Hinek A, Mecham RP 1988 Kinetics of receptor-mediated binding of tropoelastin to ligament fibroblasts. J Biol Chem 263:2280–2284

## DISCUSSION

*Pierce:* There have been a number of elastin binding proteins identified; could you clarify which of the elastin binding proteins you were referring to?

*Hinek:* I referred to the 67 kDa elastin/laminin binding protein which is homologous to spliced variants of $\beta$-galactosidase (Hinek 1994, Hinek et al 1993). Our 67 kDa protein was the only one which we could detect with so-called 'elastin affinity' columns. When we immobilize tropoelastin, $\kappa$-elastin, $\alpha$-elastin or we use insoluble elastin, the really strong band we see is the 67 kDa protein, which we call the elastin binding protein. We think the 67 kDa protein is the only one which binds to tropoelastin intracellularly (Hinek & Rabinovitch 1994). What happens on the cell surface or extracellularly is the big question, because, of course, microfibril-associated glycoprotein (MAGP) is also an elastin binding protein.

*Robert:* In pharmalogical experiments where we refer to elastin receptors, the main component seems to be the 67 kDa protein.

*Pierce:* Is there a specific name for this protein? Is this the S-GAL that you have talked about before?

*Hinek:* For a long time we had problems characterizing the 67 kDa protein. We knew that its interaction with galactosugars actually disrupted its ability to bind elastin, laminin and collagen type IV. Most of our problems with characterizing this protein were to do with sequencing it, because the N-terminus is always blocked. By a stroke of luck we realized that the indigenous protein is homologous to a splice variant of $\beta$-galactosidase. This really speeded up the whole process. We identified the unique elastin-binding motif on this protein which was encoded by a frame shift-generated fragment, which is not present in the active enzyme. We are convinced that the 67 kDa protein is β-galactosidase which has lost its active site, the lysosome-driven domain, but has gained a new unique sequence which contains the elastin/laminin-binding domain (Hinek 1994, Hinek et al 1993).

*Mecham:* In your internalization studies did you have a positive control? For instance, did you use the mannose-6-phosphate receptor to make certain that you were observing an internalization pathway?

*Hinek:* No. I used other antibodies to eliminate the possibility of non-specific immunoreaction, but I didn't use mannose-6-phosphate receptor, because it goes to lysozomes and does not recycle as quickly as the 67 kDa receptor.

For a long time we also tried to identify the other two proteins which were always attached to the cell-surface 67 kDa protein. We knew from the very first studies with Bob Mecham (Hinek et al 1988, Mecham et al 1989a) that these proteins could only be released from the cell membrane fraction by detergent, so we assumed that 61 and 55 kDa proteins which attach to the 67 kDa protein must be integral membrane proteins. We now have some evidence for the identity of these proteins; one of them is 61 kDa sialidase, which is a membrane-bound protein, and the second one is the so-called 'protective' protein, which always forms a complex with active β-galactosidase (unpublished results).

*Ooyama:* Are the smooth muscle cells synthesizing elastic fibres? Are they in a synthetic state?

*Hinek:* Yes.

*Ooyama:* So this elastin binding protein could be used as a marker for changes in the phenotype of the smooth muscle cells?

*Hinek:* No; this protein is present in many other cell types that do not produce elastin. One of its possible functions in elastin-producing cells is to serve as a molecular chaperone for tropoelastin (Hinek & Rabinovitch 1994), but it's also present in other cell types, such as monocytes, neutrophils and several cancer cells (Senior et al 1989, Mecham et al 1989b, Hinek 1994).

*Robert:* White blood cells express the elastin receptor (Fülöp et al 1986) and they don't produce elastin, as far as we can tell. A few years ago, it was suggested that the elastin receptor might be a heat shock protein. Subsequently it has been shown that a heat shock protein is an integral part of the oestrogen receptor (Baulieu 1987). Do you have any evidence that, for instance, heat shock increases the expression of the 67 kDa protein?

*Hinek:* No, I don't have direct evidence. Of course, our protein is not a 'classical' heat shock protein, but it plays the role of a molecular chaperone, as many heat shock proteins do. The only difference is the mechanism by which this complex between the ligand and the binding protein can be dissociated. We think that the galactosugars play a role in dissociation; with Hsp70, ATPase and other enzymes are needed to dissociate the complex. But I think that your comment is still very interesting because we recently observed the significant up-regulation of the deposition of insoluble elastin in cultures of aortic smooth muscle cells maintained at 41 °C. Perhaps this rise in temperature also up-regulates production of the 67 kDa protein, but so far we haven't detected a striking difference in this respect between cells cultured at 37 °C or 41 °C.

*Davidson:* In your inhibitor studies you implied that you were getting sequestration of elastin binding protein in a compartment where it couldn't participate in transport. Does the anti-binding protein antibody produce the same kind of effect, while being a more specific reagent for elastin binding protein?

*Hinek:* Yes, because this antibody was to a synthetic peptide, which is located in the middle of the frameshift-generated sequence which is very specific to the 67 kDa splice variant of $\beta$-galactosidase. We also immobilized this peptide on the resin, and we noted that the sequence to which we raised the antibody can only bind laminin and elastin, and nothing else. We tried fibronectin and many types of collagen without success.

*Davidson:* Does this antibody alter the partitioning of elastin between cell layer and medium, or the efficiency of incorporation of elastin?

*Hinek:* Yes; if cells are cultured in the presence of this antibody, elastic fibre assembly is completely inhibited, because the particular sequence on the 67 kDa protein that normally binds tropoelastin is blocked by the antibody.

*Davidson:* Do you see parallel effects on laminin production using these kinds of inhibitor?

*Hinek:* I haven't studied this.

*Davidson:* Does the elastin binding protein interfere with lysyl oxidase-mediated cross-linking?

*Hinek:* Probably not. I think the role of elastin binding protein is to present elastin at the cell surface. Later on, this touches some galactosugars, probably protruding from the microfibril proteins, and under the influence of galactosugars tropoelastin is released from this so-called chaperone protein and is subsequently anchored to the microfibrillar scaffold. The next event is probably cross-linking, but this is not affected by elastin binding protein.

*Davidson:* Do these data reconcile with the concept that elastin binding protein provides a cell recognition mechanism for elastin? We know that there are biological responses of cells to elastin: is this molecule participating in that reaction?

*Hinek:* Together with Fred Keeley and Kevin Fosket, I used the video microscope and the fura-2 label to study the response of the smooth muscle cells to elastin. When we treated our cells with $\kappa$-elastin or tropoelastin we observed a very quick influx of $Ca^{2+}$. This looks like a receptor-mediated response. Once you find a good responder you can play for half an hour up and down. But if you flush your chamber with lactose (i.e. if you strip your receptor from the cell surface), you can't do this again. This was quite an interesting observation, but we did not publish the results because we couldn't explain why not all the cells responded to the elastin.

*Keeley:* The problem with these observations was that although cells that responded were very consistent, showing a sharp $Ca^{2+}$ spike, it was often very difficult to find a cell that responded.

*Robert:* In studies on $Ca^{2+}$ influx mediated by the elastin receptor, we first used smooth muscle cells and fibroblasts in culture and monitored $^{45}Ca^{2+}$ influx (Jacob et al 1987). As soon as we found the elastin receptor and the $Ca^{2+}$ response triggered by elastin peptides in white blood cells, we did much of the pharmacological work on these cells, and part of it also on cancer cells with Dr Lapis. We didn't study differences between individual cells; we used cell suspensions and spectrophotometric monitoring.

*Hinek:* I think you are right; smooth muscle cells are not the best model to use for studying this kind of signal, because they produce lots of laminin. When cells are cultured, even after 24 h most cells are already surrounded by laminin, which consequently occupies many of the 67 kDa elastin receptors. I think that our responder cells were cells which were recently detached and thus didn't have receptors saturated with laminin, so they could still respond to elastin.

*Davidson:* Is the leukocyte response to elastin blocked by lactose?

*Hinek:* Yes. We have shown in chemotactic chambers that leukocytes show a beautiful chemotactic response to elastin fragments which can be totally abolished by flushing with lactose (Hinek 1994).

Another interesting receptor-mediated phenomenon is that treating smooth muscle cells with lactose very quickly up-regulates the deposition of fibronectin (Hinek et al 1992). This is a strange phenomenon which we didn't understand until recently, when we learnt that it can be abolished in the presence of the soluble inhibitor of the interleukin (IL)-1$\beta$ receptor. We found that the elastin binding protein is located very close to the IL-1 receptor on the cell surface, and somehow interferes sterically with the approach of IL-1 to its receptor. Once we strip this receptor from the cells or saturate it with elastin we can observe this IL-1-driven phenomenon.

*Mungai:* Have there been any studies to explain the relationship between the cellular mechanisms for the production of elastin and the local forces to which elastic fibres respond? Have there been any experimental attempts to imitate those forces and see how they affect the cellular mechanisms?

*Robert:* That's a very good question. Leung et al (1976) showed that if you pulse smooth muscle cell cultures on an elastin sheet, you can increase extracellular matrix biosynthesis as well as cell proliferation.

Although smooth muscle cells have a basement membrane, they still show a beautiful $Ca^{2+}$ intake in the presence of elastin peptides (Jacob et al 1987). As we had to work quickly and wanted to explore the elastin receptor in clinical experiments and ageing studies, we chose the white blood cells.

*Davidson:* A number of years ago we published a study showing that by blocking the secretory pathway with monensin we could down-regulate both the production and the release of elastin, and we also seemed to reduce elastin mRNA levels (Frisch et al 1985). It would seem from your data that the secretory pathway that you're obstructing doesn't interfere with overall expression. What was the time course of these studies?

*Hinek:* 24 h. I think the monensin study is very interesting, since you also showed some influence of monensin on elastin deposition. The interesting thing is that you can never block it entirely. Monensin blocks mostly the Golgi-dependent pathway of secretion. Elastin is not glycosylated, so why does it have to go to the Golgi? Our understanding is that it is actually the elastin binding protein, which is glycosylated, that goes to the Golgi. Once you block this pathway, you end up with unglycosylated elastin binding protein which may block the final secretion of elastin. But a little bit of elastin is always secreted; we think this doesn't go through the Golgi.

## References


Baulieu EE 1987 Steroid hormone antagonists at the receptor level: a role for the heat shock protein MW 90 000 (Hsp90). J Cell Biol 35:161–174

Frisch SM, Davidson JM, Werb Z 1985 Blockage of tropoelastin secretion by monensin represses tropoelastin synthesis at a pretranslational level in rat smooth muscle cells. Mol Cell Biol 5:253–258

Fülöp T, Jacob MP, Varga Z, Foris G, Leovey A, Robert L 1986 Effect of elastin peptides on human monocytes: a $Ca^{2+}$ mobilization, stimulation of respiratory burst and enzyme secretion. Biochem Biophys Res Comm 141:92–98

Hinek A 1994 Nature and multiple functions of the 67 kDa elastin/laminin binding protein. Cell Adhes Commun 2:185–193

Hinek A, Rabinovitch M 1994 67-kD elastin binding protein is a protective "companion" of extracellular insoluble elastin and intracellular soluble tropoelastin. J Cell Biol 126:563–574

Hinek A, Wrenn DS, Barondes SH, Mecham RP 1988 The elastin receptor–galactoside binding protein. Science 239:1539–1541

Hinek A, Boyle J, Rabinovitch M 1992 Vascular smooth muscle cells' detachment from elastin and migration through elastic laminae is promoted by chondroitin sulfate-induced shedding of the 67 kDa cell surface elastin binding protein. Exp Cell Res 203:344–353

Hinek A, Rabinovitch M, Keeley F, Callahan J 1993 The 67 kD elastin/laminin-binding protein is related to an alternatively spliced β-galactosidase. J Clin Invest 91:1198–1205

Jacob MP, Fülöp T, Foris G, Robert L 1987 Effect of elastin peptides on ion fluxes in mononuclear cells, fibroblasts, and smooth muscle cells. Proc Natl Acad Sci USA 84:995–999

Leung DYM, Glagoc S, Mathews MB 1976 Cyclic stretching stimulates synthesis of matrix components by arterial smooth muscle cells *in vitro*. Science 191:475–477

Mecham RP, Hinek A, Entwistle R, Wrenn DS, Griffin G, Senior R 1989a Elastin binds to a multifunctional 67 kD peripheral membrane protein. Biochemistry 28:3716–3722

Mecham RP, Hinek A, Griffin G, Senior RM, Liotta LR 1989b 67 kD elastin binding protein is homologous to the tumor cell 67 kD laminin receptor. J Biol Chem 264:16652–16657

Senior RM, Hinek A, Griffin G, Crauch E, Mecham RP 1989 Neutrophils are chemotactic to type IV collagen and its 7S domain and contain 67 kDa type IV collagen binding protein with galactolectin properties. Amer J Resp Cell Molec Biol 1:479–487

# General discussion III

*Kimani:* Dr Boyd, what is the significance of the finding that in the human elastin gene, exon 34 is missing?

*Boyd:* Joel Rosenbloom and his colleagues showed some time ago that exons 34 and 35 were missing in all of the human tropoelastin cDNA and genomic DNA recombinants that they had studied (Bashir et al 1989). We followed this up and have never been able to find exons 34 and 35 in any human transcripts. We looked at transcripts in this region of the tropoelastin gene in primates, and we were able to show that every single primate we looked at (from mandrels to rhesus monkeys to gorillas to chimps) had exon 34 but lacks exon 35 (Levi-Minzi 1993). This is in contrast to the very conserved maintenance of these two exons in everything from chickens to cows. Those two exons are certainly present in the mouse and rat tropoelastin genes as well. There's a striking gradation of loss of these two exons during primate evolution which takes place over a relatively short period of time, if you compare this with the maintenance of these exons between chickens and cows, who last shared a common ancestor some 150–200 million years ago. This suggests that the loss of exon 35 in primates and the complete loss of 34 and 35 in humans is of some physiological consequence to the elastin that is produced in these different primates.

*Mariani:* It's interesting how the genes encoding different microfibrillar proteins seem to be clustering in different regions of the genome. Joel Rosenbloom mentioned before that the genes encoding the two microfibril-associated proteins he described co-localize with the genes encoding the two fibrillin proteins. Furthermore, the genes for lysyl oxidase and the lysyl oxidase-related protein are also very close to the loci for the two fibrillin genes (Lee et al 1991, Hamalainen et al 1991, Mariani et al 1992, Kenyon et al 1993). The clustering of these genes must be taken into account when performing linkage analyses for disease loci.

*Keeley:* Twice now, once from Herb Kagan and once from Lynn Sakai, we have seen evidence that dibasic sites in fibrillin and lysyl oxidase might be important for processing of these proteins. Is there any evidence that smooth muscle cells produce convertases that would cleave at those dibasic sites?

*Kagan:* I don't know if they produce the classical dibasic-site cleaving enzyme. They certainly secrete an enzyme that will process the prolysyl oxidase to lysyl oxidase, although the site of cleavage has yet to be identified.

*Rosenbloom:* In microfibril-associated protein 1 (MFAP1; which is the same as Judy Foster's AMP) there is also a similar site. Judy showed in her paper

(Horrigan et al 1992) that this protein is processed from a 52 kDa precursor down to a 31 kDa protein which appears in the mature aorta. She postulated that the protein may be cleaved after the sequence RKK. This is becoming more of a general rule; it's likely that there is an enzyme and it may be involved with all of these proteins.

*Keeley:* The prohormone forms of several pituitary hormones can be processed in in many different ways to produce different peptide products. Formation of the final peptide product depends on which convertases are available. Defects in production of specific proteins or forms of proteins could therefore also be due to failure to make the approriate processing enzymes.

*Pasquali-Ronchetti:* Along different lines, pseudoxanthoma elasticum is a genetic disorder in which elastin undergoes progressive mineralization. We had the opportunity to cultivate dermal fibroblasts from a certain number of patients and we observed that the great majority of these cells retract collagen gels with less efficiency than controls, exhibit a different pattern of integrin expression, and produce a large amount of matrix products. It could be that mineralization of elastic fibres depends not on elastin *per se*, but might instead be the consequence of the deposition within the elastic fibres of molecules not produced by normal cells. Immunocytochemical studies in progress in our laboratory clearly point to the presence of molecules with high affinity for $Ca^{2+}$ ions in the regions of heavy mineralization of the pathological elastin fibres (Baccarani-Contri et al 1994).

## References

Baccarani-Contri M, Taparelli F, Boraldi F, Pasquali-Ronchetti I 1994 Mineralization in PXE is induced by proteins with high affinity for calcium. XIVth FECTS Meeting, Lyon, Abstr. N3

Bashir M, Indik Z, Yeh H et al 1989 Characterization of the complete human elastin gene. J Biol Chem 264:8887–8891

Hamalainen ER, Jones TA, Sheer D, Taskinen K, Pihlajaniemi T, Kivirikko KI 1991 Molecular cloning of human lysyl oxidase and assignment of the gene to chromosome 5q23.3-31.2. Genomics 11:508–516

Horrigan SK, Rich CB, Streeten BW, Li Z-Y, Foster JA 1992 Characterization of an associated microfibrillar protein through recombinant DNA techniques. J Biol Chem 267:10087–10095

Kenyon K, Modi WS, Contente S, Friedman RM 1993 A novel human cDNA with a predicted protein similar to lysyl oxidase maps to chromosome 15q24-q25. J Biol Chem 268:18435–18437

Lee B, Godfrey M, Vitale E et al 1991 Linkage of Marfan syndrome and a phenotypically related disorder to two different fibrillin genes. Nature 352:330–334

Levi-Minzi S 1993 The role of repetitive sequences in the evolution of the tropoelastin gene. PhD Thesis, Rutgers University

Mariani TJ, Trackman PC, Kagan HM et al 1992 The complete derived amino acid sequence of human lysyl oxidase and assignment of the gene to chromosome 5. Extensive sequence homology with the RAS recision gene. Matrix 12:242–245

# Elastin in lung development and disease

Richard A. Pierce*, Thomas J. Mariani* and Robert M. Senior†

*Divisions of *Dermatology and †Respiratory and Critical Care, Department of Medicine, Washington University School of Medicine at Jewish Hospital, St. Louis, MO 63110, USA*

*Abstract.* Elastic fibres are present in lung structures including alveoli, alveolar ducts, airways, vasculature and pleura. The rate of lung elastin synthesis is greatest during fetal and neonatal development, and is minimal in the healthy adult. We have determined that glucocorticoids up-regulate fetal lung tropoelastin expression while concomitantly accelerating terminal airspace maturation. Because there is minimal turnover of elastin in healthy adult lung, the elastin incorporated into the lung early in development supports lung function for the normal lifespan. However, in the adult lung, in pathological circumstances such as emphysema or pulmonary fibrosis there may be reactivation of elastin expression. We have found in silica-induced pulmonary fibrosis that expression of tropoelastin is primarily increased in the walls and the septal tips of the alveolus, with modest increases in other compartments which normally express tropoelastin during development. This finding suggests that the mesenchymal cell of the alveolar wall increases tropoelastin expression during fibrotic disorders. In emphysema and fibrosis, elastin is present in abnormal-appearing, probably non-functional, elastic fibres, suggesting that the adult lung cannot recapitulate the elastic fibre assembly mechanisms operative during normal lung growth.

*1995 The molecular biology and pathology of elastic tissues. Wiley, Chichester (Ciba Foundation Symposium 192) p 199–214*

Elastic fibres, composed of elastin and microfibrillar proteins, are responsible for the resilient properties of respiratory structures (including alveoli, alveolar ducts, airways, vasculature and pleura) at low levels of stress. At higher levels of stress, collagenous fibres provide the resistance to further stretch and generate tissue recoil. Evidence from animal studies indicates that most lung elastin is deposited during late fetal and neonatal development; lung elastin synthesis is terminated in childhood. Both human and animal data show that there is minimal turnover of elastin in healthy adult lung (Rucker & Dubick 1984, Shapiro et al 1991). Thus, the elastin incorporated into the lung early in life supports lung function for the normal lifespan.

Within the lung parenchyma, elastic fibres loop around the alveolar ducts and form rings at the mouths of alveoli. Elastin is more prevalent than collagen at the alveolar mouths and in fact constitutes most of the tissue at these sites (Kuhn & Oldmixon 1990, Mercer & Crapo 1990). Concentrations of elastic fibres are also found where two alveolar septa meet in a tent-like configuration. Otherwise, elastic fibres are sparse in alveolar septa. For information about lung elastin generally, see reviews by Starcher (1986) and Mecham et al (1991).

In certain lung diseases, particularly emphysema, there is turnover of lung elastin as well as reactivation of lung elastin expression. Pulmonary fibrosis and states of chronic pulmonary hypertension are other situations in which elastin may be synthesized after the normal period of lung elastogenesis is over. In these circumstances, elastin is assembled into abnormal-looking, probably dysfunctional, elastic fibres (Kuhn et al 1976, Starcher 1986, Starcher et al 1978). Alterations in human lung elastic fibres consistent with destruction and resynthesis have also been noted in other settings, including pulmonary Langerhans' cell granulomatosis (Fukuda et al 1990a) and lympangiomyomatosis (Fukuda et al 1990b). Production of aberrant-looking elastic fibres in these situations suggests that the adult lung is able to make elastin and elastic fibres but cannot recapitulate the elastic fibre assembly mechanisms that operate during normal lung growth. This incapacity may also occur in early life because abnormal lung parenchymal elastic fibres have been observed in premature infants who received mechanically assisted ventilation (Bruce et al 1992).

In this paper, we will comment about elastin expression and turnover in the normal lung and in pulmonary emphysema and pulmonary fibrosis. Changes in elastin metabolism associated with pulmonary hypertension are reviewed by Keeley & Barboszewicz (1995, this volume).

## Elastin during lung development

In the lung, as in other elastic tissues, elastin is primarily produced during late fetal, neonatal and early postnatal growth. To understand better the expression of elastin in interstitial lung diseases, we have first focused on its normal patterns of expression. In the rat, tropoelastin mRNA in the lung is first detectable in the canalicular stage of fetal lung development, peaks during alveolarization in the neonate and is barely detectable in the adult. Because of difficulties in obtaining tissue, an organized study of the normal pattern of tropoelastin expression in the human lung has not been done. It is likely, however, that the pattern of tropoelastin expression in all mammals is similar during equivalent stages of lung development.

To determine the spatial patterns of tropoelastin expression in developing lung, we examined sections of fetal and neonatal rat lung by *in situ* hybridization, using $^{35}$S-labelled antisense cRNA probes. In the fetal rat thorax at 19 d of gestation, tropoelastin signal is detected in the aorta, the pleura, the pericardium of the heart, the periosteum of the vertebra and ribs, as

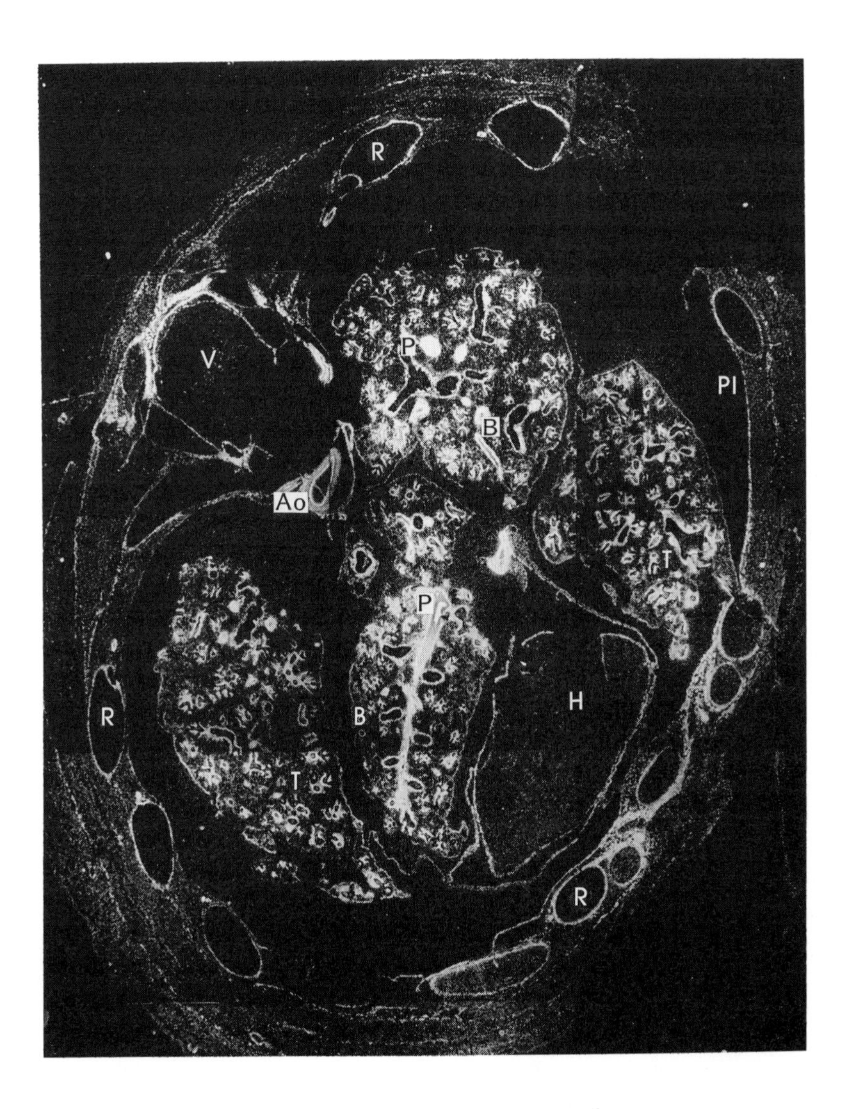


FIG. 1. Expression of tropoelastin in the fetal thorax. Shown is a dark-field view of a cross-section of a 19 d fetal rat thorax hybridized *in situ* for tropoelastin mRNA, using a $^{35}S$-labelled antisense tropoelastin cRNA probe. In this 60 d exposure, tropoelastin mRNA is detected as bright white silver grains in several thoracic structures, including the aorta (Ao), the periosteum of the vertebra (V) and ribs (R), the pericardium of the heart (H), and the parietal pleura (Pl). In the lobes of the lung, prominent signal for tropoelastin is detected in the intralobar pulmonary arteries (P), in the smooth muscle surrounding bronchi and bronchioles (B), and surrounding developing terminal airways (T).

well as in several lung compartments (Fig. 1). During this canalicular stage of lung development, tropoelastin signal is detected in the intralobar pulmonary arteries, the smooth muscle surrounding the bronchioles and in mesenchyme surrounding developing airways.

To determine the onset and pattern of tropoelastin expression in the fetal rat lung, we compared tropoelastin expression by *in situ* hybridization in lungs from 17, 18 and 19 d fetuses. At the cellular level, tropoelastin expression is first detectable during the pseudoglandular stage of fetal lung development in the smooth muscle of the intralobar pulmonary arteries and veins at 17 d (Fig. 2A, B). At this stage of lung development, signal is absent in the mesenchyme. As the fetal lung progresses through the canalicular stage of development, tropoelastin mRNA is also detected in the smooth muscle layer subadjacent to the epithelium of bronchioles and developing airways at 18 and 19 d (Fig 2C–F). During alveolarization, which occurs during the neonatal period in the rat, prominent tropoelastin expression is observed in the parenchyma, associated with cells expressing smooth muscle markers such as α-smooth muscle actin. The highest levels of tropoelastin mRNA are detected at the septal tips and bends in the alveolar wall, locations which are rich in elastic fibres in the mature lung (Fig. 3A, B). Therefore, at different stages of lung development, different cell types contribute in varying degrees to the total lung elastin production. As we will demonstrate, in interstitial pulmonary disease caused by silica, tropoelastin production is reinitiated primarily in the bends of the alveolar wall and in alveolar septal tips.

In spite of recent advances in the elucidation of the structure of the tropoelastin gene and promoter, the factors responsible for the temporally regulated, tissue-specific expression of tropoelastin in the lung are largely unknown. In particular, little is known of the signals that induce tropoelastin expression during the differentiation of the various elastogenic cell types in the lung. Because glucocorticoids have been used clinically to accelerate lung development and induce surfactant production in premature neonates, we reasoned that these agents may also accelerate fetal lung tropoelastin expression. Timed-pregnant female rats were treated daily for 3 d with 1 mg/kg of dexamethasone. Northern blot analysis of fetal lung RNA demonstrated that glucocorticoids increase fetal lung tropoelastin expression *in vivo* (Pierce et al 1995) (Fig. 4). The accelerated tropoelastin expression in dexamethasone-treated fetal rat lung is accompanied by accelerated development of terminal airspaces, suggesting elastin may affect lung development (Fig. 4, compare A and C). Increased tropoelastin expression is pronounced in mesenchymal cells surrounding the more complex airspaces in the dexamethasone-treated lung (Fig. 4, compare B and D). These data suggest that glucocorticoids may accelerate lung development, in part, by accelerating elastin expression in the developing terminal airspaces. Several other physiological mediators of tropoelastin expression, such as IGF-1 (Foster et al 1990), TGF-β

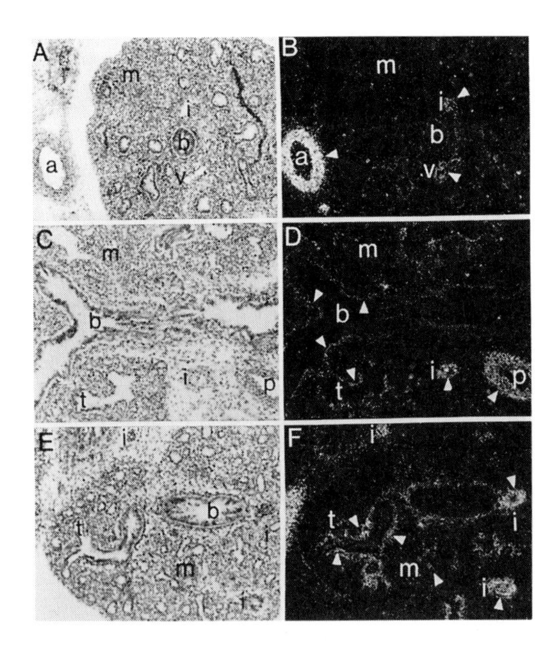


FIG. 2. Progression of tropoelastin expression during fetal lung development. Shown are paired bright- and dark-field views of 17 d (A and B), 18 d (C and D) and 19 d (E and F) fetal lung sections hybridized *in situ* for tropoelastin mRNA. (A,B) Pseudoglandular 17 d fetal rat lung. Intense signal for tropoelastin is detected in the adjacent aorta (a). In contrast, weak signal for tropoelastin (arrowheads) is detected in the intralobar arteries (i) and veins (v), and no signal is detected in a bronchus or the undifferentiated mesenchyme (m). (C,D) 18 d fetal rat lung. Intense signal for tropoelastin (arrowheads) is detected in the main intralobar artery (p) and in a smaller intralobar artery (i). Tropoelastin mRNA is also now detected in the mesenchymal cells surrounding developing terminal airways (t), and in the smooth muscle subadjacent to the cuboidal epithelial cells of the bronchioles (b). Regions of mesenchyme (m) remain relatively undifferentiated and condensed. (E,F) 19 d canalicular fetal rat lung. Intense signal for tropoelastin is detected in the intralobar arteries (i). Tropoelastin mRNA is now prominent in the mesenchymal cells surrounding dividing terminal airways (t), and in the smooth muscle subadjacent to the cuboidal epithelial cells of the bronchiole (b). The mesenchyme (m) is becoming organized into interstitial regions between developing terminal airspaces.

(McGowan 1992), retinoids (Liu et al 1993) and bFGF (Brettell & McGowan 1994), which have been shown to modulate tropoelastin production in lung cells *in vitro*, may modulate lung tropoelastin *in vivo*, but their roles in controlling the normal patterns of expression of tropoelastin are not established.

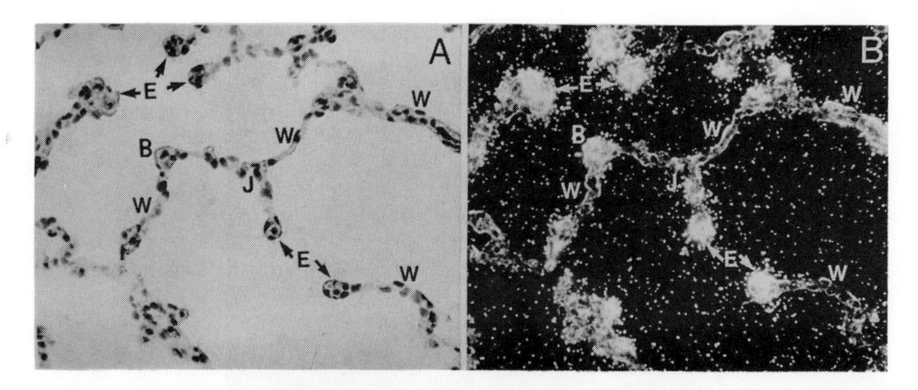


FIG. 3. Expression of tropoelastin during alveolarization of the fetal rat lung occurs primarily in alveolar septal tips and bends in the alveolar wall. Shown are paired bright-field (A) and dark-field (B) views of a 10 d neonatal rat lung section hybridized *in situ* for tropoelastin mRNA. (A) Alveolar structures include the ends of septa (E), bends where two alveolar walls meet in a tent-like fashion (B), junctions of two alveolar walls at right angles (J), and walls (W). In this bright-field view, the plane of focus was below the silver grains to reveal precisely the morphology. (B) Intense signal for tropoelastin (*arrowheads*) is detected primarily in the ends of septa and in the bends of the alveolar wall. At this stage of development, the majority of tropoelastin expression in the lung is detected in the parenchyma.

## Turnover of elastin in the normal lung

In contrast to the developing lung, studies in animals and humans indicate that the normal adult lung displays minimal synthesis or turnover of elastin. Specifically, dosing adult rodents *in vivo* with [$^3$H]valine has revealed little incorporation of radiolabel into lung elastin, in contrast to young animals with growing lungs (Dubick et al 1981). Similarly, the specific activity of lung elastin that is radiolabelled *in vivo* in young animals decreases very slowly as the animals age, indicating that relatively little synthesis of new elastin occurs beyond the early stages of life (Pierce et al 1967).

By two methods, aspartic acid racemization and carbon dating, Shapiro et al (1991) found that there is minimal turnover of human lung elastin. Because L-aspartic acid racemizes to the D-enantiomer at a rate of about 0.1% per year, and because only L-aspartic acid is incorporated into newly synthesized elastin, these workers reasoned that the percentage of D-aspartic acid in lung elastin should reflect the average age of the elastin. In lung elastin from individuals aged 6–78 years who died without lung disease, they found that the percentage of D-aspartic acid corresponded to the subjects' ages, indicating that the lung elastin was all synthesized around the same time, and that this time was near birth. An interesting additional finding was that the total content of aspartic acid in the elastin samples was higher than expected for pure elastin, suggesting the presence of small amounts of microfibrillar protein in the elastin preparations. Because the correlations between the subjects' ages and percent racemization of aspartic acid were consistent despite the presumed presence of microfibrillar proteins, the investigators speculated that elastin-associated microfibrils in the lung have the same slow physiological turnover as elastin. Radiocarbon dating revealed increased $^{14}C$ prevalence in the lung elastin of individuals whose fetal and postnatal periods of life corresponded to times of elevated atmospheric $^{14}C$ due to testing of nuclear weapons. Together, these data support the findings in rodents that the lung elastin synthesized around birth persists throughout life and that relatively little elastin is normally added to the lungs after that period of life.

## Lung elastin in emphysema

The discovery of an association between emphysema and hereditary $\alpha_1$-antitrypsin deficiency in 1963, together with the development of animal models of emphysema that involved putting proteinases into lungs in 1964, led to the idea that destruction of lung elastin is critical to the pathogenesis of emphysema (Janoff 1985). These early developments focused attention on elastin because $\alpha_1$-antitrypsin is the main circulating inhibitor of neutrophil elastase and because only proteinases that degrade insoluble elastin proved capable of causing emphysema in animals' lungs. Over the ensuing years, these ideas have been supported by numerous observations so that it is now widely accepted that destruction of lung elastin, caused by a relative excess of elastases over elastase inhibitors in the lung tissue, is pivotal to the development of emphysema. Leukocytes and macrophages can release a variety of elastases and are the likely sources of the proteolytic activity that destroys lung elastin (Janoff 1985, Reilly et al 1991, Senior et al 1991, Shapiro et al 1993, Shi et al 1994). Smokers' lungs contain much larger numbers of these inflammatory cells than do the lungs of non-smokers (Reynolds & Newball 1974).

### *Elastin content*

Although destruction of lung elastin appears to be essential for the development of emphysema, it remains unknown how the breakdown of lung elastic fibres actually translates into the deformity recognized as emphysema. Moreover, it has been difficult to establish that emphysematous lung tissue is actually deficient in elastin or that its elastin has a faster turnover than elastin in normal lung tissue. Widely variable results have been reported regarding the elastin content of the emphysematous lung, but as pointed out recently (Cardoso et al 1993), problems arise in interpreting the results of many of these studies, particularly because of the tissue samples used for analysis. Samples have included whole lungs, or portions of lung that

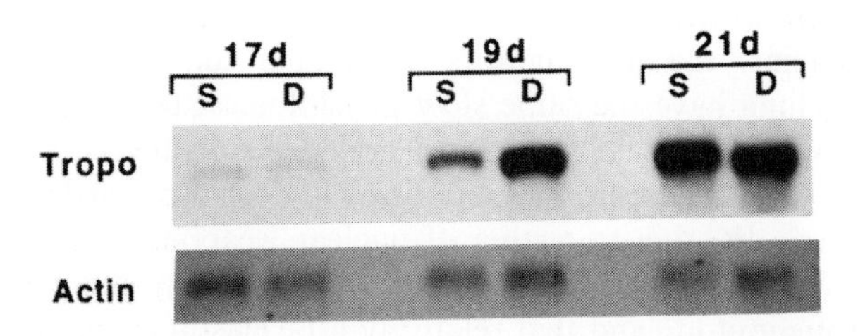
17d
S
D
19d
S
D
21d
S
D
Tropo
Actin

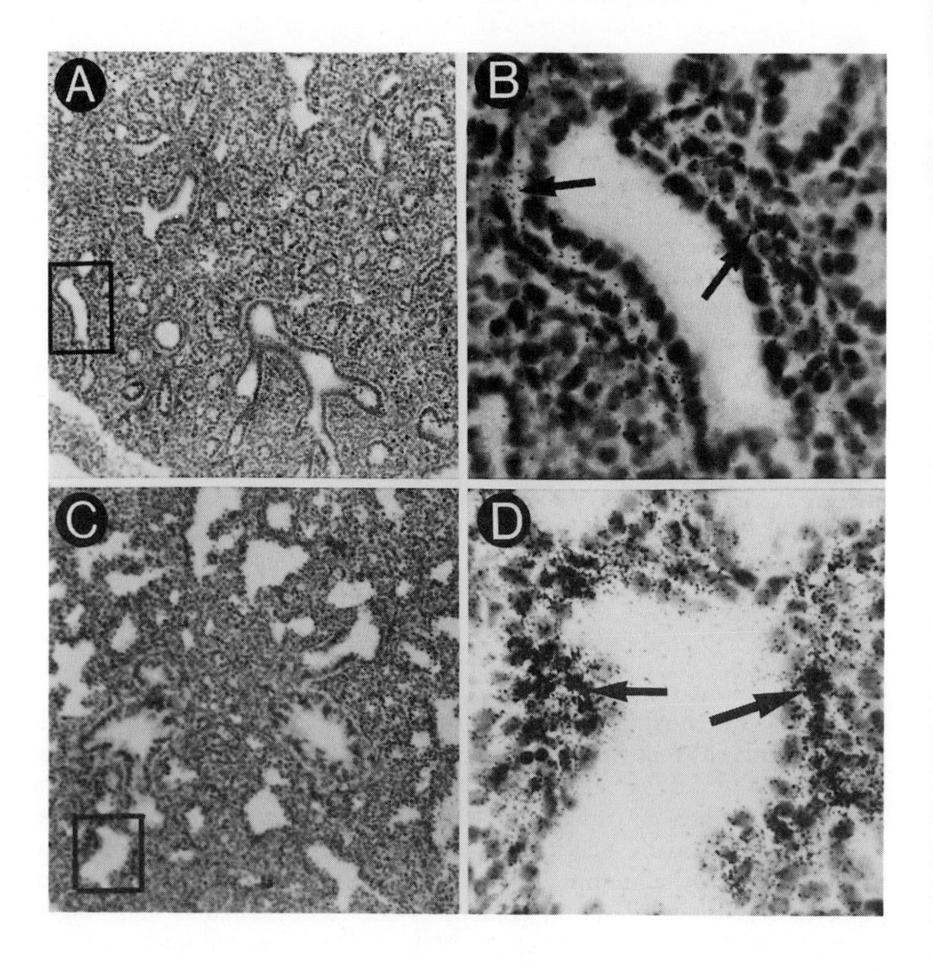
A
B
C
D

contained blood vessels, airways and pleura, or portions of lung that include both normal and emphysematous tissue. Such samples could dilute or obscure what has happened to the elastin content of the emphysematous lung parenchyma, which is the relevant lung tissue.

Cardoso et al (1993) used a dissecting microscope to characterize lung tissue for the type and severity of emphysema and took adjacent slices of the tissue for analysis of elastin content. In this way, virtually the same tissue was used for both the biochemical and anatomical studies. In multiple specimens obtained from upper lobes of 25 individuals ranging in age from 50–79 years they found that emphysematous tissue had significantly less elastin (expressed as $\mu$g per mg dry weight of tissue) than normal lung parenchyma, and that the elastin content of the tissue correlated inversely with the severity of the emphysematous pathology. The lowest levels of elastin occurred in panacinar emphysema. An important finding was that within emphysematous lungs those parts of the tissue that appeared normal had normal amounts of elastin whereas parts of the tissue showing emphysema had low levels of elastin. Accordingly, elastin depletion was specifically restricted to the sites showing emphysema, rather than being a global deficiency of the lung that contained regions of emphysema.

*Elastin degradation*

There is no direct information about the synthesis of new elastin in emphysematous human lung or about the turnover of lung elastin in emphysema. However, the discovery that elastic fibres appear abnormal in emphysema (Fukuda et al 1989) and measurements of elevated elastin peptides in the circulation of individuals with chronic obstructive lung disease suggest that there is synthesis of new lung elastin and that lung elastin is turning over at an accelerated rate in individuals with emphysema.

Increased concentrations of elastin peptides in the circulation and urine have been reported from individuals with chronic obstructive lung disease (Akers et

FIG. 4. Glucocorticoid treatment accelerates fetal lung development and tropoelastin expression. *Top*: Northern blot analysis of total RNA isolated from saline- (S) and dexamethasone-treated (D) fetal rat lungs. Increased tropoelastin mRNA, relative to $\beta$-actin mRNA, is detected in 19 d dexamethasone-treated lungs compared with saline-treated control. *Bottom*: (A,B) Canalicular 19 d saline control lung hybridized *in situ* for tropoelastin mRNA. Note there is little development of terminal airspaces. (B) Higher magnification of the area boxed in A. Scant positive signal for tropoelastin is detected by the presence of small dark silver grains in mesenchymal cells subadjacent to the cuboidal epithelium of a developing terminal airway (*arrows*). (C,D) Saccular 19 d dexamethasone-treated lung hybridized *in situ* for tropoelastin mRNA. Note the greater development and complexity of terminal airspaces. (D) Higher magnification of area boxed in C. Abundant positive signal for tropoelastin is detected by the presence of small dark silver grains in mesenchymal cells subadjacent to the flattened epithelium of a developing terminal airway.

al 1992, Dillon et al 1992, Shriver et al 1992). Because spirometry is notoriously unreliable for ascertaining and quantifying emphysema, Gelb et al (1993) and Dillon et al (1992) characterized subjects for extent and severity of emphysema by computed axial tomography of the chest and by measuring the elastic properties of the lungs. Using a polyclonal antibody to oxalic acid-solubilized human lung elastin in a competitive ELISA, Dillon et al (1992) found that plasma from emphysematous subjects had elastin peptide concentrations of approximately 50 ng/ml in contrast to values of about 3 ng/ml or less among individuals with normal lung function, including smokers. Although there were highly significant correlations between plasma elastin-derived peptide levels and both radiographic and functional indices of emphysema, in a number of subjects these correlations were not found. An implication from studies of elastin peptides in emphysema is that excessive degradation of lung elastin appears to persist in advanced emphysema, consistent with the progressive nature of the problem in most of these individuals even after they have stopped smoking. Attempts to demonstrate elevated elastin peptide concentrations in bronchoalveolar lavage fluid (BALF) in association with chronic obstructive lung disease by Shriver et al (1992) were inconclusive, but these workers estimated that BALF elastin peptide concentrations, corrected for dilution associated with the lavage procedure, were higher than plasma, suggesting that the lung is an important source of plasma and urinary elastin peptides.

Studies of elastase-induced emphysema in animals have illustrated the capacity of the adult lung to synthesize elastin. After intratracheal instillation of pancreatic elastase, lung elastin content is depleted by one-third or more at 24 h. However, a large fraction of the elastin that has been lost is replenished within 72 h (Goldstein & Starcher 1978) and after about two months total lung elastin is normal (Kuhn et al 1976). These measurements of total lung elastin fit with studies using incorporation of radiolabel into elastin. Interestingly, studies of lung elastin after intratracheal elastase treatment antedated current molecular biological approaches so that there are no reports of tropoelastin mRNA content or localization in lung tissues in animal lungs after exposure to elastases or in emphysematous human lung tissue that might be obtained at the time of resectional lung surgery.

The elastic fibres in the lungs of animals that have recovered lung elastin after intratracheal elastase treatment do not look normal (Kuhn et al 1976). The fibres are tortuous, clumped and frayed. Abnormal elastic fibres are also seen in human emphysema (Fukuda et al 1989) leading to the suggestion, on the basis of the animal model, that elastogenesis occurs in the human disease.

## Lung elastin in pulmonary fibrosis

Elastic fibres are often observed in areas of airspace fibrosis in humans, and in this disease alterations in the distribution of elastic fibres are also seen (Crouch 1990). In bleomycin-induced pulmonary fibrosis in rats, lung tropoelastin expression is increased markedly, resulting in increased total lung elastin (Starcher et al 1978, Raghow et al 1985, Last et al 1993).

Recently, we examined the expression of tropoelastin in a silica-induced model of experimental pulmonary granulomatous disease (R. A. Pierce, T. J. Mariani, unpublished results). Sterile silica (40 mg in 0.4 ml of sterile saline), was instilled into the tracheas of adult male 250 g rats and the animals' lungs were harvested 3, 7, 14 and 21 d after instillation. In this experimental model, under baseline conditions, low levels of tropoelastin expression were sometimes detected by *in situ* hybridization in the lung vasculature or surrounding the smooth muscle of bronchi, but not in the parenchyma. Seven days after instillation, tropoelastin expression was markedly up-regulated in silicotic lungs compared with saline controls. Surprisingly, expression of tropoelastin was detected as early as 3 d after instillation of silica, before significant accumulation of matrix was evident. Abundant tropoelastin mRNA was detected by *in situ* hybridization in the tips of the alveoli and in the bends of the alveolar wall (Fig. 5, compare B and D). The expression of tropoelastin co-localized with immunostaining for α-smooth muscle actin indicating that mesenchymal cells were the source of the tropoelastin expression. Interestingly, tropoelastin was primarily found in areas of normal-looking parenchyma distinct from the granulomatous tissue. This indicates that soluble mediators released by resident or inflammatory cells may induce tropoelastin expression during pulmonary granulomatous disease.

## Concluding perspective

In addition to its importance in providing resilience critical for lung function, elastin is an important architectural component of the lung and likely influences lung morphogenesis. During late fetal and neonatal periods, the expression of tropoelastin is induced in several elastogenic cell types at different stages of development, then declines after alveolarization. The factors regulating this developmental program are largely unknown.

Under normal circumstances, lung elastin is virtually stable for a lifetime following its synthesis and assembly into elastic fibres; new elastin is not made during adult life. Thus, practically speaking, elastin is neither synthesized nor degraded in the healthy adult lung. However, in a number of pathological states affecting the lungs—notably pulmonary emphysema, various forms of pulmonary fibrosis, and chronic pulmonary hypertension—there is evidence of elastogenesis. In some lung diseases, particularly emphysema, elastogenesis might be part of a response to depletion of lung elastin, whereas in others, such as pulmonary fibrosis, elastin production may be a non-specific response to global signals to elaborate extracellular matrix. It is unlikely that newly synthesized elastin in the diseased adult lung serves the function of providing resilience because it is assembled into elastic fibres that do not look normal.

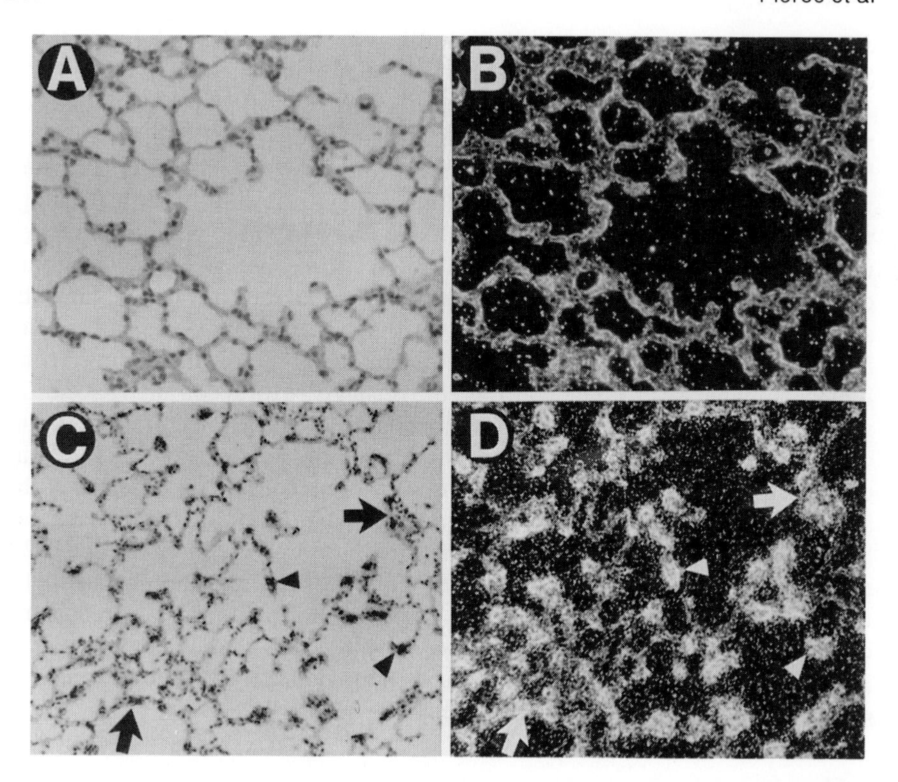


FIG. 5. Expression of tropoelastin in lung parenchyma is re-initiated in the adult rat lung in silica-induced pulmonary granulomatous disease. Shown are paired bright- and dark-field views of normal (A and B) and silicotic (C and D) rat lung sections hybridized *in situ* for tropoelastin mRNA. (A,B) Although the lung parenchyma is autofluorescent under dark-field illumination, no specific signal for tropoelastin is detected in the normal male adult rat lung parenchyma. (C,D) Intense signal for tropoelastin is detected in the normal-appearing parenchyma of silicotic lung.

The mechanisms controlling lung elastin expression in disease are not known, and the synthesis and secretion of microfibrils, which are essential for the assembly of elastic fibres, are yet to be investigated either in the normal or the diseased lung.

## *Acknowledgements*

We gratefully acknowledge the technical assistance of Stephanie Sandefur, Jill Roby and Teresa Tolley, and thank Dr John A. Pierce for helpful suggestions.

Supported by grants HL29594 and HL47328 from the National Heart, Lung and Blood Institute of the United States Public Health Service and a Grant-in-Aid from the American Heart Association. R. A. P. is the recipient of a Dermatology Foundation/ Dermik Laboratories Career Development Award.

## References


Akers S, Kucich U, Swartz M et al 1992 Specificity and sensitivity of the assay for elastin-derived peptides in chronic obstructive pulmonary disease. Am Rev Respir Dis 145:1077–1081

Brettell LM, McGowan SE 1994 Basic fibroblast growth factor decreases elastin production by neonatal rat lung fibroblasts. Am J Respir Cell Mol Biol 10:306–315

Bruce MC, Schuyler M, Martin RJ, Starcher BC, Tomashefski JF, Wedig KE 1992 Risk factors for the degradation of lung elastic fibers in the ventilated neonate. Am Rev Respir Dis 146:204–212

Cardoso WV, Sekhon HS, Hyde DM, Thurlbeck WM 1993 Collagen and elastin in human pulmonary emphysema. Am Rev Respir Dis 147:975–981

Crouch E 1990 Pathobiology of pulmonary fibrosis. Am J Physiol 259:L159–L184

Dillon TJ, Walsh RL, Scicchitano R, Eckert B, Cleary EG, Mclennan G 1992 Plasma elastin-derived peptide levels in normal adults, children, and emphysematous subjects. Am Rev Respir Dis 146:1143–1148

Dubick MA, Rucker RB, Cross CE, Last JA 1981 Elastin metabolism in rodent lung. Biochim Biophys Acta 672:303–306

Foster JA, Rich CB, Miller M, Benedict MR, Richman RA, Florini JR 1990 Effect of age and IGF-I administration on elastin gene expression in rat aorta. J Gerontol 45:B113–B118

Fukuda Y, Masuda Y, Ishizaki M, Masugi Y, Ferrans VJ 1989 Morphogenesis of abnormal elastic fibres in panacinar and centriacinar emphysema. Hum Pathol 20:652–659

Fukuda Y, Basset F, Soler P, Ferrans VJ, Masugi Y, Crystal RG 1990a Intraluminal fibrosis and elastin fiber degradation lead to lung remodeling in pulmonary Langerhans cell granulomatosis (histiocytosis). Am J Pathol 137:415–424

Fukuda Y, Kawamoto M, Yamamoto A, Ishizaki M, Basset F, Masugi Y 1990b Role of elastic fiber degradation in emphysema-like lesions of pulmonary lymphangiomyomatosis. Hum Pathol 21:1252–1261

Gelb AF, Schein M, Kuei J et al 1993 Limited contribution of emphysema in advanced chronic obstructive pulmonary disease. Am Rev Respir Dis 147:1157–1161

Goldstein RA, Starcher BC 1978 Urinary secretion of elastin peptides containing desmosine after intratracheal injection of elastase in hamsters. J Clin Invest 61:1286–1290

Janoff A 1985 Elastases and emphysema. Am Rev Respir Dis 132:417–433

Keeley FW, Bartoszewicz LA 1995 Elastin in systemic and pulmonary hypertension. In: The molecular biology and patholgy of elastic tissues. Wiley, Chichester (Ciba Found Symp 192) p 259–278

Kuhn C, Oldmixon EH 1990 The interstitium of the lung. In: Schraufnagel DE (ed) Electron microscopy of the lung. Marcel Dekker, NY, p 177–214

Kuhn C, Yu SH, Chraplyvy M, Linder HE, Senior RM 1976 The induction of emphysema with elastase. II. Changes in connective tissue. Lab Invest 34:372–380

Last JA, Gelzleichter TR, Pinkerton KE, Walker RM, Witschi H 1993 A new model of progressive pulmonary fibrosis in rats. Am Rev Respir Dis 148:487–494

Liu R, Harvey CS, McGowan SE 1993 Retinoic acid increases elastin in neonatal rat lung fibroblast cultures. Am J Physiol 265:L430–L437

McGowan SE 1992 Influences of endogenous and exogenous TGF-$\beta$ on elastin in rat lung fibroblasts and aortic smooth muscle cells. Am J Physiol 263:L257–L263

Mecham RP, Prosser IW, Fukuda Y 1991 Elastic fibers. In: Crystal RG, West JB (eds) The lung: scientific foundations. Raven Press, New York, p 389–398

Mercer RR, Crapo JD 1990 Spatial distribution of collagen and elastin fibers in the lungs. J Appl Physiol 69:756–765

Pierce JA, Resnick H, Henry PH 1967 Collagen and elastin metabolism in the lungs, skin, and bones of adult rats. J Lab Clin Med 69:485–493

Pierce RA, Mariencheck WI, Sandefur S, Crouch EC, Parks WC 1995 Glucocorticoids upregulate tropoelastin expression during late stages of fetal lung development. Am J Physiol 268:L491–L500

Raghow R, Lurie S, Seyer J, Kang AH 1985 Profiles of steady-state levels of messenger RNAs coding for type I procollagen, elastin, and fibronectin in hamster lungs undergoing bleomycin-induced interstitial fibrosis. J Clin Invest 76:1733–1739

Reilly JJ, Chen P, Sailor LZ, Wilcox D, Mason RW, Chapman HA 1991 Cigarette smoking induces an elastolytic cysteine proteinase in macrophages distinct from cathepsin L. Am J Physiol 261:L41–L48

Reynolds HY, Newball HH 1974 Analysis of proteins and respiratory cells from human lungs by bronchial lavage. J Lab Clin Med 84:559–573

Rucker RB, Dubick MA 1984 Elastin metabolism and chemistry: potential roles in lung development and structure. Environ Health Perspect 53:179–191

Senior RM, Griffin GL, Fliszar CJ, Shapiro SD, Goldberg GI, Welgus HG 1991 Human 92-kilodalton and 72-kilodalton type IV collagenases are elastases. J Biol Chem 266:7870–7875

Shapiro SD, Endicott SK, Province MA, Pierce JA, Campbell EJ 1991 Marked longevity of human lung parenchymal elastic fibers deduced from prevalence of D-aspartate and nuclear weapons-related radiocarbon. J Clin Invest 87:1828–1834

Shapiro SD, Kobayashi DK, Ley TJ 1993 Cloning and characterization of a unique elastolytic metalloproteinase produced by human alveolar macrophages. J Biol Chem 268:23824–23829

Shi G-P, Webb AC, Foster KE et al 1994 Human cathepsin S: chromosomal localization, gene structure, and tissue distribution. J Biol Chem 269:11530–11536

Shriver EE, Davidson JM, Sutcliffe MC, Swindell BB, Bernard GR 1992 Comparisons of elastin peptide concentrations in body fluids from healthy volunteers, smokers, and patients with chronic obstructive pulmonary disease. Am Rev Respir Dis 145:762–766

Starcher BC 1986 Elastin and the lung. Thorax 41:577–585

Starcher BC, Kuhn C, Overton JE 1978 Increased elastin and collagen content in the lungs of hamsters receiving an intra-tracheal injection of bleomycin. Am Rev Respir Dis 117:229–306

## DISCUSSION

*Uitto:* The arguments about transcriptional versus post-transcriptional or even translational control often revolve around the quantification of these changes. You didn't give any specific numbers which would indicate the magnitude of changes in mRNA and pre-mRNA levels.

*Pierce:* What we saw varied, depending on whether we were working *in vivo*, whether were working *in vitro* with lung minces, or whether we were working *in vitro* with highly differentiated cells already expressing tropoelastin at a reasonable level. *In vivo* we saw a two- to threefold increase in tropoelastin expression in dexamethasone-treated lungs at 19 days. In lung mince cultures, we had a very low baseline of tropoelastin expression. With the addition of

$10^{-8}$ M dexamethasone to those minces for 24 h, we observed increases of up to 10-fold. Cortisol increased tropoelastin expression to about the same level as $10^{-10}$ M dexamethasone. When we added dexamethasone with cortisol we observed a further increase which was probably about 15-fold above baseline. These remarkable increases might be the result of the artificially low baseline in this serum-free system. Alternatively, the increases that we saw might be due to continued differentiation of the lung minces as well as direct effects by glucocorticoids on tropoelastin expression.

In the *in vitro* system which we used initially, Northern blots showed a high basal level of tropoelastin mRNA in these cells; in them, we were able to get only a twofold increase with $10^{-8}$ M dexamethasone. In transfection studies, we found that the increase we saw at the promoter level was very similar to the increase that we saw at the mRNA level. In other words, we were able to get about two- to threefold increase in CAT activity using the human tropoelastin promoter–CAT construct, pEP52-CAT. So we see quite different results depending on whether we look *in vivo*, in the mince culture, or we're working with highly differentiated cells that are already producing a significant amount of tropoelastin.

*Uitto:* We have some very recent data which indicate that glucocorticosteroids up-regulate human elastin promoter activity in transgenic mice, as well as in cultures established from these animals (Ledo et al 1994). For example, subcutaneous injection of 100 ng triamcinolone acetonide enhanced the promoter activity by about 30-fold. Even more dramatically, incubation of skin fibroblast cultures established from the transgenic animals with 10 ng/ml of triamcinolone acetonide enhanced the promoter activity almost 80-fold (Ledo et al 1994). Thus, there is clearly evidence for transcriptional regulation of the human elastin gene expression *in vivo*. In support of this mode of regulation by glucocorticosteroids is the fact that human elastin promoter contains three glucocorticoid-responsive elements (Bashir et al 1989).

*Pierce:* What do you see at the mRNA level?

*Uitto:* We have not addressed the elastin mRNA levels in our transgenic mouse model. However, incubation of human skin fibroblasts in culture with similar concentrations (10 ng/ml) of triamcinolone acetonide resulted in about four- to sixfold increases in the elastin mRNA levels (S. D. Katchman, J. Uitto, unpublished results). I thought that such changes were comparable with your results.

*Starcher:* In addition to fibrosis, your mice appear to have severe emphysema. A point I would like to make is that you can have animals with lungs that are falling apart, and they survive. Also, I would like to draw attention to the septal retractions where you have the elastin synthesis. One theory is that when the septa are forming, elastin synthesis is a driving force in the formation of new alveoli. By putting elastase in the lung we have artificially created septal balls where the alveolar walls have retracted after being broken.

After 6–8 weeks of recovery, we find that these areas are rich in elastin and the damaged lungs have more elastin than normal lungs. We felt that the animals have tried to re-make the septa, but can't rebuild the architecture of the lung once it has been destroyed. Have you looked at your neonates around 8–10 days of age where they actually form these septa? Is this a period of great elastin synthesis?

*Pierce:* We see in the four and 10 day neonate almost precisely the same pattern of *in situ* hybridization as we see in the re-initiated adult. It appears, as you say, that in pulmonary silicosis the animals are attempting in some way to compensate for stresses induced by the loss of air space, and are attempting to re-create alveoli under those conditions. My understanding is that these septal tips are actually the rings which are the entrances to the alveoli.

*Starcher:* I have always thought that when the alveolus was damaged by enzymic cleavage, the broken wall retracts into a ball of elastin, and these are the septal tips that you are speaking of.

*Pierce:* But we see similar structures in the neonate. I know that under emphysematous conditions, where you have walls collapsing because of cleavage or breakage of the alveolar wall, you do obtain these balls. However, I think these are different structures: they are the rings (or entrances) to the alveolar sac.

*Starcher:* Why don't you see these balls in the normal lung?

*Pierce:* You do; however, we clearly need to assess these morphological structures more carefully in the silicotic lungs and determine whether emphysematous changes are present.

## References

Bashir MM, Indik Z, Yeh H et al 1989 Characterization of the complete human elastin gene: delineation of unusual features in the 5′ flanking region. J Biol Chem 264:8887–8891

Ledo I, Wu M, Katchman SD et al 1994 Glucocorticosteroids up-regulate human elastin gene promoter activity in transgenic mice. J Invest Dermatol 103:632–636

# Elastin and mechanoreceptor mechanisms with special reference to the mammalian carotid sinus

James Kirumbi Kimani

*Department of Human Anatomy, University of Nairobi, PO Box 30197, Nairobi, Kenya*

*Abstract.* Light microscopic studies reveal that the carotid baroreceptor region in mammals, located at the origin of the internal carotid artery, has a preponderantly elastic structure and a thick tunica adventitia. Electron microscopy discloses the presence of sensory nerve endings within the parts of the tunica adventitia adjoining the preponderantly elastic zone of the internal carotid artery. Bundles of collagen fibres in the tunica adventitia form convolutions or whorls around the nerve terminals and often terminate on the surface of the elastic fibres or into the basement membranes of the neuronal profiles. It is concluded that the large content of elastic tissue in the tunica media of the baroreceptor region renders the vessel wall highly distensible to intraluminal pressure changes, and thereby facilitates transmission of the stimulus intensity to sensory nerve terminals. However, a change in the geometrical configuration of the bundles of collagen under the influence of elastic fibres may provide a better insight into the mechanisms of distortion of the baroreceptors related to and/or in contact with collagen fibres. In support of this is the demonstration of contact sites between collagen and elastic fibres.

*1995 The molecular biology and pathology of elastic tissues. Wiley, Chichester (Ciba Foundation Symposium 192) p 215–236*

The dilated part of the internal carotid artery close to the bifurcation of the common carotid artery forms the carotid sinus (Adams 1958, Heymans & Neil 1958). This part of the arterial wall has a baroreceptor or pressoreceptor function (Adams 1958, Heymans & Neil 1958). In ruminants, however, the arteries that supply the brain arise from the internal maxillary artery, one of the two terminal branches of the external carotid artery. In these animals, the origin of the occipital artery is the site of a dense sensory innervation which corresponds to that of the carotid sinus (sinus occipitalis) in mammals with a functional internal carotid artery (Adams 1958, Rees 1968, Kimani 1979, 1987, Kimani & Mungai 1983).

Since the publication of the classical monograph on comparative morphology of the carotid body and carotid sinus by Adams (1958), numerous studies have been carried out on the ultrastructure of the presumptive baroreceptor nerve terminals (Knoche & Schmitt 1964, Rees 1967, Chiba 1972, Bock & Gorgas 1976, Knoche & Addicks 1976, Krauhs 1979, Knoche et al 1980, Kimani & Mungai 1983, Kimani 1987, 1992). These studies have confirmed earlier observations based on light microscopy that the arteries of the carotid bifurcation subserving a baroreceptor function are characterized by a largely elastic structure within the tunica media.

This review discusses the significance of the functional relationships between the presumptive sensory nerve endings and the mural structural components comprising collagen, elastin and smooth muscle cells. The review also attempts to shed more light on the functional relationships between collagen and elastin and, in turn, the sensory nerve endings.

## Elastin and the carotid baroreceptor function

Evidence has accumulated during the last 100 years or so to show that the mammalian carotid sinus wall or its occipital equivalent is specialized in that it contains a thinner and largely more elastic tunica media (Fig. 1) than the adjacent walls of the common carotid or external carotid arteries (Binswanger 1879, De Castro 1928, Sunder-Plassmann 1930, de Boissezon 1936, Addison 1944, Adams 1955, 1957, Rees 1967, 1968, Rees & Jepson 1970, Aumonier 1972, Bock & Gorgas 1976, Knoche & Addicks 1976, Knoche et al 1980). The largely elastic structure of the carotid sinus wall is associated with a striking loss of muscularity of the tunica media (Binswanger 1879, De Castro 1928, Sunder-Plassmann 1930, de Boissezon 1936, Addison 1944, Adams 1955, 1957, Rees 1967, 1968, Bagshaw & Fischer 1971). Stereological analysis of the elastin and smooth muscle content has revealed an elastin : smooth muscle ratio of $3.943 \pm 1.726$ for the carotid sinus tunica media, and $0.915 \pm 0.417$ and $0.576 \pm 0.222$ ($\pm$SD, $n = 3$) for the adjacent external and common carotid arterial walls, respectively (Kimani 1987). This raises the question of whether the high elastin content and low smooth muscle content are functionally related to the baroreceptor function of the carotid sinus area.

It has been suggested that the specialization of the carotid baroreceptor area results from its location at an arterial branching site (Boss & Green 1956). The structural aspects of the baroreceptor area owing to its location at a branching site can, however, be distinguished from those that are related to the pressoreceptor function. A corollary to this view is the fact that, as demonstrated in the giraffe, the carotid–occipital junction as well as showing an elastic intimal thickening commonly associated with branching sites (Berry 1973, Yohro & Burnstock 1973, Gorgas & Bock 1975), has in addition, an elastic tunica media which is specifically coextensive with the densely innervated parts of the arterial wall. In this regard, the baroreceptor requirement for an elastic structure of the arterial wall appears to have taken advantage of the modified tunica intima while the elevated mural stresses at branching sites in combination with the elastic structure constitute a stimulus amplifier and multiplier system which, in turn, renders the baroreceptor areas highly sensitive to intraluminal pressure changes.

Another difficulty has been that most previous workers have used material from animals in which the transition of the arterial wall at the junction of the common carotid and the internal carotid sinus is more gradual and sometimes not distinct (Rees 1967, 1968, Bagshaw & Fischer 1971, Bock & Gorgas 1976, Knoche et al 1980). In their study, Kimani & Mungai (1983) demonstrated, however, that the giraffe carotid artery has a largely muscular structure which is replaced by an elastic structure in the carotid sinus area located at the carotid–occipital junction. Similar observations have been made in other long-necked animals, such as the camel, goat and sheep (J. Kimani, unpublished results). These findings lend support to the suggestion that the high content of elastin in the carotid sinus wall and concomitant diminution of smooth muscle is a morphological adaptation of the arterial wall to the baroreceptor function. In this regard, elastin might render the arterial wall readily distensible to intraluminal pressure changes with subsequent stimulation of the adjunct sensory nerve endings. Experimental studies have revealed that factors that tend to diminish the distensibility of the arterial wall, such as hypertrophy of the arterial wall in chronic hypertension and/or atheromatous changes in the human carotid sinus, lead to a decrease in the pressoresponse of the baroreceptors to a given intrasinus pressure change (McCubbin et al 1956, Aars 1968, 1969, 1971, Angell James 1971, 1973, Heath et al 1973, Winson et al 1974, Pickering & Sleight 1977). Thus, as demonstrated by Bergel et al (1975) and Pickering & Sleight (1977), the mechanical properties of the sinus region are such that sudden changes in pressure produce a more rapid change in diameter in the sinus region than in the less distensible common carotid or external carotid arteries. This is in conformity with the physiomechanical properties of elastin, and can be explained, at least in part, on the basis of the large elastin content in the tunica media of the carotid sinus wall.

## Cytoarchitecture of baroreceptor nerve endings and mural tissue components

Numerous axonal varicosities, rich in mitochondria and containing glycogen granules and sometimes lamellar bodies, occur in the mammalian carotid sinus area (Fig. 2). These varicosities occur largely in the tunica adventitia; here they are associated with both collagen and elastic fibres, many of these approaching the media–adventitia junction (Rees 1967, 1968, Chiba 1972, Bock & Gorgas 1976, Knoche et al 1980, Yates & Thoren 1980, Kimani & Mungai 1983, Kimani 1992). The results available suggest that the stimulus of

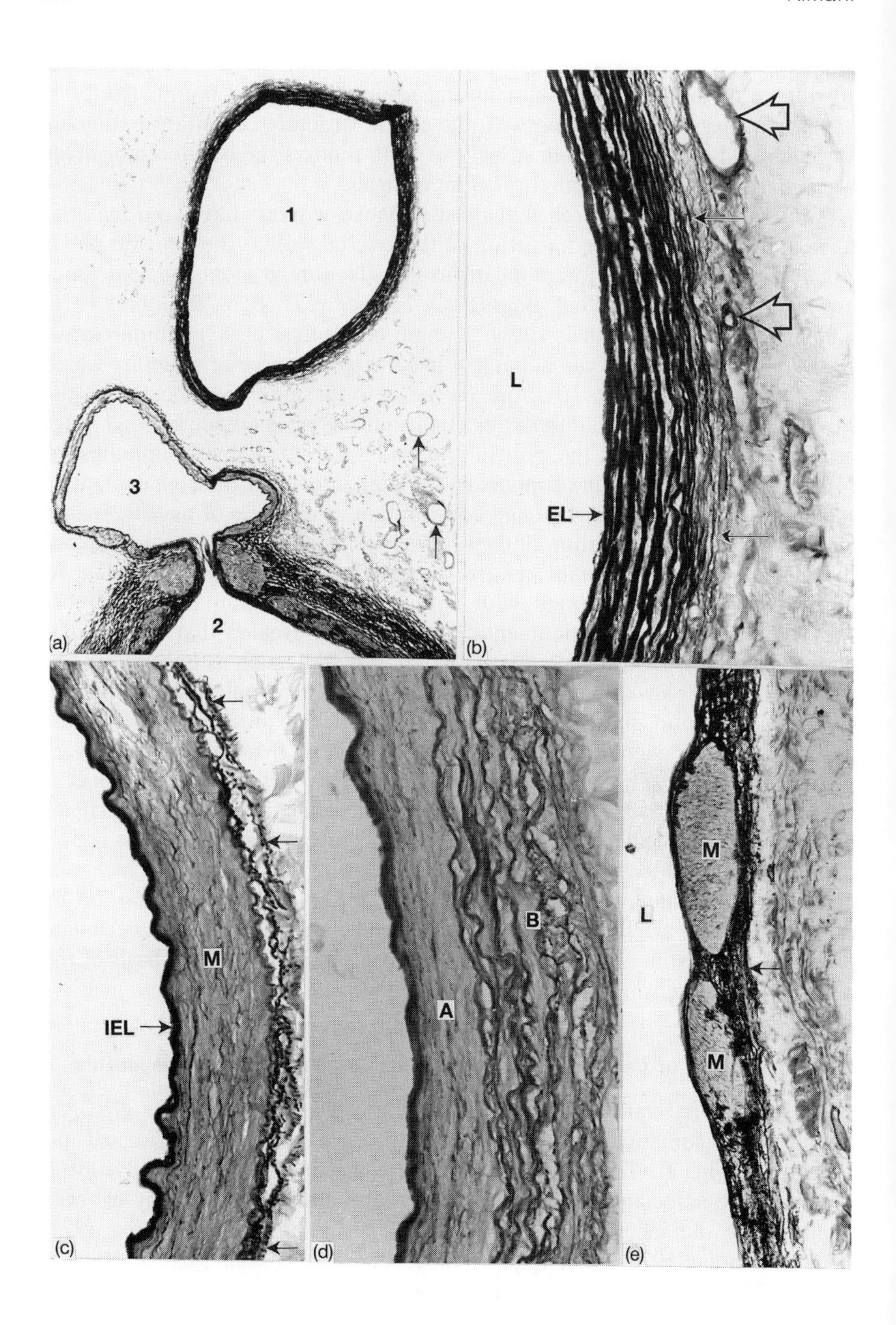
1
3
2
(a)
L
EL
(b)
M
IEL
(c)
A
B
(d)
M
L
M
(e)

the sensory nerve terminals within the walls of the carotid sinus is a mechanical deformation of the juxtaposed fibrous connective tissue components by intraluminal pressure changes (Heymans & Neil 1958, Peterson 1962).

The first electron microscopic studies of baroreceptors were carried out by Knoche & Schmitt (1964) and Rees (1967). These and subsequent ultrastructural investigations (Chiba 1972, Bock & Gorgas 1976, Knoche & Addicks 1976, Krauhs 1979, Knoche et al 1980, Abdel-Magied et al 1982, Kimani & Mungai 1983, Kimani 1987, 1992, 1993) have considered, in detail, the structure of the carotid sinus wall and the relationships between its components and nerve endings. The view generally held is that the axonal endings form mitochondrion-rich varicosities of different sizes which are located in the tunica adventitia and media–adventitia border. In some animals, axonal profiles have been followed through the tunica adventitia to the outer layers of the tunica media (Bock & Gorgas 1976, Knoche & Addicks 1976). The axonal terminals demonstrated with the use of an electron microscope conform to the so-called loops, compact and/or fibrillar expansions demonstrated in the earlier light microscopic studies. Few studies have presented evidence of neuronal profiles that resemble the ring terminals, particularly in sections cut longitudinally with respect to the vessel wall (Krauhs 1979, Kimani & Mungai 1983).

The data obtained from our studies, based on a number of animals, provide interesting features regarding the carotid sinus wall. Firstly, it has been demonstrated that the elastic and collagen fibres in the tunica adventitia are closely associated with evidence of multiple areas of contact between them (Fig. 2b).

---

FIG. 1. (a) Photomicrograph of a transverse section through the internal (1) and external (2) carotid arteries of porcupine close to their origin from the common carotid artery. Note also the common origin of the ascending pharyngeal and occipital arteries (3). The section passes through the carotid sinus portion of the internal carotid artery characterized by a largely elastic structure. Blood vessels in the intercarotid region are arrowed. Weigert's–Van Gieson's stain (×20). (b) High power photomicrograph of the carotid sinus wall as seen in (a). Note the presence of closely packed elastic lamellae (EL) in the tunica media, a fine network of elastic fibres in the media–adventitia junction (solid arrows) and vasa vasorum (open arrows). L, lumen. Weigert's–Van Gieson's stain (×125). (c) Photomicrograph of a transverse section of the porcupine internal carotid artery away from the carotid sinus area. Elastic tissue occurs as internal elastic lamina (IEL) and in the tunica adventitia (arrows) while the tunica media (M) is largely muscular. Weigert's–Van Gieson's stain (×125). (d) Transverse section of porcupine internal carotid artery wall showing the mode of the arterial transition from the elastic structure of the carotid sinus wall to the muscular structure of its cranial portion. Note the differentiation of the tunica media into a luminal muscular (A) and an outer elastic (B) zone. Weigert's–Van Gieson's stain (×200). (e) Same as (d) but seen from a longitudinal section. Note muscular nests (M) interconnected by an elastic network (arrow). L, lumen. Weigert's–Van Gieson's stain (×50). (Reproduced with permission from Kimani 1987.)

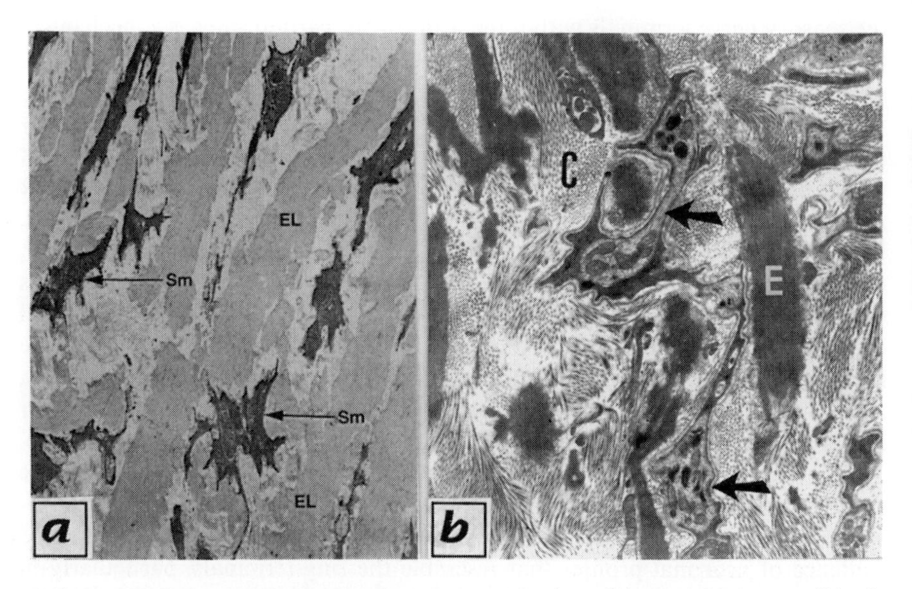


FIG. 2. (a) Electron micrograph of a transverse section of the carotid sinus wall in the porcupine showing the tunica media. Note scattered smooth muscle cells (Sm) between elastic layers (EL) (×3900). (Reproduced with permission from Kimani 1987.) (b) Axonal varicosities (arrows) in the rock hyrax carotid sinus embedded in connective tissue rich in collagen (C) and elastic fibres (E) (×5760). (Reproduced with permission from Kimani 1992.)

Secondly, apart from the close apposition of nerve terminals to the elastic fibres, further evidence has been presented on insertion of collagen fibres into the surface of elastic fibres closely associated with the nerve terminals (Fig. 4). These observations have rendered further support to the studies of Krauhs (1979) and Kimani & Mungai (1983) indicating that some neuronal profiles form sleeves on bundles of collagen and/or elastic fibres (Figs 5, 6).

Regarding the relationship between collagen and elastic fibres, a direct association between collagen and elastic fibres has been denied in most earlier studies on the vascular wall (Keech 1960, Karrer 1961). Later findings based on polarizing and electron microscopy (Oakes & Bialkower 1977, Kimani 1992, 1993), indicate that collagen and elastin may be physically linked through multiple sites of contact.

A renewed interest in the geometrical configuration of the collagen fibres and their association with elastic fibres in the carotid sinus wall has arisen from the need to explain the mechanical components of the arterial wall associated with the transduction of a mechanical force into a neural impulse. Electron microscopic studies have revealed that the sensory nerve endings are located in the tunica adventitia where they are closely related to both elastic and collagen fibres (Rees 1967, Chiba 1972, Knoche et al 1980, Kimani and

Mungai 1983, Kimani 1982, 1992, 1993). It has been shown that collagen fibrils and microfibrils of the elastic fibres terminate directly upon the basement membrane surrounding the presumptive baroreceptor terminals and Schwann cells. This close association between the connective tissue stroma and the basement membrane implies that stretching of the collagen and elastic fibres tugs against the basement membrane thereby causing a mechanical deformation of the presumptive baroreceptor terminals. In this regard, Rees (1967) noted that the receptors, having an intimate relationship with the more extensible elastic fibres, may be stimulated more readily than those related to the less extensible collagen fibres, the relative threshold of stimulation of the individual nerve terminal being dependent on the particular collagen/elastin admixture of its surrounding. Accordingly, this would be in keeping with the wide spectrum of thresholds of stimulation between individual baroreceptors demonstrated by Bronk & Stella (1932).

Both polarized light and electron microscopical studies show, however, that collagen bundles exist in a slightly folded or convoluted state in relaxed tissues and that elongation results in their unfolding and re-alignment in the direction of strain (Glagov & Wolinsky 1963, Wolinsky & Glagov 1964, Oakes & Bialkower 1977). In the carotid sinus area, it has been reported that the collagen fibres surrounding receptors are either convoluted or randomly orientated (Chiba 1972, Knoche et al 1980, Kimani 1992, 1993). As this convoluted state does not coincide with any of the thermodynamically accepted configurations of the collagen fibre, it can be assumed that a comprehensive pre-stress is required to maintain collagen in this configuration. The findings that in the aorta (Coulson et al 1965) and in the elastic wing tendon of the domestic fowl (Oakes & Bialkower 1977) administration of elastase (an enzyme known to have no collagenolytic activity) causes spontaneous unfolding of the collagen fibres imply, therefore, that the network of elastic fibres and/or an elastase-susceptible component which binds either collagen fibres together, or collagen fibres to elastic fibres, is responsible for maintaining collagen in a folded configuration as well as modulating its unfolding when it is stretched. It would appear, therefore, that the elastic structure of the carotid sinus wall provides the presumed compressive pre-stress that maintains collagen bundles in a folded conformation while also controlling its intermittent unfolding and folding in each cardiac cycle.

We recently reported a structural linkage between collagen and elastic fibres in which collagen fibrils are embedded and/or intermingled with the outer fibrillar mantle of the elastic fibres (Kimani 1992, 1993). This structural linkage may form part of the mechanism by which elastic fibres, during their recoil, tug on the collagen fibres thereby bringing about their folding and unfolding as they elongate. Much remains to be discovered, however, about the actual mechanism of adhesion (if any) between collagen and elastic fibres. A possibility may also exist that the 2D picture seen in transmission electron

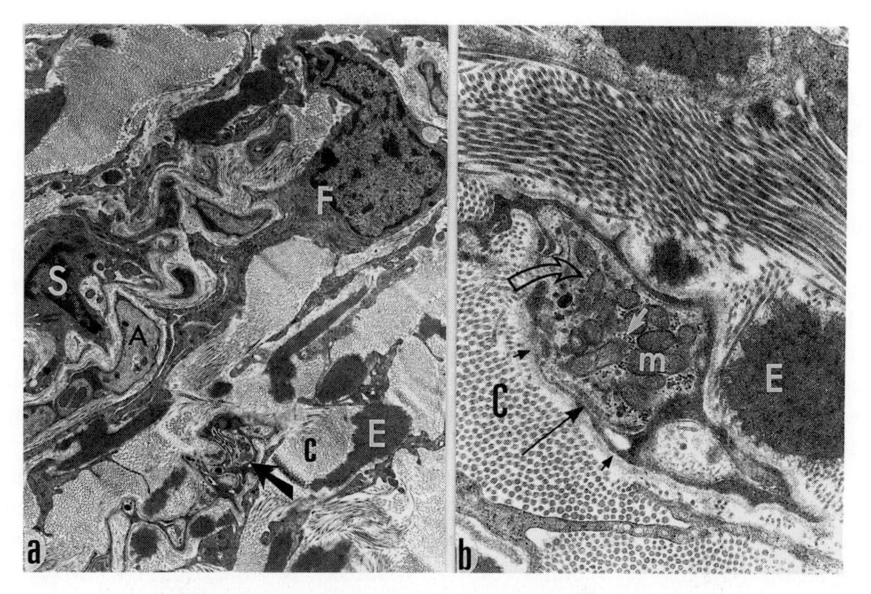


FIG. 3. Electron micrographs of the carotid sinus wall in the hyrax showing the tunica adventitia. (a) Neuronal profile in the tunica adventitia. Cellular profiles comprising Schwann cells (S), fibroblasts (F), and axonal profiles (arrow). C, collagen; E, elastic fibres ($\times 2560$). (b) Axonal varicosity (open arrow) embedded in connective tissue rich in collagen (C) and elastic fibres (E). M, mitochondria; tip of the white arrow, glycogen granules; long black arrow, Schwann cell processes; small black arrow, basement membrane ($\times 9600$). (Reproduced with permission from Kimani 1992.)

microscopy does not represent an actual insertion of collagen fibrils into elastic fibres, but rather an overlap of the latter as they run over the surface of an elastic fibre. This does not explain, however, why bundles of collagen unfold when elastin is removed by enzymic digestion even under a relaxed state unless they are bound to, or maintained in the folded state, by elastin or by an elastase-susceptible component (Oakes & Bialkower 1977). It remains to be shown, therefore, whether or not adhesion molecules occur on elastin that binds collagen and/or cellular elements onto it and how such molecules, if they exist, are genetically encoded.

The above findings indicate that the elongation and recoil of the elastic fibres modulates the mechanical events leading to the distortion of the sensory nerve endings located within the carotid sinus wall. In this scheme, elastic fibres, as they elongate, may cause the stimulation of the nerve terminals applied on them, as well as those related to collagen, by influencing the unfolding of the bundles of collagen tugged on them. Therefore, a change in the geometrical configuration of bundles of collagen under the modulating influence of elastin may provide a better insight into the mechanisms of deformation of the

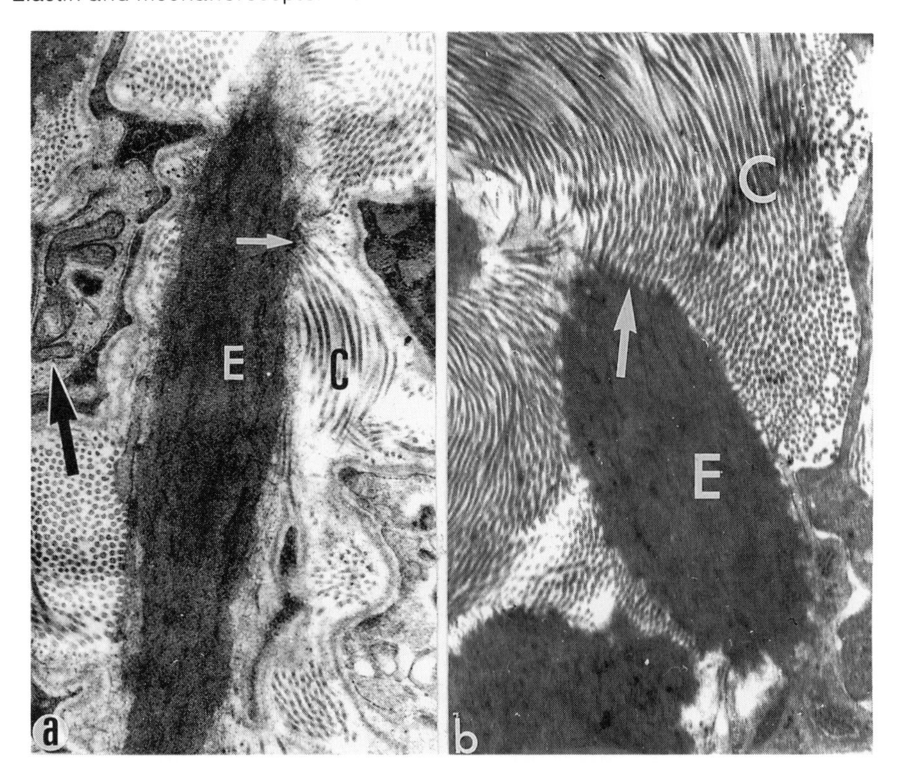


FIG. 4. Electron micrographs illustrating the presumptive physical linkage between elastic (E) and collagen (C) fibres in the carotid sinus wall of the rock hyrax. (a) Bundle of collagen fibres (C) converging on the surface of the elastic fibre (E) (white arrow). Black arrow, axonal varicosity (longitudinal section) ( × 12 000). (b) Sites of contact (arrow) between collagen fibrils (C) and elastic fibres (E) (transverse section; × 12 000). (Reproduced with permission from Kimani 1992.)

receptors or their surfaces in contact with fibrils of collagen. Thus, axonal varicosities closely related to collagen fibres can be deformed as easily as those in contact with the elastic fibres during re-alignment of the collagen fibrils. Pertinent to this is the fact that in the Golgi tendon organ, sensory nerve terminals are sandwiched between spiral bundles of collagen (Schoultz & Swett 1972). This implies, therefore, that a squeezing action, rather than elongation of the collagen fibrils, is the most likely mechanical stimulus for the related nerve endings.

There is a substantial body of evidence showing that sensory nerve terminals in the carotid sinus wall are also closely applied onto elastic fibres. This relationship was demonstrated initially by Knoche & Schmitt (1964) and was subsequently confirmed by Rees (1967), Bock & Gorgas (1976) and Knoche et al (1980). The mode of attachment of the neuronal profiles onto the elastic

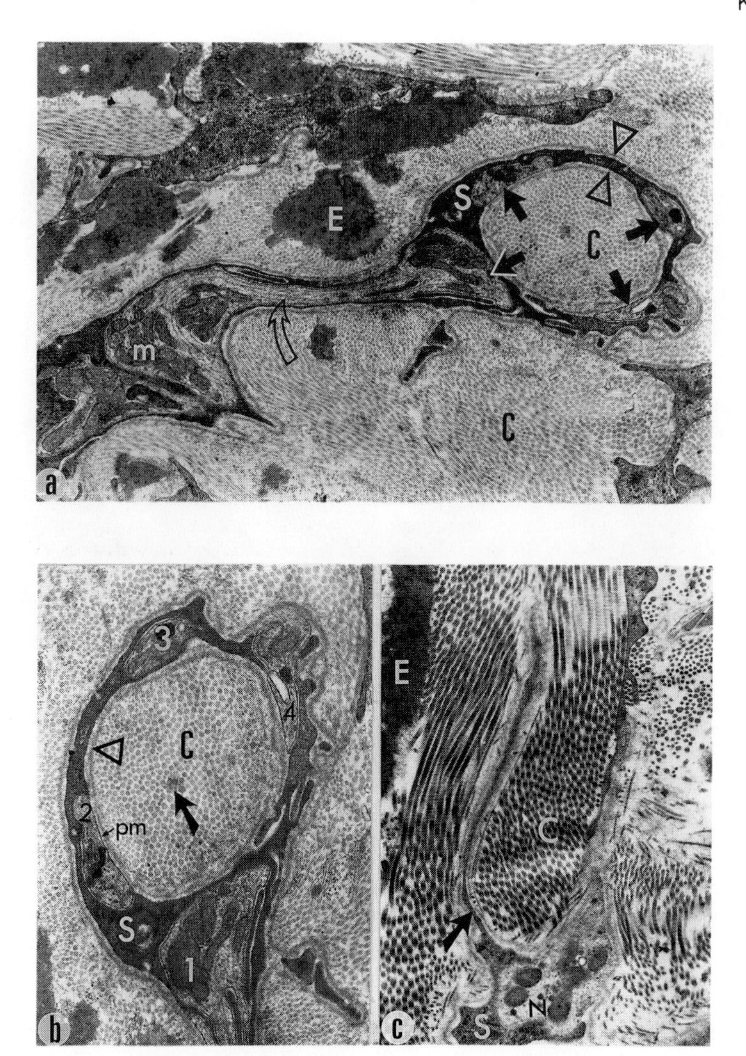


FIG. 5. Electron micrographs of the hyrax carotid sinus wall (longitudinal section). (a) Bundle of collagen in a sleeve of neuronal profile. Outer and inner basement membrane layers (open arrowheads), Schwann cell process (S), axonal varicosities (solid arrows) and an intervaricose segment (open arrow) are shown. C, collagen; E, elastic fibres; m, mitochondria ( × 9600). (b) Part of neuronal sleeve shown above. Schwann cell processes (S) and axonal varicosities (1–4) containing mitochondria are shown. Note juxtaposition of the axolemma (pm) to the basement membrane (open arrowhead). Arrow, elastic fibre; C, collagen fibres ( × 16 000). (c) Neuronal profile forming an open ring. Bundle of collagen (C) extends out through the open end of the ring. S, Schwann cell and thin extensions over the bundle of collagen (arrow); N, axonal varicosity ( × 16 000). (Reproduced with permission from Kimani 1992.)

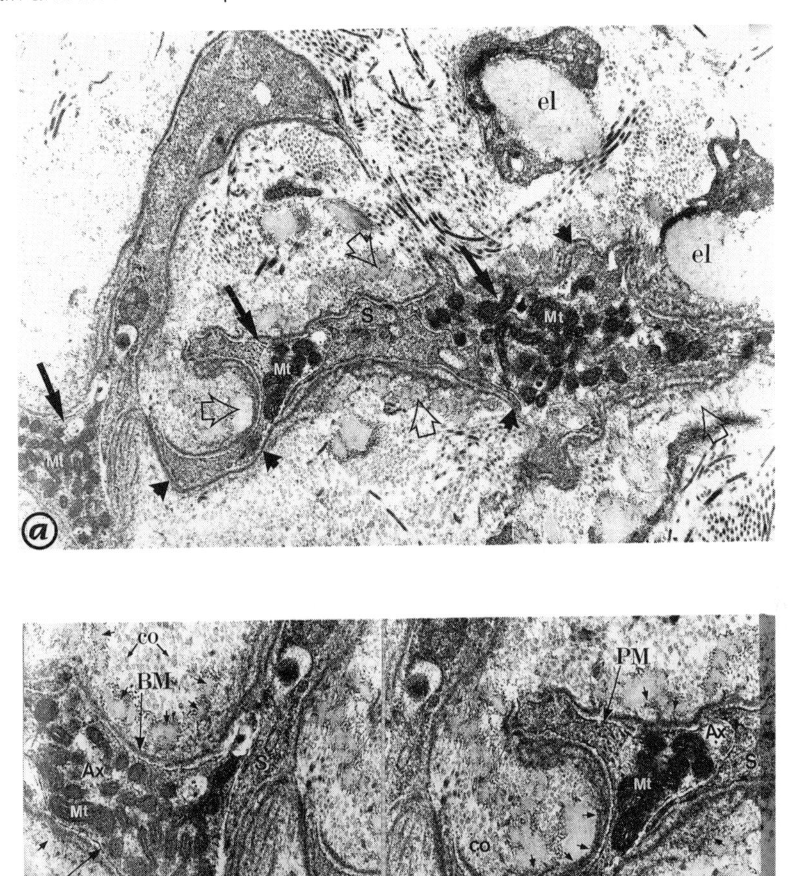


FIG. 6. Transmission electron micrographs showing neuronal profiles in the giraffe carotid sinus wall in longitudinal section. (a) A large neuronal profile in the deep layers of the tunica adventitia. The Schwann cell (S) and mitochondria (Mt)-rich axonal varicosities (solid arrows) are completely enveloped by a basement membrane (solid arrowheads). The connective tissue stroma contains numerous elastic fibres (open arrowheads) that form a fibrous basket around the nerve terminals. Note a large elastic fibre (el) to which the axonal varicosity is attached (×18000). (Reproduced with permission from Kimani & Mungai 1983.) (b,c) Higher magnification micrographs of two axonal profiles from the sensory profile shown in (a), demonstrating the topographic organizational relationship between the axonal varicosities (Ax), Schwann cell processes (S), basement membrane complex (BM), axonal plasma membrane (PM) and the surrounding connective tissue stroma which is rich in elastic fibres, to which the basement membrane is often anchored (arrows). MT, mitochondria; co, collagen (×34000). (Reproduced with permission from Kimani & Mungai 1983.)

fibres has not been fully investigated. However, as demonstrated in Fig. 3a, neuronal profiles may run along the surface of elastic fibres and/or form sleeves on them. The basement membrane or limiting lamina of the axonal profile may be applied on the surface of the elastic fibre. Beneath this, there may be a cytoplasmic process of a Schwann cell, or sometimes none, thus exposing the plasmolemma to the limiting membrane (Fig. 3b). Thus, the main structural barrier between the sensory nerve endings and the connective tissue stroma seems to be the limiting membrane whose main component is collagen type IV. In such a mechanoreceptor unit, the stimulus energy could deform the limiting membrane thereby causing a response of the sensory nerve terminal.

That the apparent relationship between the axonal profiles and the connective tissue elements of the arterial wall (and elastin in particular) may be coupled funtionally to mechanoreception was demonstrated by Krauhs (1979) and recently by Kimani (1992, 1993). Krauhs (1979) described 'ring receptor' fields in which the nerve terminals and their associated Schwann cell processes are threaded around bundles of collagen and/or elastic fibres. Kimani & Mungai (1983) described similar structures in the giraffe carotid sinus, but their functional significance remained obscure in view of the poor fixation of the material used in that study. Kimani (1992) produced for the first time a transmission electron micrograph that clearly demonstrates this relationship (Fig. 5), thereby confirming the earlier studies. A similar structural relationship has been demonstrated in the Golgi tendon organ (Schoultz & Swett 1972).

It may be argued, however, that this structural relationship supports the contention by Rees (1967) that the relative collagen/elastin admixture determines the thresholds of stimulation of the baroreceptors. In such a scheme, the receptor units forming sleeves on bundles of collagen would be less sensitive than those on elastic fibres. In further support of this is the fact that such bundles of collagen have a preferential circular orientation with respect to the vessel wall.

The picture is, however, complicated by the fact that the axolemma of the contained axonal profiles may be exposed to the inner and outer sides of the sleeve. Thus, whereas the collagen fibrils contained within the sleeve show a regular arrangement, those outside the sleeve form a 3D network (Figs 5a, 6a–c). Furthermore, although an attempt to follow these structures serially proved difficult because of the quality of the microtome used in these studies, it is likely that after leaving the sleeve, the bundles of collagen fibres merge with the surrounding convoluted bundles (Fig. 5c).

The available data indicate that it may not be easy to speculate on the structural basis for the differences in thresholds of stimulation of the individual axons of the sinus nerve on the evidence derived from the associated connective tissue fibres. Physiological studies have revealed, however, that the non-myelinated afferent axons from the carotid and aortic arch/brachiocephalic baroreceptor areas have thresholds of 30–50 mmHg higher than those of their medullated counterparts, and much lower sensitivities and maximum frequencies (Thoren & Jones 1977, Coleridge & Coleridge 1980, Yao & Thoren 1983). Yet, at present, there is no corresponding evidence to show that the receptor nerve terminals of the non-myelinated axons are related preferentially to the less distensible collagen fibres, and the myelinated receptors to the more extensible elastic fibres. Rather, as demonstrated in this paper and elsewhere (Krauhs 1979, Kimani & Mungai 1983, Kimani 1987, 1992, 1993), a complex nerve terminal region is formed and both myelinated and non-myelinated axons become varicose. Whereas part of the surface of each terminal region is ensheathed by Schwann cells, the remaining surface is in direct contact with the basement membrane. This implies that the thresholds of stimulation of the receptors may be intrinsic properties of the individual nerve fibres and their terminal arborizations irrespective of the elastin/collagen admixture. Instead, the connective tissue acts largely as a medium through which the stimulus energy is transmitted to the receptors. This, it seems, can be achieved through stretching of the elastic fibres and/or a change in the geometrical configuration of the bundles of collagen fibres.

## Conclusion

One of the main morphological adaptations of the mammalian carotid sinus to its baroreceptor function is an elastic structure of the tunica media. It is conceivable that elastic fibres, because of their continuous nature and rubber-like physiomechanical properties, render the arterial wall readily distensible to intraluminal pressure change. This in turn renders the sinus wall more responsive to a given pressure change.

Physiomechanical studies have demonstrated that a mechanical distortion of the arterial wall, rather than a pressure change *per se*, forms the stimulus energy for the sensory nerve endings within the tunica adventitia. This implies, therefore, that, as shown in previous studies, factors that decrease the distensibility of the arterial wall, such as atherosclerosis, may lead to a decrease in pressoresponse of the baroreceptors to a given intrasinus pressure change.

It is further suggested that collagen fibres provide rigid constraints that limit the overal distention of the arterial wall and conceivably therefore the sensitivity of the receptors. It seems probable, however, that mechanical distortion and unfolding of collagen fibres, rather than their elongation, is the most likely mechanical event causing the stimulation of the related nerve terminals. There is evidence that this unfolding of collagen fibres is modulated by elastic fibres and/or an elastase-susceptible component that binds collagen to elastin. Our studies have shown multiple contact sites between collagen and elastin.

Our studies have also shown the presence of numerous concentric and/or encapsulating receptor terminals in which an admixture of collagen and elastin is ensleaved by Schwann cell processes containing axonal varicosities. The elastic fibres contained within the hook-like axoplasmatic processes or the membranous sheaths may modulate the distension of the other tissues, including the nerve terminals, while collagen fibres provide constraints that check the overall distention of these mechanoreceptors.

*Acknowledgements*

The earlier work referred to in this review was carried out under the supervision of Professor Joseph Maina Mungai, to whom the author is greatly indebted for guidance, support and assistance. Electron microscopic studies relating to the rock hyrax were carried out in the Institute of Anatomy, University of Lübeck, under the sponsorship of the German Agency for Academic Exchange (DAAD). Illustrations have been reproduced with permission from Zoological Society (Figs 1,2a), Wiley-Liss, Inc. (Figs 2a,3,4,5,) and Karger, Basel (Fig. 6) whom I thank. I thank the technical staff, Department of Human Anatomy, University of Nairobi, for technical support.

## References

Aars H 1968 Aortic baroreceptor activity in normal and hypertensive rabbits. Acta Physiol Scand 72:298–309

Aars H 1969 Relationship between aortic diameter and baroreceptor activity in normal and hypertensive rabbits. Acta Physiol Scand 75:406–414

Aars H 1971 Effect of noradrenaline on activity in single aortic baroreceptor fibres. Acta Physiol Scand 83:335–343

Abdel-Magied EM, Taha AAM, King AS 1982 An ultrastructural investigation of a baroreceptor zone in the common carotid artery of the domestic fowl. J Anat 135:463–475

Adams WE 1955 The carotid sinus complex 'parathyroid' III and thymo-parathyroid bodies, with special reference to the Australian opossum, *Trichosurus vulpecula.* Am J Anat 97:1–57

Adams WE 1957 The carotid sinus complex in the hedgehog, *Erinaceus europaeus.* J Anat 91:207–227

Adams WE 1958 The comparative morphology of the carotid body and carotid sinus. Thomas, Springfield, IL

Addison WHF 1944 The extent of the carotid pressoreceptor areas in the cat as indicated by its special elastic tissue wall. Anat Rec 48:418–419

Angell James JE 1971 The effects of changes of extramural 'intrathoracic' pressure on aortic arch baroreceptors. J Physiol 214:89–103

Angell James JE 1973 Characteristics of single aortic and right subclavian fibre activity in rabbit with chronic renal hypertension. Circ Res 32:149–161

Aumonier FJ 1972 Histological observations on the distribution of baroreceptors in the carotid and aortic regions of the rabbit, cat and dog. Acta Anat 82:1–16

Bagshaw RJ, Fischer GM 1971 Morphology of the carotid sinus in the dog. J Appl Physiol 31:198–202

Bergel DH, Bertram CD, Brooks DE, MacDermott AJ, Robbinson JL, Sleight P 1975 Simultaneous recording of the carotid sinus dimensions and the baroreceptor nerve in the anaesthetized dog. J Physiol 252:15–16

Berry CL 1973 The establishment of the elastic structure of arterial bifurcation and branches. Its relevance to medial effects of cerebral arteries. Atherosclerosis 18:117–127

Binswanger O 1879 Anatomische Untersuchungen uber die Ursprungstelle und den Anfangstheil der Carotis interna. Arch Psychiatr Nervenkr 9:351–368

Bock P, Gorgas K 1976 The fine structure of baroreceptor terminals in the carotid sinus of guinea-pigs and mice. Cell Tissue Res 170:95–112

Boss J, Green JH 1956 The histology of the common carotid baroreceptor area of the cat. Circ Res 4:12–17

Bronk DW, Stella G 1932 Afferent impulses in the carotid sinus nerve. J Cell Comp Physiol 1:113–130

Chiba T 1972 Fine structure of the baroreceptor nerve terminals in the carotid sinus of the dog. J Electron Microsc 21:139–148

Coleridge HM, Coleridge JCG 1980 Cardiovascular afferents involved in regulation of peripheral vessels. Annu Rev Physiol 42:413–427

Coulson RWN, Weisman N, Carnes WH 1965 Cardiovascular studies on copper deficient swine. VII. Mechanical properties of aortic and dermal collagen. Lab Invest 14:303–309

de Boissezon P 1936 La trifurcation carotidienne et le corpuscle intercarotidien du cheval. Anns Anat Path 13:733–747

De Castro F 1928 Sur la structure et l'innnervation du sinus carotidien de l'homme et des mammiferes. Nouveaux faits sur l'innervation et al fonction du glomus caroticum. Etud Anat Physiol Trab Lab Invest Biol Univ Madrid 25:331–3380

Glagov S, Wolinsky H 1963 Aortic wall as a two-phase material. Nature 199:606–608

Gorgas K, Bock P 1975 Studies on nitra-arterial cushions. I. Morphology of the cushions at the origins of intercostal arteries in mice. Anat Embryol 148:59–72

Heath D, Smith P, Winson M 1973 The atherosclerotic human carotid sinus. J Pathol 110:49–58

Heymans C, Neil E 1958 Reflexogenic areas in the cardiovascular system. Churchill, London

Karrer HE 1961 An electron microscope study of the aorta in young and ageing mice. J Ultrastruct Res 5:1–27

Keech MK 1960 Electron microscope study of normal rat aorta. J Biophys Biochem Cytol 7:533–538

Kimani JK 1979 Some aspects of the structural organisation of the carotid arterial system of the giraffe (*Giraffa camelopardalis*) with special reference to the carotid baroreceptor equivalent. PhD thesis, University of Nairobi, Nairobi, Kenya

Kimani JK 1987 Observations on the structure and innervation of the carotid sinus complex in the African porcupine (*Hystrix cristata*). J Zool 213:569–590

Kimani JK 1992 Electron microscopic structure and innervation of the carotid baroreceptor region in the rock hyrax (*Procavia capensis*). J Morphol 212:201–211

Kimani JK 1993 Structural linkage between collagen and elastic fibres in the mammalian carotid sinus wall with special reference to baroreceptor mechanisms. Discovery Innovation 5:51–56

Kimani JK, Mungai JM 1983 Observations on the structure and innervation of the presumptive carotid sinus area in the giraffe (*Giraffa camelopardalis*). Acta Anat 115:117–133

Knoche H, Addicks K 1976 Electron microscopic studies of the pressoreceptor fields of the carotid sinus of the dog. Cell Tissue Res 173:77–94

Knoche H, Schmitt G 1964 Beitrag zur Kenntnis des Nervengewebes in der Wand des Sinus caroticus. I. Mitteilung Z Zellforsch Mikrosk Anat 63:22–36
Knoche H, Wiesner-Menzel L, Addicks K 1980 Ultrastructure of baroreceptors in the carotid sinus of the rabbit. Acta Anat 106:63–83
Krauhs JM 1979 Structure of rat aortic baroreceptors and their relationship to connective tissue. J Neurocytol 8:401–414
McCubbin JW, Green JH, Page IH 1956 Baroreceptor function in chronic renal hypertension. Circ Res 4:205–210
Oakes BW, Bialkower B 1977 Biomechanical and ultrastructural studies on elastic wing tendon from domestic fowl. J Anat 123:369–387
Peterson LH 1962 Properties and behaviour of living vascular wall. Physiol Rev 42:309–325
Pickering TG, Sleight P 1977 Baroreceptors and hypertension. Prog Brain Res 47:43–60
Rees PM 1967 Observations on the fine structure and distribution of presumptive baroreceptor nerves at the carotid sinus. J Comp Neurol 131:517–548
Rees PM 1968 Electron microscopical observations on the architecture of the carotid arterial walls with reference to the sinus portion. J Anat 103:35–45
Rees PM, Jepson P 1970 Measurements of arterial geometry and wall composition in the carotid baroreceptor area. Circ Res 26:461–467
Schoultz TW, Swett JE 1972 The fine structure of the Golgi tendon organ. J Neurocytol 1:1–26
Sunder-Plassmann P 1930 Untersuchungen uber den Bulbus carotidis bei Mensch und Tier in Hinbick auf die 'Sinursreflexe' nach H. E. Hering; ein Vergleich mit anderen Gefasstrecken; die Histopathologie des Bulbus carotidis; des Glomus caroticum. Z Anat Entwicklungsgesch 93:567–622
Thoren P, Jones JV 1977 Characteristics of aortic baroreceptor C-fibres in the rabbit. Acta Physiol Scand 99:448–456
Winson M, Heath P, Smith P 1974 Extensibility of the human carotid sinus. Cardiovasc Res 8:58–64
Wolinsky H, Glagov S 1964 Structural basis for the static mechanical properties of the aortic media. Circ Res 14:400–413
Yao T, Thoren P 1983 Characteristics of brachiocephalic and carotid sinus baroreceptors with non-medullated afferents in rabbits. Acta Physiol Scand 117:1–8
Yates T, Thoren P 1980 An electron microscopic study of the baroreceptors in the internal carotid artery of the spontaneously hypertensive rat. Cell Tissue Res 205:473–483
Yohro T, Burnstock G 1973 Fine structure of intimal cushions at branching sites in coronary arteries of vertebrates. Z Anat Entwicklungsgesch 140:187–202

## DISCUSSION

*Rosenbloom:* Did I understand correctly that the collagen fibres in the baroreceptor are woven or twisted in some way, so that when you initially get the stretching these then unwind a little bit?

*Kimani:* Yes.

*Rosenbloom:* When elastic ligaments are treated with elastase, do they retain their folded appearance? Did you try stretching the elastase-treated material to see what would happen?

*Kimani:* We haven't done that part of the study. I only referred to the studies of Oakes & Bialkower (1977). In these studies they used unstretched elastic ligaments and found that collagen fibres unfold after enzymic digestion of elastin by elastase and without any stretching of the elastase-treated ligament. It is for this reason that they concluded that either elastin or an elastase-susceptible component that binds collagen to elastin is responsible for maintaining collagen in a folded configuration in the relaxed state.

*Rosenbloom:* So elastin is doing its usual thing; it is recoiling; bringing it back into that woven pattern.

*Kimani:* Yes. You can imagine that a change in conformation is occurring, for example, in the baroreceptor mechanisms. Collagen is not stretching, but it is changing in conformation. That change in conformation could be brought about by the stretching of the elastic fibres.

*Robert:* It may be more than that; your presentation reminded me of a very old incident which nearly broke a friendship. Professor Anna Kadar, in Budapest, viewed one of our highly purified elastin preparations, boiled in sodium hydroxide, by electron microscopy. There was a hole in one of these lamellae, and a native collagen fibre stuck out of the hole on both sides. We were sure that we had boiled the preparation for three 15 min treatments and had washed it properly but, somehow, at least one collagen fibril still survived. Perhaps Dan Urry's explanation holds true here: that is, maybe the hydrophobic effect somehow uncharged the hydroxyl ion near the site where this collagen fibril was crossing the elastic lamella. When elastin stretches it certainly has the potential to deform collagen fibrils in its immediate vicinity.

*Kimani:* That is a plausible explanation. By transmission electron microscopy it is not possible to demonstrate with any degree of certainty the presumptive contact sites between collgen and elastin. It is necessary to apply other methods, such as scanning electron and confocal microscopy. However, as I noted in my paper, other workers such as Rees (1968) have provided further morphological evidence in support of the existence of contact sites between collagen and elastin, at least in the carotid sinus area. We are reluctant to speculate that this is a unique finding in this part of the arterial wall.

*Robert:* Do the nerve endings end blindly in the carotid sinus wall or do they have synapses on the smooth muscle cell membrane?

*Kimani:* There are two main types of nerve ending in the carotid sinus wall, namely, afferent and efferent terminals of the sympathetic type. We have shown that the afferent nerve endings form ramifications within the connective tissue stroma of the tunica adventitia. In some instances, they even penetrate the outer layers of the tunica adventitia. Isolated smooth muscle cells are seen in the tunica adventitia but the afferent nerve endings, though not forming synaptic contacts, may come relatively close to the cells.

Regarding the distribution of the efferent nerve fibres, morphological studies based on fluorescent histochemical techniques and transmission electron microscopy have demonstrated noradrenergic innervation of the carotid sinus area which is restricted mainly to the tunica adventitia and away from the vasomotor control of the tunica media (Rees 1967, Reis & Fuxe 1968, Stanton & Hinrichsen 1980). This anatomical finding has led to the suggestion that sympathetic nerves may modify baroreceptor reflex through release of noradrenaline (Rees 1967, Reis & Fuxe 1968), although no generalization can be made owing to the reported species difference in the relative densities of catecholamine innervation of the sinus wall (Stanton & Hinrichsen 1980).

*Robert:* These are very important problems which are not studied carefully enough. It was shown years ago that when you make one passage in culture (or at the most two) of smooth muscle cells they loose these receptors, so this can only be studied *in vivo*.

*Davidson:* You said that the muscularization of the proximal segment of the common carotid artery in the giraffe was a postnatal event, so that an extensive remodelling of the vessel wall occurs when the neonate starts walking around. This might be an excellent model for examining the activity of the elastolytic enzymes that Leslie Robert and Alek Hinek have been studying. This is a non-pathological remodelling process that could involve this kind of smooth muscle cell-derived protease.

*Robert:* That's a very good suggestion, because if the giraffe imitates humans and up-regulates its elastases with age, it may be cutting these connections progressively.

*Kimani:* We were intrigued initially to find that the proximal part of the common carotid artery, which in the adult giraffe has a luminal elastic and outer muscular structure, is all elastic in the neonate. We could only assume then, as Dr Davidson has pointed out, that the muscularization of the outer part of the carotid artery in the giraffe is a postnatal remodelling process. Subsequent studies in our laboratory (J. Kimani, unpublished results) have shown that in a three-month-old fetus, one finds a thin zone luminally with a definite layering of the smooth muscle and elastic lamellae separated from the tunica adventitia by a zone of undifferentiated tissue. At the time of birth, this latter layer is not well defined, but one may assume that the postnatal remodelling process occurs by way of appositional growth and differentiation of bundles of smooth muscle at the media–adventitia junction.

*Winlove:* I wanted to ask a very simple physical question: if when the giraffe has its head down its blood pressure is 350 mm Hg, how does the blood ever get back to the heart?

*Kimani:* The jugular veins in the giraffe have a series of valves and a well-defined muscular cuff at the entrance of the superior vena cava to the heart (Goertz & Keen 1957). These valves face the heart, therefore preventing backflow in the head-down position. Considering that the animal has a thick skin and fasciae, the tone of the neck musculature could work in a manner similar to the 'cuff-pump' effect in the lower extremities.

It is interesting to consider the expected impact of a downward rush of blood in the caroid arteries when the giraffe lowers its head to drink. In our earlier studies (Kimani 1979), we suggested that this is accompanied by an immediate vasoconstriction of the cranial vessels leading to the brain. Our subsequent studies (Kimani et al 1991) revealed a very rich adrenergic innervation of these vessels, particularly those that feed the carotid rete mirabile. The carotid rete is, however, poorly innervated, in contrast to the moderate innervation of the intracranial vessels arising from it. We believe that the vessels feeding the carotid rete may demonstrate a localized vasoconstrictor effect when the animal bends to drink, in a manner similar to the diving reflex in the seal (White et al 1973).

*Robert:* The cranial vessels must have a very efficient coupling with the peripheral vasoreceptors, otherwise it would be very difficult to control the blood flow to the brain so quickly.

*Mungai:* In the giraffe fetus we see development of elastic tissue during gestation, and muscle develops postnatally. Now this is in a blood vessel. That would be similar to the tropoelastin expression in the rat during gestation where the vessels have a lot of development of elastic fibres. What is the relationship between this development of elastin and the actual force of the heartbeat (or, in the case of the lungs, breathing) when the young is born?

*Pierce:* It is more or less dogma that during vascular development, increases in blood pressure contribute to the increase in elastin production within, for instance, the aortic wall. There are a number of lines of evidence that indicate that during cases of hypertension—either pulmonary hypertension or systemic hypertension—this is often accompanied by either sustained high production of tropoelastin in the early neonate or increased elastin deposition in more mature animals. I don't believe anyone has done a careful study of the relationship between pressure in the lung and elastin deposition: it would be difficult to do studies where you don't allow animals to breathe! But there are a number of *in vitro* models. Jeff Davidson has some data on stretching and production of tropoelastin in cultured cells.

*Robert:* Although during development, when the embryo lays down the smooth muscle cell layers, the pressure is not very high, so other regulatory elements must be involved. But in general you might be right, the higher the blood pressure, the more there could be elastin deposition. But that probably must be regulated genetically before blood pressure rises in the embryo.

*Kimani:* There could be an interplay between both genetic and epigenetic factors. Genetic programming could determine that the definitive number of elastic layers are laid down in the tunica media, while epigenetic factors might influence the relative thickness of the individual elastic lamellae.

*Keeley:* I would agree that there are pressure and wall stress factors that drive increases in aortic elastin production during development. But developmental programmes are also important in determining these patterns of synthesis.

*Robert:* Years ago, an interesting study was carried out in London, which almost went unnoticed, by Looker & Berry (1972). They quantitated the elastic lamellae in developing rat aorta, and showed very nicely that concentric continuous lamellae are laid down very early on during development. Later on, there is a continuous synthesis of fibres which stretch in between the concentric lamellae but no more lamellae are laid down. When they introduced a lesion it was not repaired. The same is true for the discontinuities of the elastic laminae of the artery of the rat tail as shown by the interesting experiments of Mary Osborn (see Osborn-Pellegrin et al 1990). For the concentric lamellae this must somehow be genetically programmed, even if the blood pressure has some role to play. The concentric lamellae are synthesized during development in the womb, and their number seems to be genetically regulated.

*Kimani:* That is an interesting suggestion. There is some agreement that postnatal development of the tunica media in the elastic arteries consists of increases in the amounts of collagen and increases in thickness and branching of elastic lamellae (Cliff 1967). Considerable controversy exists, however, as to whether the number of elastic lamellae actually increases. Wolinsky & Glagov (1967) suggested that mammalian aortae that have 29 or fewer elastic lamellae at birth retain this number at maturity regardless of the increase in the overall thickness of the tunica media. On the other hand, mammalian aortae with more than 29 elastic lamellae at maturity already have this number at birth with further addition in the outer vascularized zone (Wolinsky & Glagov 1967).

*Robert:* I wouldn't be surprised if it turns out that different proteins are present in continuous lamellae and in the one-dimensional fibres.

*Davidson:* From cell culture studies it appears that this tendency to form lamellae is almost an intrinsic property of certain vascular cells. It might not be an environmentally or pressure-driven process, but instead may represent the embryonic differentiated aspect of vascular wall biology. Pressure or wall stress is then an important factor that modulates either the thickness of the lamellae or the number of populations that are recruited.

*Mecham:* There are some interesting differences, though, in the development of the pulmonary and systemic circulation. Elaine Davis in my laboratory has been looking at the organization of blood vessels early in embryonic development. She found that the internal elastic lamella forms first in both systemic and pulmonary vessels, and that the endothelial cell is an active player in its organization. Later, the smooth muscle cells in the medial layer of the vessel wall organize themselves and start to make elastic lamellae. But the internal elastic lamella is formed first under all circumstances.

*Robert:* It was demonstrated some time ago that the vascular endothelial and smooth muscle cells have a number of embryonal origins: ectoderm, the neural crest and mesoderm (Le Livre & Le Douarin 1975, Dieterlen-Livre et al 1993). Smooth muscle cells from each of these sources might have a completely different expression of their elastin genes. The guiding structures might just be the membranes of the smooth muscle cells, which align during development of the blood vessels. The problem is how they align and how many layers of the smooth muscle cells there are. The remodelling could come later. This doesn't exclude the interaction of different proteins during the deposition of these two-dimensional elastic lamellae. Differences in alternative exon usage might be one of the factors, but this has yet to be demonstrated.

*Kimani:* That is indeed an intruiging problem. What, for example, are the factors that determine the phenotypic expression of lamellae in one case and fibres in another within the tunica media of blood vessels?

*Keeley:* Other proteins may confer at least partial elastomeric properties on these tissues. For example, the lamprey aorta does not contain any elastin, just structures which look like microfibrils. However, this animal has a systolic blood pressure of about 40 mm Hg, so there must be some flexibility in the vessel. It may be that other proteins can supply a sufficient level of extensibility in tissues subjected to only relatively low levels of stress.

## References

Cliff WJ 1967 The aortic tunica media in growing rats studied with the electron microscope. Lab Invest 17:599–615

Dieterlen-Livre F, Paradanaud L, Godin I, Garcia-Porrer J, Cumano A, Marcos M 1993 Developmental relationships between hemopoiesis and vasculogenesis. C R Acad Sci 316:897–901

Goertz RH, Keen EN 1957 Some aspects of the cardiovascular system in the giraffe. Angiology 8:542–564

Kimani JK 1979 Some aspects of the structural organization of the carotid arterial system of the giraffe (*Giraffa camelopardalis*) with special reference to the carotid baroreceptor equivalent. PhD thesis, University of Nairobi, Nairobi, Kenya

Kimani JK, Opole IO, Ogeng'o JA 1991 Structure and sympathetic innervation of the intracranial arteries in the giraffe (*Giraffa camelopardalis*). J Morphol 208:193–203

Le Livre CS, Le Douarin NM 1975 Mesenchymal derivatives of the neural crest: analysis of chimaeric quail and chick embryos. J Embryol Exp Morphol 34:125–154

Looker T, Berry C 1972 The growth and development of the rat aorta II. Changes in nucleic acid and seroprotein content. J Anat (Lond) 113:17–34

Oakes BW, Bialkower B 1977 Biomechanical and ultrastructural studies on elastic wing tendon from domestic fowl. J Anat 123:369–387

Osborn-Pellegrin M, Farjanel J, Hornebeck W 1990 Role of elastase and lysyl oxidase activity in spontaneous rupture of internal elastic plasma lamina in rats. Atherosclerosis 10:1136

Rees PM 1967 The distribution of biogenic amines in the carotid bifurcation region. J Physiol 193:245–253

Rees PM 1968 Electron microscopical observations on the architecture of the carotid arterial walls with reference to the sinus portion. J Anat 103:35–45

Reis DJ, Fuxe K 1968 Adrenergic innervation of the carotid sinus. Am J Physiol 215:1054–1057

Stanton PD, Hinrichsen CFL 1980 Monoaminergic innervation of the carotid sinus in mammals: a histochemical survey. Acta Anat 108:293–303
White FN, Ikeda M, Elsner RW 1973 Adrenergic innervation of large arteries in the seal. Comp Gen Pharmacol 4:271–276
Wolinsky H, Glagov S 1967 A lamellar unit of aortic medial structure in mammals. Circ Res 20:99–111

# Skin elastic fibres: regulation of human elastin promoter activity in transgenic mice

Jouni Uitto, Sylvia Hsu-Wong, Stacy D. Katchman, Muhammad M. Bashir* and Joel Rosenbloom*

*Departments of Dermatology, and Biochemistry and Molecular Biology, Jefferson Medical College, and Section of Molecular Dermatology, Jefferson Institute of Molecular Medicine, Philadelphia, PA 19107 and *Department of Histology and Anatomy, University of Pennsylvania School of Dentistry, Philadelphia, PA 19104, USA*

*Abstract.* Elastic fibres form an extracellular network which provides elasticity and resilience to tissues such as the skin. To study the regulation of human elastin gene expression, we have developed a line of transgenic mice which harbour 5.2 kb of human elastin gene promoter region in their genome. This promoter is linked to the chloramphenicol acetyltransferase (CAT) reporter gene which allows determination of the expression of human elastin promoter in different tissues. The highest CAT activity was found in the lungs and aorta, tissues rich in elastin, while lower levels were detected in a variety of other tissues, including skin. Assay of CAT activity in the lungs of fetal and newborn animals revealed high activity which progressively declined during the postnatal period up to six months. Thus, there was evidence of tissue-specific and developmentally regulated expression of the human elastin promoter activity in these mice. These animals were then used to examine the expression of the elastin gene by a variety of factors which have previously shown to alter elastin gene expression, as determined at the mRNA or protein levels. First, injection of transforming growth factor $\beta 1$ (100 ng) subcutaneously into the transgenic animals resulted in a time-dependent elevation of the promoter activity up to 10-fold after a single injection. Secondly, enhancement of the human elastin promoter activity by interleukin $1\beta$ injected subcutaneously resulted in an approximately 10-fold elevation of the CAT activity. Finally, subcutaneous injection of these animals with triamcinolone acetonide or dexamethasone, two glucocorticosteroids in clinical use, resulted in marked enhancement of human elastin promoter activity. Similar changes were noted in fibroblast cultures established from the transgenic animals. These data indicate that the 5.2 kb upstream segment of the human elastin gene contains *cis*-elements which allow tissue-specific and developmentally regulated expression of the human elastin promoter. Furthermore, this segment of the gene contains responsive elements to a variety of cytokines and pharmacological agents. Collectively, these data indicate that elastin gene expression in the skin *in vivo* can be regulated at the transcriptional level.

*1995 The molecular biology and pathology of elastic tissues. Wiley, Chichester (Ciba Foundation Symposium 192) p 237–258*

## The biology of elastic fibres

Elastic fibres form an extracellular matrix network that is responsible for physiological elasticity of various connective tissues. The elastic fibres are most abundant in the aorta and other vascular connective tissues, as well as in the lungs. In normal human skin, the elastic fibres represent a relatively small fraction of the total dermal proteins (approximately 2–4%), while collagen is the predominant component. Nevertheless, several lines of evidence indicate that elastic fibres are critical for the physiological resilience of normal human skin. (For reviews on elastic fibres, see Uitto et al 1991a, Uitto & Christiano 1993, Christiano & Uitto 1994.)

Early electron microscopic examination of connective tissues demonstrated that elastic fibres consist of two morphologically distinct components. The major component has an amorphous appearance without distinct periodic structures. This component biochemically represents elastin, a well-characterized connective tissue protein (Rosenbloom et al 1991a,b, Uitto et al 1991b). The primary sequence of elastin has been deduced from full-length cDNA clones (Indik et al 1987, Fazio et al 1988a). The amorphous elastin core, when visualized by transmission electron microscopy, is surrounded by distinct fibrillar structures which have a regular diameter of 10 to 12 nm. These structures represent the so-called elastin-associated microfibrils (Cleary 1987). The microfibrils consist, at least in part, of fibrillins, which comprise a family of large glycoproteins, two of which have been extensively characterized (Sakai et al 1986). These two microfibrillar proteins, fibrillin 1 and fibrillin 2, are distinct products of genes localized to chromosomes 15 and 5, respectively. In addition to fibrillin, other microfibrillar components have been characterized (Cleary 1987, Gibson et al 1989). In particular, the human microfibril-associated protein 1 (MFAP1) has been recently cloned and shown to be structurally distinct from the fibrillins (Yeh et al 1994). Interestingly, however, the gene encoding MFAP1 has been localized to the same region of the long arm of chromosome 15 as the fibrillin gene, *FBN1* (Magensis et al 1991).

The major component of mature elastic fibres, elastin, is a well characterized connective tissue protein which is initially synthesized as a ~70 kDa polypeptide, also known as tropoelastin (Foster et al 1973). The primary sequence of elastin is encoded by a 3.5 kb mRNA, which has been detected by Northern hybridizations in a variety of cell types, including aortic smooth muscle cells and dermal fibroblasts (Olsen et al 1988, Fazio et al 1988b, Sephel et al 1989). Subsequent to translation of the elastin mRNA, several post-translational events take place. These include: (a) hydroxylation of certain prolyl residues to form *trans*-4-hydroxy-L-proline; (b) secretion of elastin polypeptides into the extracellular milieu packaged in Golgi vesicles; (c) association of the elastin polypeptides with the microfibrillar component; (d) oxidative deamination of certain lysyl residues to form allysines in a reaction catalysed by lysyl oxidase, a copper-dependent enzyme; and (e) stabilization of the elastic fibres by formation of desmosines, stable cross-link compounds derived from three allysines and an unmodified lysyl residue (Rosenbloom et al 1991a,b, Uitto et al 1991a). Thus, the pathway leading to formation of elastic fibres is a complex multistep process which involves delicate regulation of the expression of several extracellular matrix genes, coordinate translation of the corresponding mRNAs, and assembly, stabilization and maturation of the fibre structures. On the other hand, the elastic fibres are subject to degradation by elastases, a group of proteolytic enzymes. Some elastases are released from inflammatory cells, such as polymorphonuclear leukocytes, while others are secreted by connective tissue cells, including dermal fibroblasts (Banda & Senior 1990). The activities of these enzymes can be inhibited by a variety of inhibitor molecules, including $\alpha_1$-antitrypsin and $\alpha_2$-macroglobulin. Thus, the steady-state levels of functional elastic fibres in the skin are dependent on the balance between the multistep biosynthetic fibrillogenesis and the degradative pathways reflecting the presence of active proteolytic enzymes.

## Pathology of elastic fibres in cutaneous diseases

Considering the complexity of elastin fibrillogenesis and the presence of protease/antiprotease systems in the skin, distinct levels can be identified at which errors could be introduced into the structure and/or quantity of the elastic fibres. Abnormalities manifesting as functional deficiencies could occur in the structure of elastin or the microfibrillar proteins, in the activity of enzymes involved in the synthesis and degradation of these molecules, and in the balance between the fibrillogenesis and degradation resulting in excessive accumulation or loss and fragmentation of elastic fibres. In fact, histopathological and ultrastructural observations of elastic fibres in the skin have allowed identification of a variety of pathological conditions demonstrating abnormalities in the elastic structures. In some of these cases, specific defects in the biochemistry or molecular biology of elastin and the microfibrillar proteins have been identified (Uitto et al 1991b, Christiano & Uitto 1994).

## The human elastin gene: structure, chromosomal location and regulation of expression

As indicated above, the elastin polypeptides are encoded by an mRNA of approximately 3.5 kb, which consists of about 2.2 kb of coding segment and a relatively large, 3′-untranslated region of about 1.3 kb. The human elastin gene contains 34 exons spanning a total of about 45 kb of genomic DNA (Bashir et al 1989, Rosenbloom et al 1991b). This gene has been mapped by chromosomal *in situ* hybridizations to the long arm of chromosome 7, within the q11.1–21.1 region (Fazio et al 1991). This chromosomal assignment was supported by hybridizations with a panel of human–rodent cell hybrids which showed concordance with human chromosome 7, and excluded other human chromosomes.

The regulation of elastin gene expression has been demonstrated to take place at several different levels. Initial data suggested that a major control mechanism for elastin gene expression resides at the post-transcriptional level through mechanisms which influence the stability of elastin mRNA. However, recent cloning of the entire human elastin gene, including 5.2 kb of the 5′-flanking DNA, has provided information to suggest that transcriptional regulation of the elastin gene expression at the promoter level may be a significant component of the overall regulatory process (Bashir et al 1989, Fazio et al 1990, Kähäri et al 1990). Sequencing of 2.6 kb of the promoter region of the human elastin gene has revealed several interesting features. These include the lack of a canonical TATA box and a functional CAAT motif. This may explain the presence of multiple transcription initiation sites at the 5′ end of the human elastin gene (Bashir et al 1989). In addition to multiple SP-1 and AP-2 binding sites, we have identified several additional putative regulatory *cis*-acting elements (Kähäri et al 1990). These include three putative glucocorticoid responsive elements, two cAMP responsive elements and two 12-*O*-tetradecanoylphorbol-13-acetate (TPA) inducible elements.

The expression of the elastin gene in a variety of cell cultures, including aortic smooth muscle cells and dermal fibroblasts, has been shown to be modulated by a variety of factors, including transforming growth factor (TGF)-$\beta$, tumour necrosis factor (TNF)-$\alpha$, interleukin (IL)-1$\beta$, IL-10, insulin-like growth factor 1, various glucocorticosteroids, vitamin $D_3$ and cyclic nucleotides (see below). It is conceivable, therefore, that some of the putative *cis*-elements identified in the human elastin promoter region mediate the effects of these cytokine and hormone factors.

## Development of transgenic mice expressing human elastin promoter

To provide a test system to study the transcriptional regulation of human elastin gene expression at the promoter level, we have recently developed a line of transgenic mice which have incorporated a human elastin promoter/chloramphenicol acetyltransferase (CAT) reporter gene construct into their genome (Hsu-Wong et al 1994). Specifically, a 5.2 kb fragment of the 5′-flanking DNA of the human elastin gene, extending from $-5200$ to $+2$ (in relation to the ATG translation initiation codon), was ligated to a 0.7 kb CAT reporter gene, followed by 0.3 kb of DNA containing the polyadenylation signal (Fig. 1). The linearized construct, 6.2 kb in size, was injected into

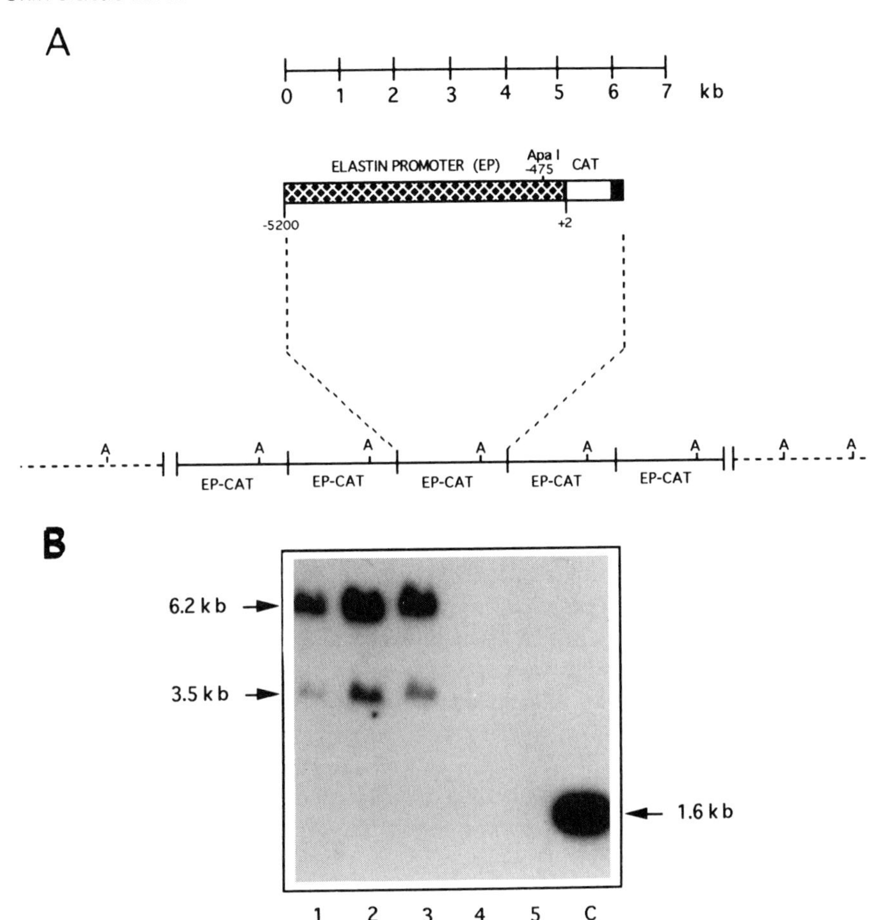


FIG. 1. Development of transgenic mice expressing a human elastin promoter–CAT construct. (A) A construct containing human elastin gene segment extending from positions −5200 to +2 (▩) was linked to the 0.7 kb fragment of DNA containing the entire coding region of CAT (□), and the 3′ end of the construct contained a 0.3 kb fragment containing the polyadenylation signal (■). Thus, the total length of the linearized construct was 6.2 kb. Note that the construct contains within the elastin promoter a single restriction enzyme site for *Apa*I endonuclease at position −475. The lower portion schematically depicts the potential integration of multiple copies of the elastin promoter–CAT construct in tandem head-to-tail arrangement. Digestion of the DNA with *Apa*I would yield multiple copies of the 6.2 kb fragment, as well as a single copy of a fragment, the size of which is dependent on the presence of *Apa*I site downstream from the integration site. (B) Southern blot analyses on three positive mice (lanes 1–3) and two negative littermates (lanes 4–5), the former ones demonstrating a strong band of 5.2 kb in size when hybridized with a 1.6 kb full-length CAT DNA. A smaller, ~3.5 kb band representing the 3′-most fragment of the tandem repeat construct, as shown in (A), was also detected. The 1.6 kb CAT DNA was used as a positive control (lane C). (Reproduced with permission from Hsu-Wong et al 1994.)

fertilized oocytes of FVB/N mice, and the eggs were implanted into pseudo-pregnant foster mothers. The offspring ($F_0$) were tested for the presence of the human elastin promoter by Southern hybridizations using DNA isolated from tail snips of the mice at three weeks of age. DNA was digested with *Apa*I restriction endonuclease, and the digestion fragments were fractionated on 0.7% agarose gels, transferred to nylon filters and hybridized with a 1.6 kb pCAT DNA. Since the elastin promoter–CAT construct contains a single *Apa*I restriction enzyme site, fragments of 6.2 kb were expected in the positive mice (Fig. 1). In addition, depending on the site of integration, a single fragment of a different size could be detected, reflecting the position of the adjacent downstream *Apa*I site.

Southern hybridization of DNA isolated from tails of pups representing five litters, initially identified one positive mouse with the characteristic band of 6.2 kb. As shown in Fig. 1, the hybridization also revealed a smaller band of 3.5 kb. However, the intensity of the 6.2 kb band was significantly stronger than that of the 3.5 kb band, suggesting that multiple copies of the transgene had integrated into the mouse genome. In fact, scanning densitometry revealed a ratio of 7:1 for the 6.2 kb and 3.5 kb bands. The positive mouse ($F_0$) was then mated with wild-type mice, and several offspring ($F_1$) positive for this 6.2 kb transgene were identified (Hsu-Wong et al 1994).

The $F_1$ generation mice were also tested for expression of the human elastin promoter activity, as reflected by CAT activity, determined by incubation of aliquots of tissue extracts with radioactive chloramphenicol. Unequivocal expression of CAT activity was detected in several organs of the $F_1$ mice (Fig. 2). The $F_1$ mice were then mated with other positive mice, and the homozygous animals strongly expressed the human elastin promoter–reporter gene construct in a variety of tissues.

## Evidence for tissue-specific and developmentally regulated expression of the human elastin promoter

The homozygous line of transgenic mice was subsequently used to examine the tissue-specific and developmentally regulated expression of the human elastin promoter activity (Hsu-Wong et al 1994). For this purpose, different organs were isolated from mice of varying ages, and CAT activity was determined. The initial analyses were performed with tissues isolated from mice of approximately 2 months of age, a stage when the organs are well developed enough to allow their precise dissection. The highest level of expression was noted in the lungs and aorta, two organs known to be rich in elastic fibres. Lower, yet clearly detectable levels of CAT activity were also noted in the skin, kidneys and brain (Fig. 2). This pattern of CAT expression was reproducible among a total of 25 individual mice representing eight separate litters. An

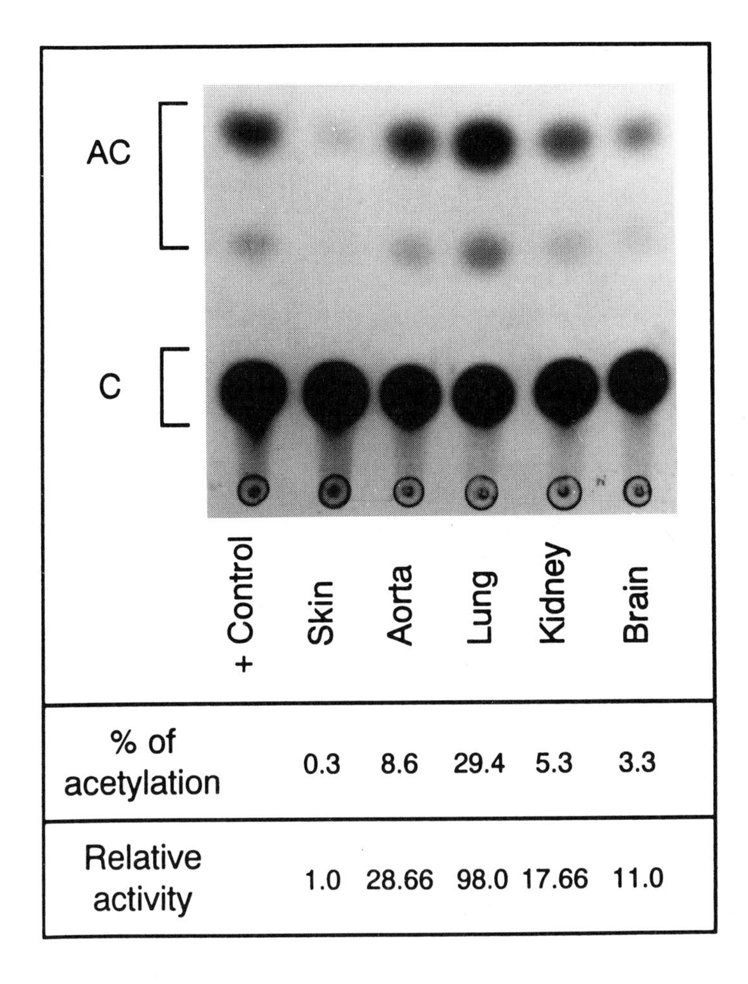


| | Skin | Aorta | Lung | Kidney | Brain |
|---|---|---|---|---|---|
| % of acetylation | 0.3 | 8.6 | 29.4 | 5.3 | 3.3 |
| Relative activity | 1.0 | 28.66 | 98.0 | 17.66 | 11.0 |

FIG. 2. Expression of human elastin promoter in different tissues of transgenic mice, as determined by assay of CAT activity. Different tissues were dissected from a 2-month-old transgenic mouse and CAT activity was determined by incubation of tissue extracts containing 100 $\mu$g of protein, with [$^{14}$C]chloramphenicol, as described elsewhere (Hsu-Wong et al 1994). The figure is an autoradiograph depicting the separation of the acetylated forms (AC) from the non-acetylated form (C) of chloramphenicol. The percentage of acetylated forms of total radioactivity is shown below the figure, and the relative activity in different organs, in relation to the skin (1.0), is also shown. The positive control (+) was an extract of a cell transiently transfected with the $SV_2$–CAT promoter construct. (Reproduced with permission from Hsu-Wong et al 1994).

exception to this consistent tissue-specific expression pattern was the brain, which in some animals demonstrated extremely high CAT activity. Although the reason for this is not clear, the variable activity of CAT in the brains of different animals may reflect the accuracy of dissection of the brain tissue from animals of different ages.

To examine the age-dependent alterations in human elastin promoter activity, we measured CAT activity in organs isolated from animals of different ages. First, assay of CAT activity of the lungs demonstrated an age-dependent decrease from 5 days to 6 months of postnatal age (Fig. 3). In fact, the levels noted in the lungs of 6-month-old animals were only about 1.6% of the values noted in the lungs of mice that were 5 days of age. Furthermore, comparison of the activity in the lungs of the 5-day-old mice with the activity in fetal lung at the 15th day of gestation indicated that the activity in the fetal lungs was about fourfold higher. These results suggest that there is a relatively rapid decline in the transcriptional expression of the human elastin gene in the lungs of transgenic animals as a function of age.

Similar experiments were performed in the skin and the aorta at various stages of postnatal development. The results indicated that the values in the aorta remained relatively constant throughout development from 5 days to 3 months, and then gradually declined. The values in the skin initially increased from 5 days to 3 months by up to 10-fold, but the values then returned to the 5-day level. CAT activity in different preparations from heart and kidney did not demonstrate a consistent trend as a function of age during the postnatal period.

### Persistent expression of the transgene in cultured cells

To extend the transgenic *in vivo* model to *in vitro*, we established cell cultures from the skin and aorta of the transgenic mice and examined them for elastin promoter expression using CAT as the reporter gene (Hsu-Wong et al 1994). The cell cultures were established by standard procedures and the subcultures were examined at passages 2–3. Examination of cultures established from the skin indicated that cells were elongated and spindle shaped, characteristic of fibroblasts. The cells derived from the aorta were fusiform in configuration, with hill-and-valley morphology, suggesting that they were smooth muscle cells. Assay of CAT activity in cultures consisting of either type of cell revealed clearly detectable activity. Cells from control animals without the transgene did not reveal any CAT activity. Thus, the cells established from the transgenic animals retained their capacity to express the human elastin promoter–reporter gene construct in culture. In conclusion, we have developed a transgenic mouse line that expresses the human elastin promoter in a tissue-specific and developmentally regulated manner (Hsu-Wong et al 1994). These mice, or cells derived from these animals, provide a model system to examine the regulation of elastin gene expression at the promoter level *in vivo* and *in vitro*.

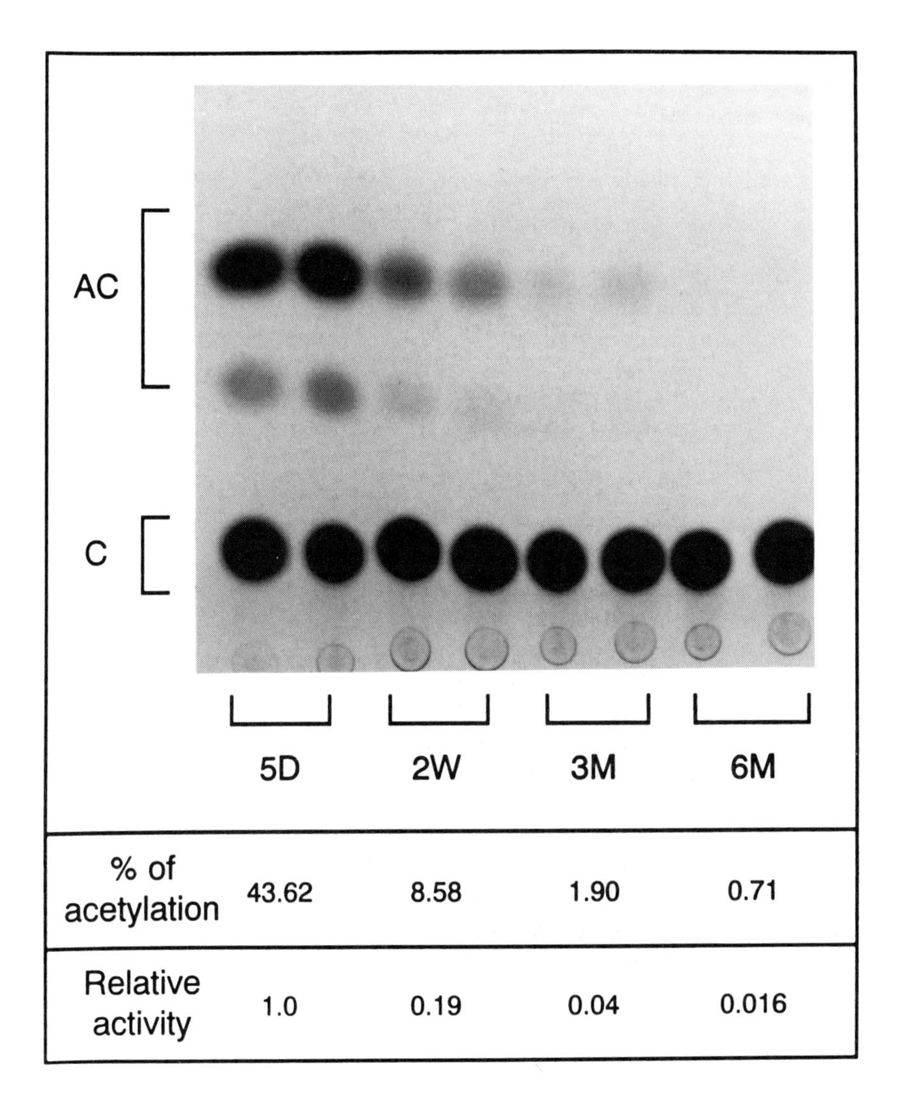


FIG. 3. Demonstration of age-dependent decline in the elastin promoter activity in the lungs of transgenic mice at five days (5D), two weeks (2W), and three and six months (3M and 6M, respectively) of age. Elastin promoter activity was determined by CAT assay as described in Fig. 2. The percentage of acetylation of radioactive chloramphenicol at each point is the mean of two parallel determinations, and the activity in relation to the 5-day-old mice (1.0) is also shown. (Reproduced with permission from Hsu-Wong et al 1994.)

**Cytokine modulation of human elastin promoter activity in transgenic mice**

As indicated above, various cytokines and hormones have been suggested to modulate elastin gene expression, yet the precise level of modulation has not been determined in most cases. One of the growth factors that has been previously shown to alter elastin gene expression is TGF-$\beta$, which was initially shown to enhance elastin gene expression in porcine aortic smooth muscle cells, as determined at the protein level by ELISA (Liu & Davidson 1989). Subsequent studies demonstrated that treatment of human skin fibroblasts in culture with TGF-$\beta$ elevated elastin mRNA steady-state levels in a dose- and time-dependent manner (Kähäri et al 1992a). In the latter study, additional experiments using transient cell transfections in fibroblast cultures with elastin promoter–CAT reporter gene constructs demonstrated essentially no effect by TGF-$\beta$ on elastin promoter activity, while mRNA stability assays suggested that the half-life of elastin mRNAs was prolonged in the presence of TGF-$\beta$. These observations were interpreted to suggest that the TGF-$\beta$ effect was primarily post-transcriptional.

To examine possible transcriptional regulation of elastin gene expression by TGF-$\beta$, we used the transgenic mice which were shown to express human elastin promoter in a tissue-specific and developmentally regulated manner as a model system to examine the effects of TGF-$\beta$1 (Katchman et al 1994). In the initial studies, TGF-$\beta$1 was injected subcutaneously into 5-day-old mice, and the treated areas of the skin were biopsied at different time points, up to 72 hours. Assay of CAT activity in the TGF-$\beta$1-treated mice demonstrated a marked (up to 10-fold) increase in the elastin promoter activity (Fig. 4). The maximum activity was noted at 24 hours, with the values levelling off between 24 and 48 hours and significantly decreasing at the 72 h time point (Katchman et al 1994). These observations clearly suggested that TGF-$\beta$1 is capable of up-regulating human elastin promoter activity in the skin *in vivo*.

We have extended these studies to other cytokines, namely, TNF-$\alpha$, IL-1$\beta$ and IL-10. These cytokines play an important modulatory role in a variety of inflammatory processes. In the first study, we also examined the modulation of elastin gene expression by TNF-$\alpha$. In contrast to TGF-$\beta$, this cytokine markedly suppressed the elastin mRNA levels in a time- and dose-dependent manner (Kähäri et al 1992b). TNF-$\alpha$ also suppressed the expression of the elastin promoter–CAT construct in transiently transfected cultured human skin fibroblasts and rat aortic smooth muscle cells, indicating regulation at the transcriptional level. This suppression of the promoter activity was temporally preceded by rapid and transient up-regulation of c-*jun* and c-*fos* genes. Several lines of evidence, including co-transfections with synthetic double-stranded AP-1 oligomer, 5′ deletion constructs, and gel mobility shift assays implicated an AP-1 *cis*-element in the position $-229$ to $-223$ within the human elastin promoter (Kähäri et al 1992b).

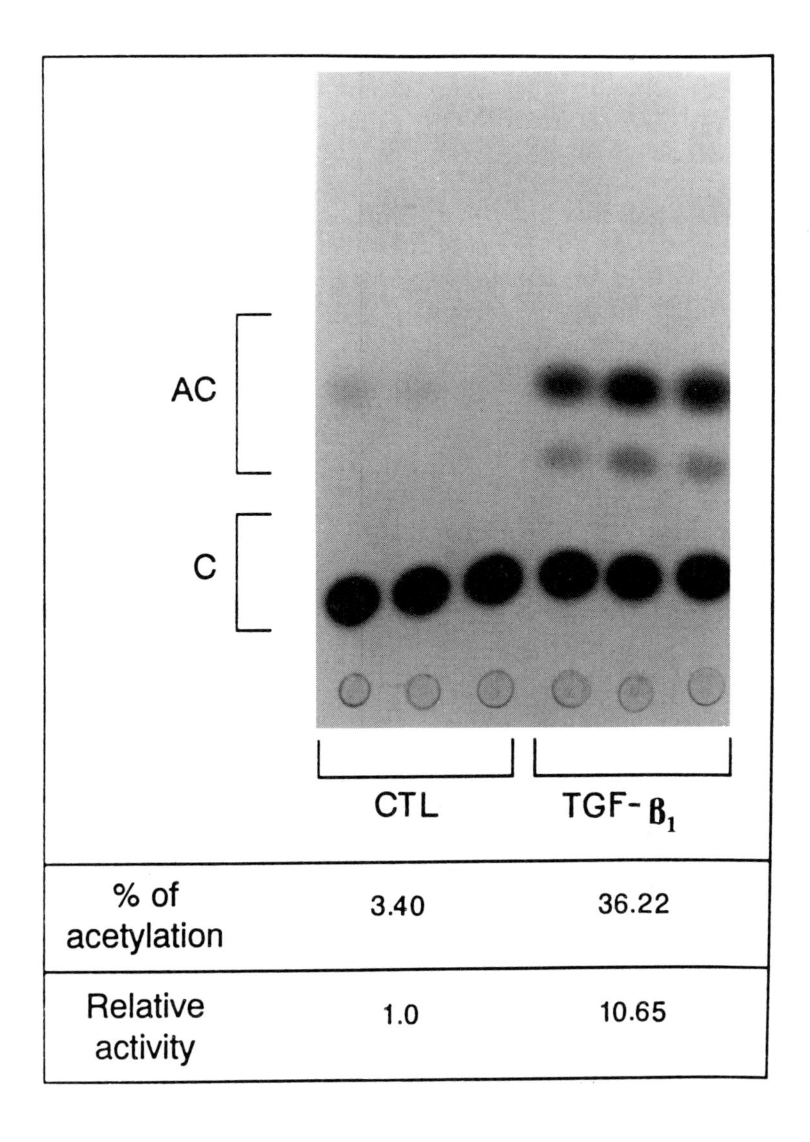


| | CTL | TGF-B1 |
|---|---|---|
| % of acetylation | 3.40 | 36.22 |
| Relative activity | 1.0 | 10.65 |

FIG. 4. Demonstration that TGF-$\beta$1 increases human elastin promoter activity in the skin of transgenic mice, as determined by the CAT assay. A single dose of TGF-$\beta$1 (100 ng) was injected subcutaneously, and the control animals (CTL) received saline. After 24 h, a full-thickness 6 mm biopsy of the skin at the site of injection was obtained, and CAT activity was determined, as described in Fig. 2. The percentage of acetylation of radioactive chloramphenicol is expressed as the mean of three individual animals injected in parallel in both groups, each animal representing the same litter of transgenic mice. (Reproduced with permission from Katchman et al 1994.)

The effects of human recombinant IL-1$\beta$ on elastin gene expression were studied in human skin fibroblast cultures by Northern hybridization and transient transfection experiments (Mauviel et al 1993). Incubation of the cells with IL-1$\beta$ elevated elastin mRNA levels by three- to fourfold, and similar increases were noted at the protein level when estimated semiquantitatively by indirect immunofluorescence of cultured cells. Transient transfections of the dermal fibroblasts with a human elastin promoter–CAT reporter gene construct suggested transcriptional regulation, since the CAT activity in cells incubated with IL-1$\beta$ was similarly increased by about threefold. Enhancement of human elastin promoter activity by IL-1$\beta$ was also noted in fibroblast cultures established from the skin and lungs of the transgenic mice described above. Furthermore, subcutaneous injection of IL-1$\beta$ to these mice resulted in an approximately fourfold elevation of the CAT activity in the skin assayed after 30 h, as compared with the CAT activity in the skin of control animals (Mauviel et al 1993). Thus, these data indicated that IL-1$\beta$ up-regulates elastin gene expression in fibroblast cultures as well as in the skin *in vivo*, and the activation occurs at the transcriptional level.

A similar study was performed utilizing recombinant human IL-10 at physiological concentrations (Reitamo et al 1994). In this study, subcutaneous injection of IL-10 (1 to 100 ng) resulted in an up to 3.5-fold increase in CAT activity 24 hours after injection. When human dermal fibroblasts were studied by Northern analyses, elastin gene expression was markedly enhanced by IL-10, and incubation of IL-10, together with TGF-$\beta$2, synergistically enhanced the elastin mRNA levels. Again, this study suggested that IL-10 has an up-regulatory effect on elastin gene expression, both *in vivo* and *in vitro*.

Collectively, these studies indicate that various inflammatory mediators—including TGF-$\beta$, TNF-$\alpha$, IL-1$\beta$ and IL-10—all modulate human elastin gene expression, as determined at the mRNA steady-state and protein levels. Furthermore, TGF-$\beta$, IL-1$\beta$ and IL-10 exert their up-regulatory effects, at least in part, through activation of the human elastin promoter. These observations may have implications for our understanding of the pathological changes in elastin noted in various inflammatory diseases.

## Glucocorticosteroids up-regulate human elastin gene promoter activity in transgenic mice

As indicated above, sequencing of 2.6 kb of the 5′-flanking DNA of the human elastin gene revealed the presence of several putative regulatory *cis*-elements (Kähäri et al 1990). These included three putative glucocorticoid responsive elements (GREs), with a partial sequence of TGTTCC, at positions −[1437 to 1433], −[1315 to 1311] and −[1023 to 1019] (counting upstream from the translation initiation site). Previous studies utilizing animals and cell cultures as models to study elastin gene expression have suggested that these GREs may be functional. Specifically, glucocorticoids have been shown to stimulate elastin synthesis in chick embryo aorta and in differentiated bovine fetal ligamentum nuchae fibroblasts in culture, as determined at the protein and mRNA levels (Keeley & Johnson 1987, Mecham et al 1984).

To examine further the effects of glucocorticosteroids on elastin gene expression at the transcriptional level, we have recently utilized the transgenic mice that express the human elastin promoter–reporter gene construct (Ledo

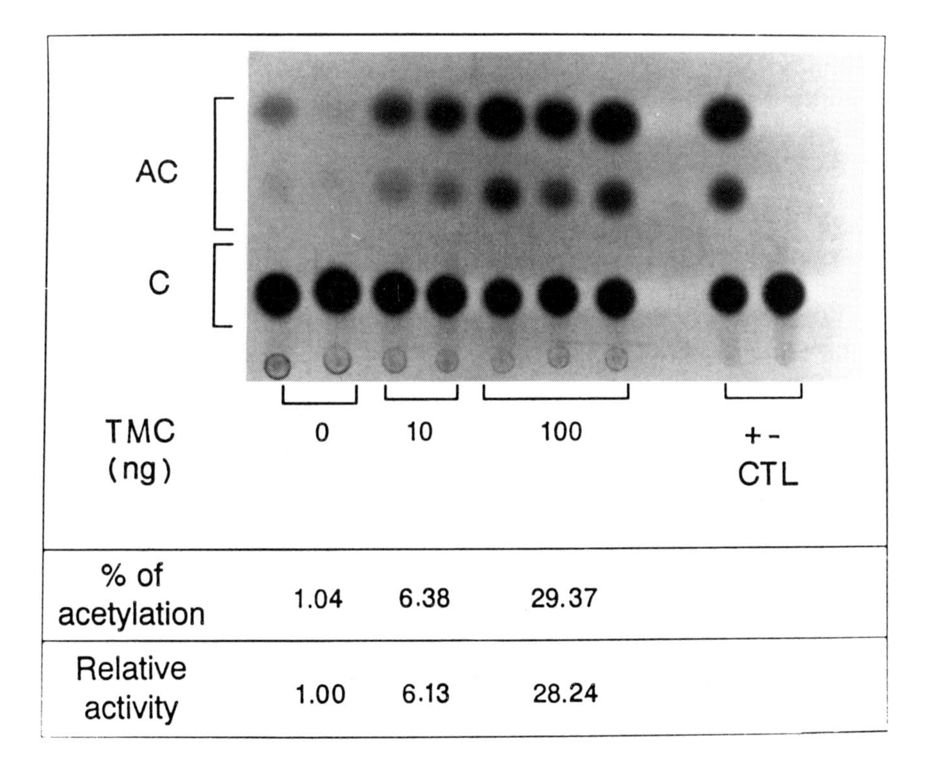


| | 0 | 10 | 100 |
|---|---|---|---|
| % of acetylation | 1.04 | 6.38 | 29.37 |
| Relative activity | 1.00 | 6.13 | 28.24 |

FIG. 5. Up-regulation of human elastin promoter activity in the skin of transgenic mice injected subcutaneously with triamcinolone acetonide (TMC). Five-day-old transgenic mice expressing the human elastin promoter-CAT construct were injected with 10 or 100 ng of TMC subcutaneously, and CAT activity was determined 24 h later as described in Fig. 4. The controls (CTL) included an extract of lung of a homozygous untreated animal as a positive control (+), and the negative control (−) consisted of the radioactive substrate incubated with an extract from a non-transgenic wild-type mouse. The relative activity in animals receiving 10 or 100 ng TMC was calculated in relation to control animals which received saline alone (1.0). Each lane represents an assay from a separate animal within a litter of homozygous transgenic animals. (Reproduced with permission from Ledo et al 1994.)

et al 1994). In the initial study, two glucocorticoids in clinical use, triamcinolone acetonide (TMC) and dexamethasone (DEX), were tested. These two glucocorticosteroids were injected subcutaneously into 5-day-old transgenic mice. Injection of as little as 10 ng of TMC resulted in a sixfold increase in the promoter activity, and injection of 100 ng of TMC increased the activity by 28-fold at the site of injection (Fig. 5). Similar injection of 100 ng of DEX resulted in a 6.6-fold increase of CAT activity, whereas 1 μg resulted in a 21.3-fold increase. Time-course experiments indicated that the enhancement of CAT activity by TMC was maximal 12–24 hours after the initial injection of 100 ng of TMC. A subsequent decline in the activity was noted, but values at 72 h were still 4.6-fold higher than those noted in the control skin.

To assess the effects of the same glucocorticosteroids on the elastin promoter activity in cell cultures, we incubated dermal fibroblasts derived from the skin of the transgenic mice with varying concentrations of TMC (Ledo et al 1994). As indicated above, these cells retained their capacity to express the human elastin promoter, as determined by CAT activity in passage 2 or 3 (Hsu-Wong et al 1994). Incubation of these cells with TMC in concentrations from 0 to 10 ng/ml for 24 h resulted in a dose-dependent increase in the CAT activity, and the relative activity in cultures incubated with 10 ng/ml of TMC showed a marked, up to 73-fold increase, as compared with control cultures incubated in parallel without TMC. Similar incubations of fibroblasts with DEX demonstrated dose-dependent enhancement of human elastin promoter activity. However, higher concentrations of DEX were again needed to achieve a similar degree of up-regulation.

Collectively, these data indicate that glucocorticosteroids clearly enhance elastin gene expression at the promoter level. Enhancement of promoter activity is probably mediated by the putative GREs identified in the human elastin promoter region. It should be noted, however, that the three putative GREs identified in the human elastin promoter region do not entirely conform with the proposed consensus sequence for GRE, GGTACAnnnTGTTCT (Beato 1989). In fact, all three GREs identified in the elastin gene promoter depict the sequence TGTTCC, a variant with the downstream half-site of the consensus GRE, whereas the upstream sequences have a lower degree of homology. However, experimental evidence from a variety of mammalian genes has suggested that the proposed consensus sequence for GRE may not be universal (Petersen et al 1988, Su & Pitot 1992). It is conceivable, therefore, that the glucocorticosteroids up-regulate the elastin promoter through binding of the ligand–receptor complexes to GRE.

An interesting observation in these studies was that TMC is consistently about a 10-fold more powerful up-regulator of elastin gene expression than DEX, when the same concentrations of these two steroids are tested in parallel. This difference could be explained by differential tissue availability of these glucocorticosteroid preparations, TMC being a suspension of particles, whereas DEX is in soluble form. These observations suggest the possibility that the transgenic mouse line used in this study may provide a biological *in vivo* model to test the relative pharmacological potency of various glucocorticosteroid preparations, using a GRE-containing promoter (the human elastin gene) as a target for ligand–receptor-mediated action. In fact, recent preliminary data by our group have indicated that topical application of high-potency glucocorticosteroids up-regulates human elastin promoter activity in the skin of the transgenic mice to a considerably higher degree than relatively weak steroids, such as 1% hydrocortisone (S. D. Katchman, J. Uitto, unpublished results). Thus, this animal model may provide a novel biological *in vivo* test system to determine the relative efficacies of topically applied corticosteroids in clinical use.

*Acknowledgements*

We thank numerous colleagues who contributed to the original studies cited in this overview. Tamara Alexander provided expert secretarial help. This study was supported in part by USPHS, NIH grants RO1-AR28450, PO1-AR20553 and T32-AR07561, and by the Dermatology Foundation.

## References

Banda MJ, Senior RM 1990 Elastin degradation mediated by the inflammatory processes. In: Tamburro AM, Davidson JM (eds) Elastin: chemical and biological aspects. Congedo Editore, Galatina, Italy, p 277–291

Bashir MM, Indik Z, Yeh H et al 1989 Characterization of the complete human elastin gene: delineation of unusual features in the 5′ flanking region. J Biol Chem 264:8887–8891

Beato M 1989 Gene regulation by steroid hormones. Cell 56:335–344

Christiano AM, Uitto J 1994 Molecular pathology of the elastic fibers. J Invest Dermatol 103:53S–57S

Cleary EG 1987 The microfibrillar component of the elastic fibers. In: Uitto J, Perejda AJ (eds) Connective tissue diseases: the molecular pathology of the extracellular matrix. Marcel Dekker, New York, p 55–81

Fazio MJ, Olsen DR, Kauh EA et al 1988a Cloning of full-length elastin cDNAs from a human skin fibroblast recombinant cDNA library: further elucidation of alternative splicing utilizing exon-specific oligonucleotides. J Invest Dermatol 91:458–464

Fazio MJ, Olsen DR, Kuivaniemi H et al 1988b Isolation and characterization of human elastin cDNAs, and age-associated variation in elastin gene expression in cultured skin fibroblasts. Lab Invest 58:270–277

Fazio MJ, Kähäri VM, Bashir MM, Saitta B, Rosenbloom J, Uitto J 1990 Regulation of elastin gene expression: evidence for functional promoter activity in the 5′-flanking region of the human gene. J Invest Dermatol 94:191–196

Fazio MJ, Mattei M-G, Passage E et al 1991 Human elastin gene: new evidence for localization to the long arm of chromosome 7. Am J Hum Genet 48:696–703

Foster JA, Bruenger E, Gray WR, Sandberg LB 1973 Isolation and amino acid sequences of tropoelastin peptides. J Biol Chem 248:2876–2879

Gibson MA, Kumaratilake JS, Cleary EG 1989 The protein components of the 12-nanometer microfibrils of elastic and nonelastic tissues. J Biol Chem 264:4590–4598

Hsu-Wong S, Katchman S, Ledo I et al 1994 Tissue-specific and developmentally regulated expression of human elastin promoter activity in transgenic mice. J Biol Chem 269:18072–18075

Indik Z, Yeh H, Ornstein-Goldstein N et al 1987 Alternative splicing of human elastin mRNA indicated by sequence analysis of cloned genomic and complementary DNA. Proc Natl Acad Sci USA 84:5680–5684

Kähäri V-M, Fazio MJ, Chen YQ, Bashir MM, Rosenbloom J, Uitto J 1990 Deletion analyses of 5′-flanking region of the human elastin gene: evidence for a functional promoter and regulatory *cis*-elements. J Biol Chem 265:9485–9490

Kähäri V-M, Olsen DR, Rhudy RW, Carrillo P, Chen YQ, Uitto J 1992a Transforming growth factor-β upregulates elastin gene expression in human skin fibroblasts: evidence for post-transcriptional modulation. Lab Invest 66:580–588

Kähäri V-M, Chen YQ, Bashir MM, Rosenbloom J, Uitto J 1992b Tumor necrosis factor-α downregulates human elastin gene expression. Evidence for the role of AP-1 in the suppression of promoter activity. J Biol Chem 267:26134–26141

Katchman SD, Hsu-Wong S, Ledo I, Wu M, Uitto J 1994 Transforming growth factor-β up-regulates human elastin promoter activity in transgenic mice. Biochem Biophys Res Commun 203:485–490

Keeley FW, Johnson DJ 1987 Age differences in the effect of hydrocortisone on the synthesis of insoluble elastin in aortic tissue of growing chicks. Connect Tissue Res 16:259–268

Ledo I, Wu M, Katchman S et al 1994 Glucocorticosteroids upregulate human elastin gene promoter activity in transgenic mice. J Invest Dermatol 103:632–636

Liu J, Davidson JM 1989 The elastogenic effect of recombinant transforming growth factor-beta on porcine aortic smooth muscle cells. Biochem Biophys Res Commun 154:895–901

Magensis RE, Maslen CL, Smith L, Allen L, Sakai LY 1991 Localization of the fibrillin (FBN) gene to chromosome 15, band q21. Genomics 11:346–351

Mauviel A, Chen YQ, Kähäri V-M et al 1993 Human recombinant interleukin-1β upregulates elastin gene expression in dermal fibroblasts. Evidence for transcriptional regulation both *in vitro* and *in vivo*. J Biol Chem 268:6520–6524

Mecham RP, Morris SL, Levy BD, Wrenn DS 1984 Glucocorticoids stimulate elastin production in differentiated bovine ligament fibroblasts but do not induce elastin synthesis in undifferentiated cells. J Biol Chem 259:12414–12418

Olsen DR, Fazio MJ, Shamban AT, Rosenbloom J, Uitto J 1988 Cutis laxa: reduced elastin gene expression in skin fibroblast cultures as determined by hybridizations with a homologous cDNA and an exon 1-specific oligonucleotide. J Biol Chem 263:6465–6467

Petersen DD, Magnuson MA, Granner DK 1988 Location and characterization of two widely separated glucocorticoid responsive elements in the phosphoenolpyruvate carboxykinase gene. Mol Cell Biol 8:96–104

Reitamo S, Remitz A, Tamai K, Ledo I, Uitto J 1994 Interleukin-10 upregulates elastin gene expression *in vivo* and *in vitro* at the transcriptional level. Biochem J 302:331–333

Rosenbloom J, Bashir M, Yeh H et al 1991a Regulation of elastin gene expression. Ann NY Acad Sci 624:116–136
Rosenbloom J, Bashir MM, Kähäri V-M, Uitto J 1991b Elastin genes and regulation of their expression. Crit Rev Eukaryotic Gene Expression 1:145–156
Sakai LY, Keene DR, Engvall E 1986 Fibrillin, a new 350-kD glycoprotein, is a component of extracellular microfibrils. J Cell Biol 103:2499–2509
Sephel GC, Byers PH, Holbrook KA, Davidson JM 1989 Heterogeneity of elastin expression in cutis laxa fibroblast strains. J Invest Dermatol 93:147–153
Su Y, Pitot HC 1992 Location and characterization of multiple glucocorticoid-responsive elements in the rat serine dehydratase gene. Arch Biochem Biophys 297:239–243
Uitto J, Christiano AM 1993 Elastic fibers. In: Fitzpatrick TB, Eisen AZ, Wolff K, Freedberg IM, Austen KF (eds) Dermatology in general medicine, 4th edn. McGraw-Hill, New York, p 339–349
Uitto J, Fazio MJ, Bashir M, Rosenbloom J 1991a Elastic fibers of the connective tissue. In: Goldsmith LA (ed) Physiology, biochemistry, and molecular biology of the skin, Vol I, 2nd edn. Oxford University Press, New York, p 530–557
Uitto J, Christiano AM, Kähäri V-M, Bashir MM, Rosenbloom J 1991b Molecular biology and pathology of human elastin. Biochem Soc Trans 19:824–829
Yeh H, Chow M, Abrams WR et al 1994 Structure of the human gene encoding the associated microfibrillar protein (MFAP1) and localization to chromosome 15q15-q21. Genomics 23:443–449

## DISCUSSION

*Robert:* Have you used your model system to look at the effect of retinoic acid on elastin gene expression?

*Uitto:* Yes, in fact we tested this by a twice-daily topical application of 0.1% all-*trans*-retinoic acid cream to our transgenic mice for up to two weeks. There was no change in the CAT activity, suggesting that retinoic acid may not have an effect on elastin synthesis, at least at the promoter level.

*Robert:* That's what we found, also. Retinoic acid might increase collagen synthesis, but the claims that it causes elastin to reappear have not been substantiated.

Have you ever looked at the effects of sex steroids on this system?

*Uitto:* We have a whole range of experiments in progress addressing the effects of different vitamins and hormones. These will be published in due course, but we do not have any definitive results at this point.

*Sandberg:* It could be argued from a pharmacological point of view that it is not desirable to have steroids which increase elastin production. Wouldn't you rather want ones that decrease elastin production?

*Uitto:* Not necessarily. The desired mode of steroid effect, enhancement versus down-regulation, obviously depends on the pathology of the disease one is attempting to treat. Nevertheless, from the pharmacological point of view, the major effects of the steroids in dermatological use relate to their anti-inflammatory and immunosuppressive properties. The question we addressed in our study was simply whether we could use these transgenic mice, which express the human elastin promoter–CAT construct, as a biological assay system to measure the strength of different steroids, and whether their performance in this assay correlates with other pharmacological parameters.

*Sandberg:* Do you feel there's a good correlation between the effect of steroids in your assay and with other pharmacological parameters?

*Uitto:* We have tested several topical glucocorticoid preparations, and there appears to be a fairly good correlation between their commercially established strength based on their ability to suppress skin erythema and the enhancement of CAT activity in the skin of our transgenic animals. We have also looked into issues relating to the pharmaceutical formulation of topical steroids. Some of the steroids clearly showed differences in their CAT-inducing activity depending on the type of base they had been formulated in. One of the limitations of the transgenic mouse system, though, is that the mouse skin is much thinner than human skin, and the topical steroids probably penetrate much more easily.

*Parks:* Rich Pierce and Fred Keeley have both shown in the developing rat lung and chick aorta that there is a window of time where elastin production is sensitive to steroid treatment. It is mostly during fetal development. How old were the mice you tested?

*Uitto:* For topical application, we usually test in 5-day-old mice, i.e. before the animals grow hair. However, we have clearly shown that intraperitoneal injection of the transgenic mice with DEX at the age of six months markedly enhances the expression of the human elastin promoter in the lungs and kidneys (Ledo et al 1994). It is clear that this response is not limited to fetal or neonatal mice.

*Davidson:* I find some of this information a little paradoxical, especially with regard to the skin behaviour. When glucocorticoids are applied to adult dermis they generally produce an atrophic rather than a trophic response in connective tissue, which includes collagen and elastin. In our forthcoming paper (Russell et al 1995), we show that DEX down-regulates elastin mRNA levels.

*Uitto:* This apparent discrepancy could well be explained by the fact that the physiological concentrations of glucocorticoids tested in our cell culture studies are probably several orders of magnitude smaller than the pharmacological levels used for treatment of skin diseases. For example, we observed in our transgenic cell culture system almost 50-fold enhancement of the elastin promoter–CAT activity, with as little as 1 ng/ml triamcinolone acetonide, while the concentration of triamcinolone acetonide normally used for intralesional injection of keloids is 5–40 mg/ml. Such high pharmacological concentrations of glucocorticosteroids have been shown to suppress collagen synthesis in the skin, resulting in dermal atrophy (Oikarinen et al 1986) and they could possible inhibit elastin synthesis as well.

*Davidson:* Are you suggesting that there might be a biphasic response?

*Uitto:* Yes, that is quite possible.

*Davidson:* In all of these data, it would be useful to know what's going on with the endogenous elastin gene. Do you have any data showing that the behaviour of the elastin promoter–reporter gene construct closely parallels that of the target tissue?

*Uitto:* Not in these mice, although we are in the process of setting up a PCR-based assay to test the expression of the endogenous gene in parallel.

*Davidson:* My third question regards the differential response of smooth muscle versus fibroblasts to TGF-$\beta$ *in vitro*: smooth muscle cells are responsive, but fibroblasts are not. In skin, however, you report that injection of TGF-$\beta$ strongly up-regulates CAT activity. What cell type is expressing this activity?

*Uitto:* We don't know which cells in the skin of these transgenic animals are responsive to TGF-$\beta$. However, if we transiently transfect cultured human skin fibroblasts with a type I collagen–CAT reporter gene construct and incubate these cultures with TGF-$\beta$, the collagen promoter activity goes up (Kähäri et al 1990). It is clear, therefore, that dermal fibroblasts do have the receptors and the binding proteins necessary for transcriptional activation of the collagen promoter. Why these same cells do not respond to TGF-$\beta$ with regard to the elastin promoter is not clear.

*Boyd:* If you're going to continue working on modulating the promoter activity of those cultured cells from a single founder animal, it's very important that you should include cell cultures from other founders. The only conclusion that you could finally draw in pharmacological modulation of elastin synthesis has to be taken from cultured lines from multiple founders. While what you have reported here is very seductive in terms of suggestions relating directly to the elastin gene promoter, if the site of integration of the transgene is next to an enhancer core of another gene it is possible that all you are measuring is the enhancer activity or any other *cis*-acting sequence that is unrelated to the promoter itself.

*Uitto:* We are well aware of this, and we are in the process of developing additional transgenic mice, not only with this 5.2 kb promoter construct but also with several 5′-deletion promoter constructs. With these mice, we hope to nail down the elements responsive to TGF-$\beta$ and glucocorticosteroids, as well as the elements underlying the tissue-specific and developmentally regulated expression of the elastin promoter–reporter gene construct in these mice.

*Parks:* Have you ever looked at circulating cells like peritoneal macrophages or perhaps even skin keratinocytes, or any cell type that does not make elastin, to see if there is tissue specificity conferred by the promoter?

*Uitto:* No, we have not. The only cells that we have examined in culture are aortic smooth muscle cells and dermal fibroblasts.

*Robert:* An unrelated question: have you looked at skin from patients with Werner's syndrome? The elastic fibres in Werner's skin are horizontal and totally inelastic. With age, also, as the density of the elastic fibres goes up, elasticity goes down (Robert et al 1988a). Do you have any thoughts about this?

*Uitto:* My group has not studied Werner's syndrome from the point of view of elastin, although we have examined collagen metabolism in these patients in the past (Bauer et al 1988). Maybe Jeff Davidson can address your question.

*Davidson:* We've looked at fibroblast populations from patients with Werner's syndrome, and they don't exhibit any abnormalities *in vitro* in terms of elastin production. Another premature ageing syndrome, Hutchinson–Gilford progeria, does exhibit that phenotype (Sephel et al 1988, Giro & Davidson 1993). Werner's syndrome fibroblasts have been characterized as having abnormal regulation of collagenase expression (Bauer et al 1986), which seems to be the more important connective tissue marker.

*Robert:* It is a paradox that although there is an important increase in skin elastin in Werner's syndrome and in sun-exposed and ageing skin, there is a decrease of elasticity.

*Uitto:* Regarding the issue of skin ageing, you have to separate two distinct biological processes. One is the innate, chronological ageing and the other relates to photodamage; they are clearly two different processes. In innate ageing, there is degradation of normal functional elastic fibres. Elastic fibres are lost, and we know that the ability to repair elastic fibres in elderly skin is diminished. In sun-exposed areas of skin, both these phenomena are taking place at the same time. In solar elastosis there is an accumulation of elastotic material, which we call elastotic simply because it stains positively with elastic stains. However, this material seems to be a mixture of elastin, and the microfibrillar components, decorin, fibrillin, fibronectin and type III collagen, among others. Even if there is elastin, it is not organized into normal fibrillar structures; rather it's a globular mass of elastotic material without functional elasticity.

*Robert:* The studies we carried out concern upper arm sun-protected areas; by two different morphometric procedures we demonstrated an age-dependent increase in elastic fibre density, which we confirmed by immunohistochemistry (Robert et al 1988a, Frances et al 1990). We showed an age-dependent decrease of elasticity (Robert et al 1988b). At any site we found only an increase of elastic fibre density by morphometry and only a decrease of elasticity. Werner's syndrome is an extreme example. So the paradox is still there. We don't yet have a good molecular explanation for why in these situations there is more elastin yet less elasticity.

*Tamburro:* According to my results, when you have a high concentration of soluble elastins the dominant aggregating structures are not fibres but globular, grape-like aggregates. These are similar to those Jouni Uitto has found in elastoderma (Kornberg et al 1985), where there is an accumulation of newly synthesized elastin.

*Robert:* In the upper arm skin we didn't see globules, but, rather, regular elastic fibres. The only change we saw was a change in the direction of alignment of the fibres; they were less vertical and more horizontal. But with our staining procedure and computerized image analysis we did not find a decrease of elastic fibres in ageing skin.

*Uitto:* It is clear that photodamage also involves alterations in other connective tissue components of the dermis, including collagen fibres and proteoglycan/glycosaminoglycan macromolecules, in addition to the elastic fibres (Uitto et al 1989). Thus, the overall architecture of the dermis is perturbed in photodamaged skin, resulting in age-associated degenerative changes including sagging and wrinkling of the skin.

*Robert:* During ageing, the whole skin gets thinner, with a 6% decrease per decade on average. Perhaps the whole problem has to do with the number of cells decreasing, accompanied by changes in their gene expression patterns. Some genes appear to be down-regulated; others, such as that for collagen type III (metalloendopeptidase), are clearly up-regulated (Labat-Robert & Robert 1988).

## References

Bauer EA, Silverman N, Busiek DF, Kronberger A, Deuel TF 1986 Diminished response of Werner's syndrome fibroblasts to growth factors PDGF and FGF. Science 234:1240–1243

Bauer EA, Uitto J, Tan EML, Holbrook KA 1988 Werner's syndrome: Evidence for a preferential regional expression of a generalized mesenchymal cell defect. Arch Dermatol 124:90–101

Frances C, Branchet MC, Boisnis S, Lesty CL, Robert L 1990 Elastic fibres in normal human skin: variations with age. A morphometric analysis. Arch Gerontol Geriatr 10:57–67

Giro MG, Davidson JM 1993 Familial co-segregation of the elastin phenotype in skin fibroblasts from Hutchinson-Gilford progeria. Mech Ageing Devel 70:163–176

Kähäri V-M, Chen YQ, Su MW, Ramirez F, Uitto J 1990 Tumour necrosis factor-$\alpha$ and interferon-$\gamma$ suppress the activation of human type I collagen by transforming growth factor-$\beta$1: evidence for two distinct mechanisms of inhibition at transcriptional and post-transcriptional level. J Clin Invest 86:1489–1495

Kornberg RL, Hendler S, Oikarinen AI, Matsuoka LY, Uitto J 1985 Elastoderma: disease of elastin accumulation within the skin. N Engl J Med 312:771–774

Labat-Robert J, Robert L 1988 Aging of the extracellular matrix and its pathology. Exp Gerontol 23:5–18

Ledo I, Wu M, Katchman SD et al 1994 Glucocorticosteroids up-regulate human elastin gene promoter activity in transgenic mice. J Invest Dermatol 103:632–636

Oikarinen AI, Uitto J, Oikarinen J 1986 Glucocorticoid action on connective tissue: from molecular technology to clinical practice. Med Biol 64:221–230

Robert C, Lesty C, Robert L 1988a Ageing of the skin: study of elastic fiber network. Modifications by computerized image analysis. Gerontology 34:91–62

Robert C, Blanc M, Lesty C, Dikstein S, Robert L 1988b Study of skin ageing as a function of social and professional conditions: modification of the rheological parameters. Measured with a non-invasive method—indentometry. Gerontology 34:84–90

Russell SB, Trupin JS, Kennedy RZ, Russell JR, Davidson JM 1995 Glucocorticoid regulation of elastin synthesis in human fibroblasts: down-regulation in fibroblasts from normal dermis but not from keloids. J Invest Dermatol 104:241–245

Sephel GC, Buckley A, Giro MG, Davidson JM 1988 Increased elastin production by progeria skin fibroblasts is controlled by the steady state levels of elastin mRNA. J Invest Dermatol 90:643–647

Uitto J, Fazio MJ, Olsen DR 1989 Molecular mechanisms of cutaneous aging: age-associated connective tissue alterations in the dermis. J Am Acad Dermatol (Suppl) 21:624–622

# Elastin in systemic and pulmonary hypertension

F. W. Keeley and L. A. Bartoszewicz

*Division of Cardiovascular Research, Hospital for Sick Children, 555 University Avenue, Toronto, Ontario M5G 1X8 and Departments of Biochemistry and Clinical Biochemistry, University of Toronto, Toronto, Canada*

*Abstract.* Increased elastin production and accumulation is a rapid and sensitive response to elevated vascular wall stress in both systemic and pulmonary hypertension. While initially protecting the vessel wall, these structural changes may in the longer term result in reinforcement of the hypertensive state and contribute to the persistence of the pathology of hypertension. Rapid responses apparently uncorrelated with increased elastin mRNA, at least in the case of systemic vessels, suggest novel mechanisms perhaps including increased efficiency of message translation or matrix accumulation of the protein. Investigations using *in vitro* organ and cell culture models have indicated a role for phospholipases and protein kinases, including protein kinase C, in stretch-induced elastin synthesis. In addition, tyrosine phosphorylation of membrane/sub-membrane/cytoskeletal sensors, including focal adhesion kinase and members of the lipocortin family, have been shown to be important in this transduction mechanism. Because its turnover is normally very slow, additional vascular elastin accumulated during hypertensive episodes, together with its consequences for the physical properties of the vessel wall, may persist long after blood pressure is restored to normal levels. Thus, recent interest has been drawn to the possibility of achieving regression of accumulated matrix elastin by promoting turnover of this protein through activation of endogenous vascular elastase and collagenase activities.

*1995 The molecular biology and pathology of elastic tissues. Wiley, Chichester (Ciba Foundation Symposium 192) p 259–278*

Remodelling of arterial vessels has been recognized for over a century as one of the important pathological consequences of both systemic and pulmonary hypertension. This restructuring process involves thickening of the vessel wall and includes accumulation of additional elastin, collagen and other matrix proteins. Elastin and collagen are the major constituents of the walls of large vessels, contributing up to 80% of their mass and accounting in large part for their passive physical properties. These physical properties are crucial to the physiological function of conduit vessels, allowing expansion during ventricular ejection at systole followed by recoil at diastole, propelling the

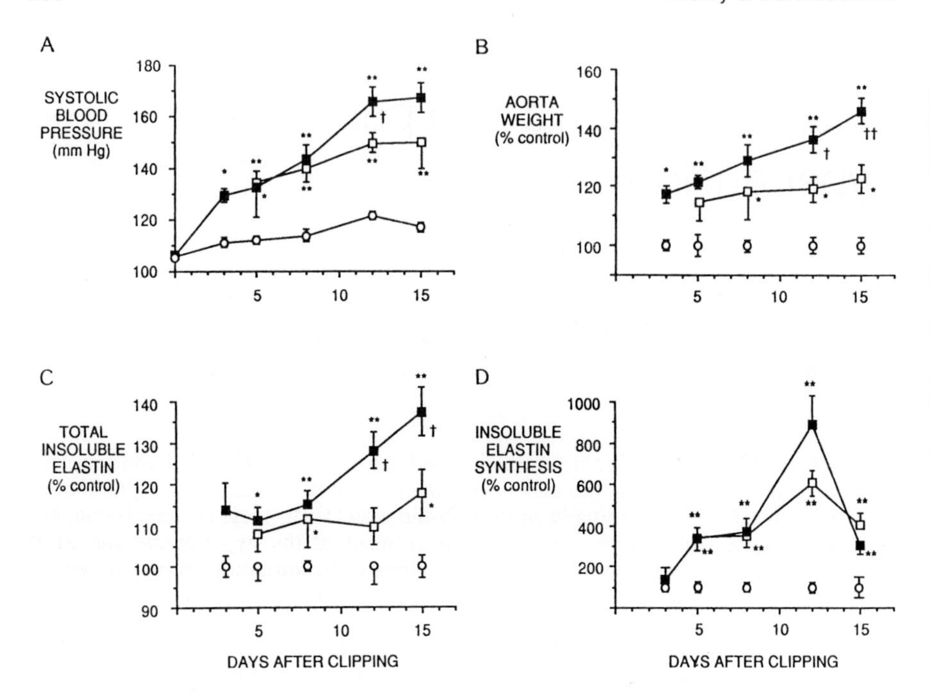


FIG. 1. Effect of developing renal clip hyptertension on (A) systolic blood pressure; (B) aorta segment weight; (C) total aortic insoluble elastin; and (D) insoluble elastin synthesis. Filled squares and open squares represent clips with slit widths of 0.2 mm and 0.25 mm, respectively. Open circles are sham-operated controls. Aorta weight, total elastin and elastin synthesis are expressed as percentage of age-matched controls. ** and * indicate $P < 0.01$ and $P < 0.05$, respectively, between treatment group and controls. †† and † indicate $P < 0.01$ and $P < 0.05$, respectively, between the two treatment groups. Reproduced with permission from Keeley & Alatawi (1991).

blood down the arterial tree and smoothing both pressure and flow waves. Even in smaller, pre-resistance and resistance vessels, in which the proportions of elastin and collagen are considerably lower, these matrix proteins make an important contribution to structural integrity, physical properties and physiological function. More than just a simple response to rising blood pressure, increased deposition of elastin and collagen during episodes of hypertension, because of the unusual stability of these proteins, may also have significance for the maintenance and progression of the hypertensive state. For these reasons it is important to understand not only the nature of the response of these vascular matrix proteins to a hypertensive stimulus, but also the mechanisms and consequences, both short and long term, of this response.

**TABLE 1 Proportions by weight of elastin and collagen in vascular tissues after 10 weeks of renal clip hypertension**

| | *Elastin* | | *Collagen* | |
|---|---|---|---|---|
| | *Control* | *Hypertensive* | *Control* | *Hypertensive* |
| Thoracic aorta | 16.6 ± 0.8 | 15.9 ± 0.6 | 11.7 ± 0.8 | 12.5 ± 0.7 |
| Abdominal aorta | 11.1 ± 0.7 | 10.7 ± 0.2 | 13.9 ± 0.5 | 14.9 ± 0.6 |
| Superior mesenteric artery | 7.8 ± 0.5 | 8.7 ± 0.4 | 12.6 ± 0.6 | 13.8 ± 0.7 |

Values are given as mg/100 mg wet weight of tissue and are means ± SEM, $n = 7$.

## Characteristics of the vascular response to increased blood pressure

According to Laplace's law, vascular wall stress is directly proportional to the distending pressure and the radius of the vessel, and inversely proportional to its wall thickness. Thus, increasing wall thickness during hypertension can be regarded as a means by which the vessel adapts to protect itself from the increased load to which it is subjected, returning its wall stress towards normal levels. Indeed, measurements of elastin synthesis and accumulation during the development of hypertension suggest that if the increase in blood pressure is limited, the response of increased elastin production is transient, returning to normal synthesis levels when sufficient new material has been added to the vessel wall (Fig. 1).

The increased mass of the arterial wall in hypertension consists of a proportional accumulation of all major extracellular matrix components, including both elastin and collagen (Cleary & Moont 1976). Thus, while absolute amounts of elastin and collagen in a given segment of vessel increase substantially (Wolinsky 1970, 1971, Keeley & Alatawi 1991), the proportions by weight of these proteins in the vessel wall generally remain unchanged (Table 1). Such a proportional response of matrix protein production appears to occur at all levels of the arterial tree (Cleary & Moont 1976) (Table 1). As a result, the response at a given location is a reflection of the normal composition of the vessel at that site.

While the elastin laid down in the vascular wall during normal development and growth is organized into concentric lamellae, additional elastin accumulated during hypertension appears to result neither in thickening of the existing lamellae nor in formation of new lamellae, but rather is deposited in interlamellar islands (Wolinsky 1971, Todorovich-Hunter et al 1988). This change in architecture may account at least in part for the fact that, in spite of unchanged proportions of elastin and collagen, physical properties of the vessels are affected, resulting in decreased distensibility (Cox 1981, Tozzi et al 1994).

Hypertrophy of smooth muscle cells also contributes substantially to increased vascular mass, especially in smaller vessels (Lee et al 1983). However, in general, at least at early stages of the hypertensive response and in the absence of injury, there appears to be no net increase in the number of smooth muscle cells in the vessel wall (Owens 1985, Keeley & Johnson 1986).

Although the hypertrophic response of cell and matrix elements in the vessel wall to increased blood pressure has been well known for some time, the rapidity and sensitivity of this response is less widely appreciated. For example, in *in vivo* models of both systemic and pulmonary hypertension, measurable hypertrophy with increases in accumulated vascular elastin and collagen occur within a few days and at elevations of systolic pressure of as little as 5 mm Hg in the pulmonary (Todorovich-Hunter et al 1988, Poiani et al 1990) and 15–20 mm Hg in the systemic circulation (Fig. 1) (Keeley & Alatawi 1991). *In vitro* models, in which increased elastin and collagen synthesis are induced by distension of explanted pulmonary or aortic tissue, are characterized by a similar rapid response, with elastin and collagen production measurably affected within a few hours of increases in vascular wall stress (Tozzi et al 1989, Keeley 1991, Belik et al 1994). The ability of these *in vitro* models to mimic the *in vivo* response also indicates that a pulsatile stimulus is not necessary, and that the capacity for the response is intrinsic to the tissue itself without a requirement for circulating factors or neurohumoral influences. However, it may be important to remember that essentially all of these *in vivo* and *in vitro* observations have been made in vessels from younger, growing animals in which matrix accumulation is still taking place at a rapid rate. These vessels may be particularly sensitive to wall stress effects as part of a normal developmental process. Similar responses, while certainly still present in vessels of older animals, may be relatively blunted.

## Mechanism of the response

One of the most interesting aspects of any consideration of vascular hypertrophy in hypertension is the question of the mechanism of the response, particularly in view of the evidence indicated above that this response is a local property of the tissue itself. What is the mechanism by which a physical force exerted on a matrix is transmitted to the cells within that matrix, relayed across the cell membrane, and transduced into an increase in protein synthesis within those cells? Effects of physical forces on cell metabolism are well known in a variety of tissues, but mechanisms underlying these effects are not yet well understood. Physical connections of elements of the extracellular matrix to cells have been described in morphological studies (Davis 1993a,b) and cell-surface receptors binding extracellular matrix proteins including elastin, collagen, fibronectin and laminin have been characterized (Ingber 1991, Hinek et al 1992, Juliano & Haskill 1993). In many cases these receptors not only are clearly linked to

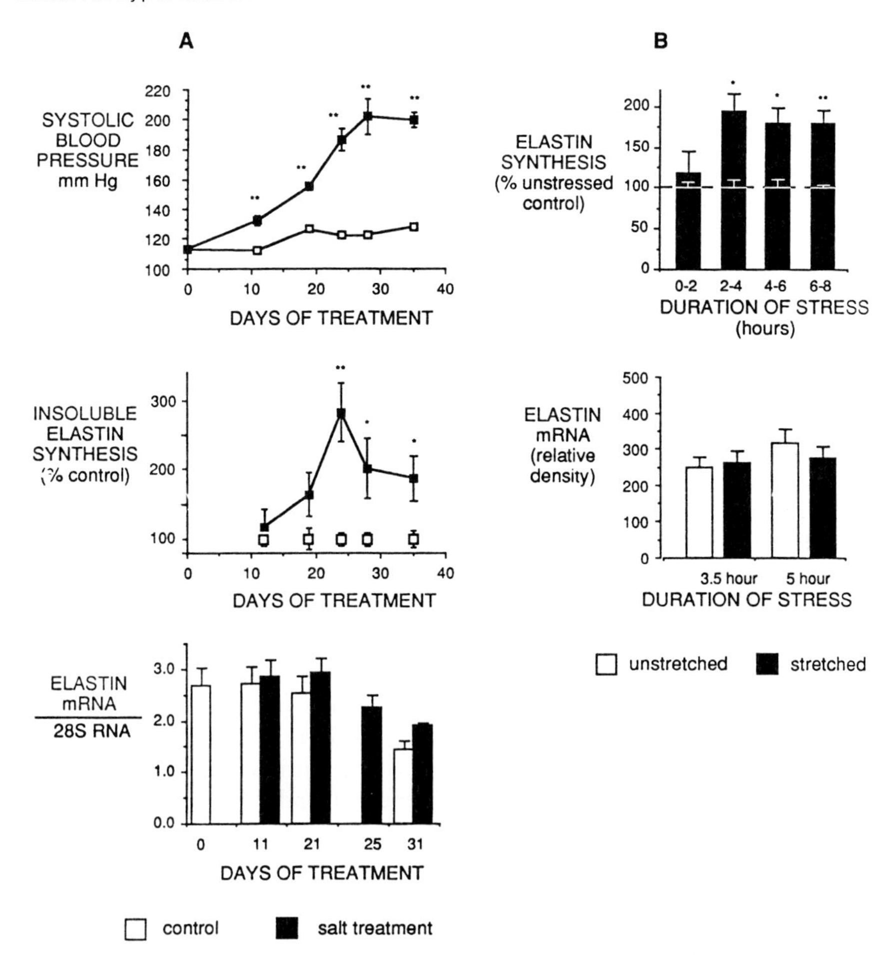


FIG. 2. Effect of *in vivo* hypertension or increased vascular wall stress *in vitro* on steady state mRNA levels for elastin. (A) Systolic blood pressure, insoluble elastin synthesis and elastin mRNA during the development of hypertension in a Dahl salt-sensitive rat model (Keeley & Alatawi 1991). Elastin mRNA was determined by Northern blotting and expressed relative to 18S RNA. (B) Synthesis of insoluble elastin and elastin mRNA levels in response to elevated wall stress by *in vitro* distension of aortic tissues. Elastin mRNA was determined by Northern blotting. Statistical significance between groups is represented as in Fig. 1. Reproduced with permission from Keeley et al (1993).

sub-membranous, cytoskeletal structures within cells, providing a plausible pathway for mechanical transmission of external stresses into the cytoplasm, but also activate other signal transduction cascades within the cell (Ingber 1991, Juliano & Haskill 1993). Some of the signal transduction pathways implicated in this response will be discussed in more detail below. For the moment, however, let us consider the final outcome of the process—increased production of matrix proteins—because evidence is mounting that mechanisms involved may not be identical in different tissues and under different situations.

There is evidence from *in situ* hybridization studies in *in vivo* models of pulmonary hypertension that increased vascular elastin and collagen production is accompanied by increased mRNAs for elastin and collagen in medial smooth muscle cells (Parks et al 1989, Prosser et al 1989). *In vitro* distension models have also reported increased steady-state levels of mRNA for collagen (Poiani et al 1990). On the other hand, our data (Fig. 2A) (Keeley et al 1993) and those of others (Takasaki et al 1990) have suggested that increased steady-state levels of mRNA for both elastin and collagen are not correlated with increased production and accumulation of aortic elastin and collagen in at least some *in vivo* models of systemic hypertension, although increased fibronectin synthesis is accompanied by elevated mRNA levels for that protein (Takasaki et al 1990). Similarly, increased elastin production in distended aortic tissue in organ culture (Keeley et al 1993) does not appear to be accompanied by increased mRNA levels for elastin (Fig. 2B). These data suggest that enhanced transcriptional activity may not be the only mechanism operating in at least some models of hypertension. Other differences between pulmonary and systemic hypertension models should also be noted. Distension models of pulmonary hypertension indicate that the connective tissue response is endothelium dependent (Tozzi et al 1989, Belik et al 1994). In contrast, in similar models of systemic hypertension using aortic tissues, this response is present even in the absence of endothelium (Keeley et al 1993). Whether these differences are species-, tissue-, or model-related, or reflect differences in characteristics of early and later responses to elevated wall stress, is not clear. Nevertheless, they serve as a caution against indiscriminate extrapolation of conclusions between pulmonary and systemic models of hypertension.

Our own studies of signal transduction mechanisms linking tissue and cell distension and increased elastin synthesis have implicated several pathways. As might be expected, mechanisms with relevance to the cytoskeleton appear to predominate. Earlier, using an *in vivo* model of systemic hypertension, we showed that elevated blood pressure could be dissociated from vascular hypertrophy and increased elastin and collagen deposition by pre-treatment of animals with doses of colchicine which did not impair either general body growth or normal development of the vascular system (Keeley & Alatawi 1991). Similar effects have been seen for both colchicine and cytochalasin D in *in vitro* distension models using aortic organ cultures, with stretch-induced insoluble elastin production abolished, but normal, background levels of synthesis of this protein unaffected. Stretch-activated $Ca^{2+}$ and $Na^{+}$ channels are not required for this distension-induced response, nor is inositol

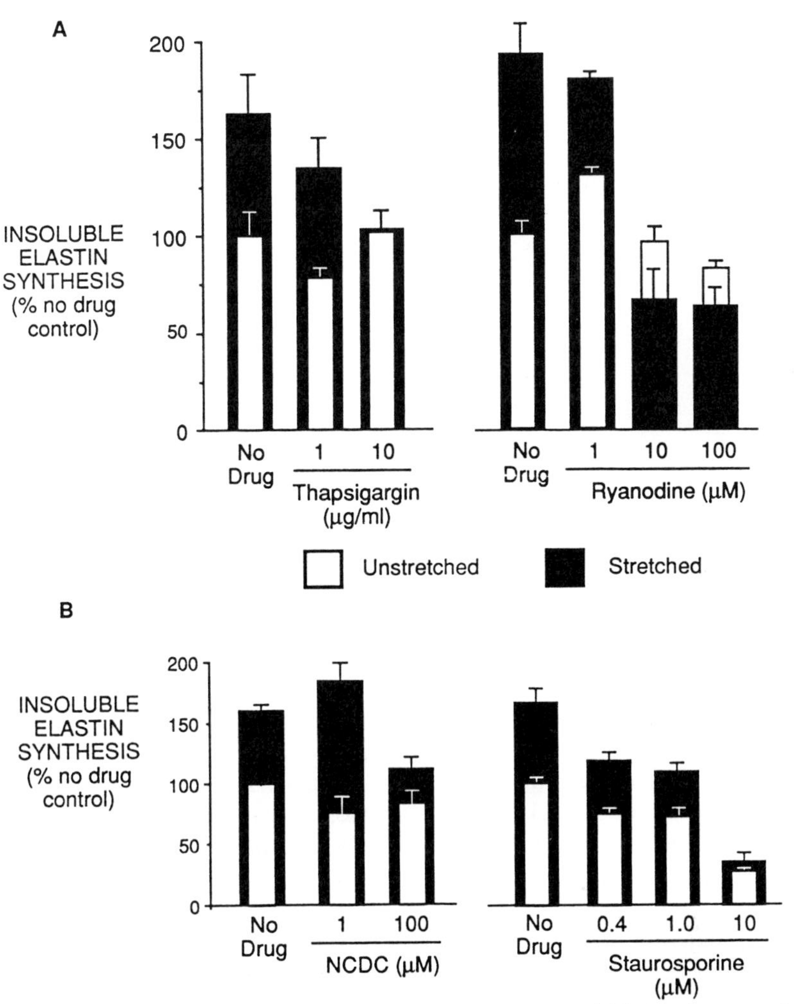


FIG. 3. Effect of (A) thapsigargin and ryanodine, agents which block re-uptake of $Ca^{2+}$ into the sarcoplasmic reticulum, and (B) NCDC (2-nitro-4-carboxyphenyl-*N*-diphenylcarbamate) and staurosporine, inhibitors of phospholipase C and protein kinase C activities, respectively, on stretch-induced insoluble elastin synthesis in an *in vitro*, organ culture model. Elastin synthesis is measured by incorporation of [$^{14}$C]-proline into CNBr-insoluble elastin (Keeley & Alatawi 1991).

trisphosphate-mediated release of $Ca^{2+}$ from intracellular stores. However, flooding of the cytoplasm with $Ca^{2+}$ by emptying of intracellular stores using ryanodine, thapsigargin, or caffeine selectively abolishes the stretch-induced response (Fig. 3A), perhaps through $Ca^{2+}$-mediated disruption of the integrity of the cytoskeleton. Inhibitors of phospholipase C and protein kinase C

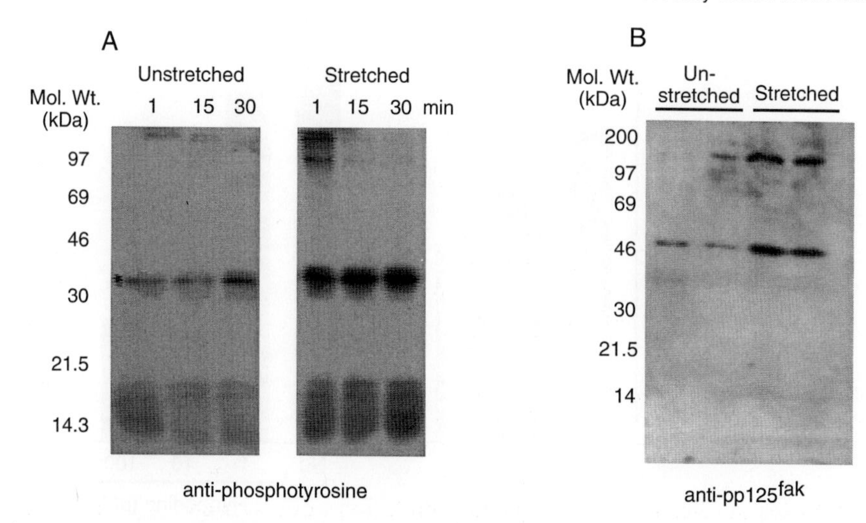


FIG. 4. Effect of stretch on tyrosine phosphorylation of proteins in aortic tissues. (A) Western blotting using an antibody to phosphotyrosine on SDS extracts of aortic tissues after *in vitro* stretching for various periods. (B) Western blotting using an antibody to pp125$^{fak}$ on SDS extracts of aortic tissues after *in vitro* stretching for 5 min.

activities also appear to abolish the stretch-induced response (Fig. 3B), implying a role for this signal transduction pathway.

Distension of aortic tissues results in a rapid increase in tyrosine phosphorylation of several cytoplasmic proteins, including a doublet at a molecular weight of 35–36 kDa (Fig. 4), which we have tentatively identified as members of the lipocortin or annexin family of proteins. These proteins have been implicated by a variety of membrane–cytoskeleton and membrane–membrane interactions (Pepinsky 1991, Creutz 1992), signals for which appear to be regulated through tyrosine phosphorylation mechanisms. Significantly, inhibition of distension-induced tyrosine phosphorylation of proteins in aortic tissues with tyrphostin 25 (but not tyrphostin 1, an inactive analogue) also selectively abolishes stretch-induced elastin production. Increased tyrosine phosphorylation of a kinase (pp125$^{fak}$) associated with focal adhesion plaques (Juliano & Haskill 1993), sites of interactions between extracellular matrix proteins and sub-membranous cytoskeletal connectors (talin, vinculin, $\alpha$-actinin) mediated through integrin and other cell-surface receptors, has also been identified as an early result of distension of aortic tissues using this model (Fig. 4). In view of the evidence discussed earlier that enhanced transcription may not be the only mechanism for increasing elastin and collagen production in hypertension, it is interesting to note that downstream consequences of such phosphorylation mechanisms have been reported to include effects both on transcription and on efficiency of translation of mRNAs.

In several respects our observations in distended aortic tissue in organ culture correspond to conclusions from similar studies of transduction methods in statically stretched cardiac myocytes, in which the end-point consequence measured was either increased expression of immediate early genes such as c-*fos*, or increased protein synthesis (Yamazaki et al 1993, Yazaki et al 1993, Sadoshima & Iumo 1993a,b).

Another plausible mechanism for alteration of rates of matrix protein accumulation in response to hypertension which should be considered involves effects on the efficiency of permanent incorporation of newly synthesized connective tissue proteins into the extracellular matrix. In the case of collagen, it has been recognized for some time that a portion of newly synthesized collagen is degraded very soon after its synthesis. While this was initially regarded as an artefact of cell culture, such rapid turnover has also been demonstrated in normal tissues (Laurent 1987), and variations in the fraction of rapidly degraded newly synthesized collagen have been shown to contribute to changes in rates of net collagen accumulation both with ageing (Mays et al 1991) and in hypertension-induced vascular and cardiac hypertrophy (Bishop et al 1990, Eleftheriades et al 1993). Similar investigations of the possibility of rapid breakdown of a portion of newly synthesized elastin in tissues have not been undertaken. Although it is clear that, once firmly incorporated into the extracellular matrix, insoluble elastin is a highly stable protein (Shapiro et al 1991, Davis 1993c), recent evidence of apparent mismatches between elastin synthesis and accumulation (Todorovich-Hunter et al 1988, Johnson et al 1993, Strauss et al 1994) suggests that such a possibility should not be entirely dismissed. Indeed, such adjustments in the proportion of newly synthesized connective tissue proteins ultimately incorporated into the extracellular matrix would provide a post-translational mechanism ideally suited for rapid responses to increased wall stress in vessels.

## Consequences of the vascular connective tissue response to hypertension

Although the hypertrophy of the vessel wall resulting from hypertension is proportional by weight, these changes are not without consequence for the physical properties of the tissue. As indicated earlier, larger conduit vessels show increased stiffness, compromising their ability to moderate pressure and flow waves. In smaller pre-resistance and resistance vessels, hypertrophy also causes encroachment of the vessel wall into the lumen. Since the resistance to flow in a vessel is related to the 4th power of its luminal diameter, even small decreases in calibre can have major effects on the resistance of the vascular bed. In both cases these effects conspire to reinforce and even exacerbate the hypertensive state, and constitute the basis for the 'vascular amplifier' effect described originally by Folkow (Folkow 1978, Keeley et al 1991). Because elastin and collagen, once permanently laid down in the extracellular matrix,

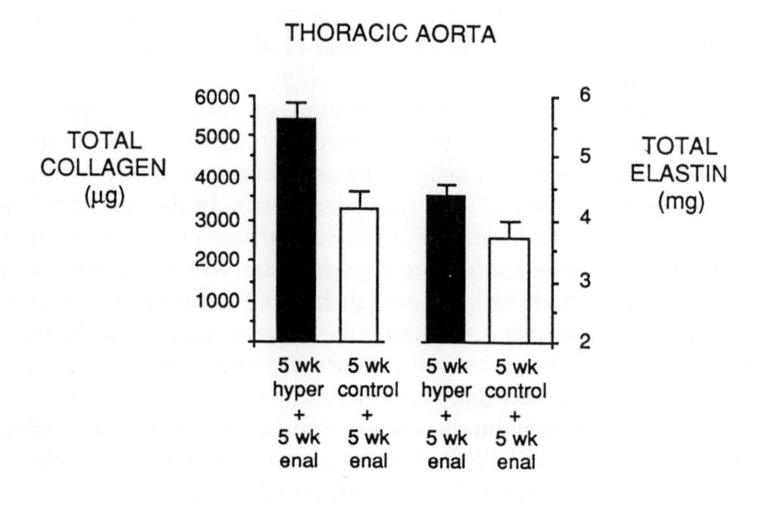


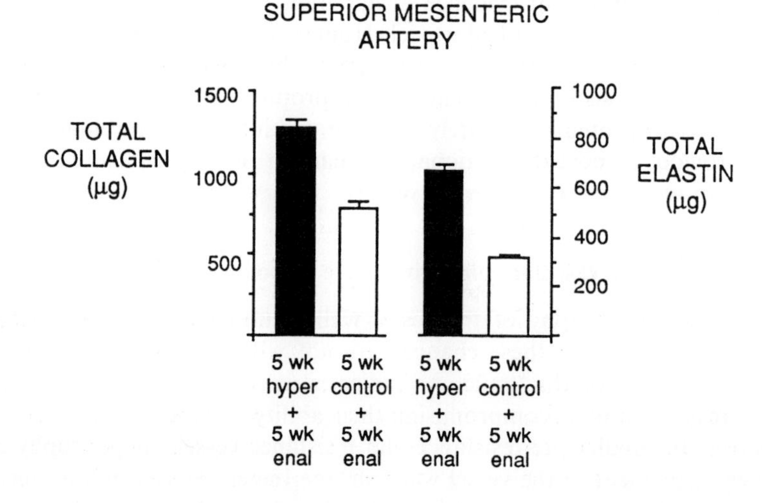


FIG. 5. Effect of normalization of blood pressure by enalapril treatment on total elastin and collagen in aortic and superior mesenteric artery segments of rats. Animals were made hypertensive (hyper) by the renal clip method. After 5 weeks of hypertension, rats were treated with enalapril (enal) for a subsequent 5 weeks. Reproduced with permission from Keeley et al (1991).

are unusually stable proteins turning over only very slowly if at all, it might be expected that additional elastin and collagen accumulated during a period of hypertension would persist long after blood pressure had been lowered to normal levels. Indeed, data from ourselves and others (Wolinsky 1971, Brecher et al 1978, Keeley et al 1991) has indicated that, unlike the effects on smooth muscle cell hypertrophy, established vascular connective tissue changes induced by hypertension are not quickly reversed (Fig. 5).

Slow regression of such connective tissue changes would therefore contribute to the clinically and experimentally observed persistence of pathologically elevated resistance in vascular beds, even after prolonged normalization of blood pressure (Mulvany 1987). Furthermore, rapid vascular connective tissue changes in response to elevated blood pressure, coupled with subsequent slow or incomplete regression of these changes provides a mechanism through which periodic or episodic bouts of hypertension may produce cumulative structural changes in the vessel wall, and may account at least in part for the slow and insidious development of essential hypertension.

While inhibition of elastin and collagen production during development of hypertension would prevent or attenuate these effects, in most cases hypertension and its accompanying changes in vascular elastin and collagen are well established before treatment can be initiated. Thus it would appear that the only feasible approach to regression is to promote turnover of these proteins in the vessel wall. Are there any reasonable prospects for such an approach?

My own response to this question has become more positive over the last several years. In spite of compelling evidence for the very slow turnover of vascular elastin and collagen once firmly incorporated into the extracellular matrix (Shapiro et al 1991, Davis 1993c), it is clear that in at least some situations, regression of connective tissue changes can and does take place. An example of this is the involuting uterus, in which large quantities of elastin and collagen laid down during pregnancy are rapidly degraded (Sharrow et al 1989). In addition, using a hypoxia model of pulmonary hypertension, Riley and his colleagues have shown that recently accumulated elastin and collagen in pulmonary artery can be subsequently lost within several days of normoxia (Poiani et al 1990, Tozzi et al 1991). Furthermore, it is clear that a mechanism must exist for developmental remodelling of arterial vessels to accommodate their growth in length, thickness and calibre. Such developmental remodelling of vessel walls must involve expression of elastase and collagenase activities which are closely regulated both temporally and spatially. Both elastases and collagenases have been identified in the vessel wall (Derouette et al 1981, Todorovich-Hunter et al 1992, Shemie & Rabinovitch 1993, Zhu et al 1994) and have been shown to be up-regulated in a variety of physiological and pathological situations (Yamada et al 1986). While the mechanisms for such remodelling processes are only now being investigated (Wong 1994), it seems plausible to suggest that manipulation of such endogenous, regulated protease activities might provide an opportunity for promoting reversal of connective tissue changes in hypertension and, in doing so, allow an important therapeutic step beyond simple control of blood pressure.

## Summary

Hypertrophy of arterial vessels in hypertension involves substantial increases in elastin and collagen in the vessel wall. While initially protecting the vessel from elevated wall stress, these structural changes also contribute to the maintenance and progression of hypertension. Using both *in vivo* and *in vitro* organ and cell culture models of hypertension, we have shown that the connective tissue response to elevated blood pressure (wall stress) is rapid and sensitive. The mechanism for this response appears to involve signal transduction pathways which require an intact cytoskeleton and which utilize phosphorylation cascades in the cell. Elements of this signal transduction may be similar to those involved in stretch-induced responses in cardiac myocytes and other tissues. Although turnover of vascular elastin and collagen is normally very slow, there is emerging evidence that manipulation of the activities of endogenous vascular elastases and collagenases may offer the possibility of promoting regression of the persistent vascular connective tissue changes associated with hypertension.

### *Acknowledgements*

This work was supported by grants from the Heart and Stroke Foundation of Ontario.

## References

Belik J, Keeley FW, Baldwin F, Rabinovitch M 1994 Pulmonary hypertension and vascular remodelling in fetal sheep. Am J Physiol 266:H2303–H2309

Bishop JE, Guerreiro D, Laurent GJ 1990 Changes in the composition and metabolism of arterial collagens during the development of pulmonary hypertension in rabbits. Am Rev Respir Dis 141:450–455

Brecher P, Chan CT, Franzblau C, Faris B, Chobanian AV 1978 Effects of hypertension and its reversal on aortic metabolism in the rat. Circ Res 43:561–569

Cleary EG, Moont M 1976 Hypertension in weanling rabbits. Adv Exp Med Biol 79:477–490

Cox RH 1981 Basis for the altered arterial wall mechanics in the spontaneously hypertensive rat. Hypertension 3:485–495

Creutz CE 1992 The annexins and exocytosis. Science 258:924–930

Davis EC 1993a Endothelial cell connecting filaments anchor endothelial cells to the subjacent elastic lamina in the developing aortic intima. Cell Tissue Res 272:211–219

Davis EC 1993b Smooth muscle cell to elastic lamina connections in developing mouse aorta. Role in aortic medial organization. Lab Invest 68:89–99

Davis EC 1993c Stability of elastin in the developing mouse aorta: a quantitative radioautographic study. Histochemistry 100:17–26

Derouette S, Hornebeck W, Loisance D, Godeau G, Cachera JP, Robert L 1981 Studies on elastic tissue of aorta in aortic dissections and Marfan syndrome. Pathol Biol 29:539–547

Eleftheriades EG, Durand JB, Ferguson AG, Engelmann GL, Jones SB, Samarel AM 1993 Regulation of procollagen metabolism in the pressure-overloaded rat heart. J Clin Invest 91:1113–1122

Folkow B 1978 Cardiovascular structural adaptation: its role in the initiation and maintenance of primary hypertension. Clin Sci Mol Med 55:3S–22S

Hinek A, Boyle J, Rabinovitch M 1992 Vascular smooth muscle cell detachment from elastin and migration through elastic laminae is promoted by chondroitin sulfate-induced 'shedding' of the 67-kDa cell surface elastin binding protein. Exp Cell Res 203:344–353

Ingber D 1991 Integrins as mechanochemical transducers. Curr Opin Cell Biol 3:841–848

Johnson DJ, LaBourene J, Rabinovitch M, Keeley FW 1993 Relative efficiency of incorporation of newly synthesized elastin and collagen into aorta, pulmonary artery and pulmonary vein of growing pigs. Connect Tissue Res 29:213–221

Juliano RL, Haskill S 1993 Signal transduction from the extracellular matrix. J Cell Biol 120:577–585

Keeley FW 1991 Dynamic responses of collagen and elastin to vessel wall perturbations. In: Gotlieb AI, Langille BL, Fedoroff S (eds) Atherosclerosis: cellular and molecular interactions in the artery wall. Plenum Press, New York, p 101–114

Keeley FW, Alatawi A 1991 The response of aortic elastin synthesis and accumulation to developing hypertension. The inhibitory effect of colchicine on this response. Lab Invest 64:499–507

Keeley FW, Johnson DJ 1986 The effect of developing hypertension on the synthesis and accumulation of elastin in the aorta of the rat. Can J Biochem Cell Biol 64:38–43

Keeley FW, Elmoselhi A, Leenen FHH 1991 Effects of antihypertensive drug classes on regression of connective tissue components of hypertension. J Cardiovasc Pharmacol (suppl 2)17:S64–S69

Keeley FW, Bartoszewicz LA, Leenen FHH 1993 The effect of wall stress on vascular connective tissue accumulation. Eur Respir Rev 30:623–628

Laurent GJ 1987 Dynamic state of collagen: pathways of collagen degradation *in vivo* and their possible role in regulation of collagen mass. Am J Physiol 252:C1–C9

Lee RMKW, Garfield RE, Forrest JB, Daniel EE 1983 Morphometric study of structural changes in the mesenteric blood vessels of spontaneously hypertensive rats. Blood Vessels 20:57–71

Mays PK, McAnulty RJ, Campa JS, Laurent GJ 1991 Age-related changes in collagen synthesis and degradation in rat tissues. Importance of degradation of newly synthesized collagen in regulating collagen production. Biochem J 276:307–313

Mulvany MJ 1987 The structure of the resistance vasculature in essential hypertension. J Hypertens 5:129–136

Owens GK 1985 Differential effects of antihypertensive drug therapy on vascular smooth muscle cell hypertrophy, hyperploidy, and hyperplasia in the spontaneously hypertensive rat. Circ Res 56:525–536

Parks WC, Mecham RP, Crouch EC, Orton EC, Stenmark KR 1989 Response of lobar vessels to hypoxic pulmonary hypertension. Am Rev Respir Dis 140:1455–1457

Pepinsky B 1991 Phosphorylation of lipocortin-1 by the epidermal growth factor receptor. Methods Enzymol 198:260–272

Poiani G, Tozzi C, Yohn S et al 1990 Collagen and elastin metabolism in hypertensive pulmonary arteries of rats. Circ Res 66:968–978

Prosser IW, Stenmark KR, Suthar M, Crouch EC, Mecham RP, Parks WC 1989 Regional heterogeneity of elastin and collagen gene expression in intralobar arteries in response to hypoxic pulmonary hypertension as demonstrated by *in situ* hybridization. Am J Pathol 135:1073–1088

Sadoshima J, Izumo S 1993a Mechanical stretch rapidly activates multiple signal transduction pathways in cardiac myocytes: potential involvement of an autocrine/paracrine mechanism. EMBO J 12:1681–1692

Sadoshima J, Izumo S 1993b Mechanotransduction in stretch-induced hypertrophy of cardiac myocytes. J Recept Res 13:777–794

Shapiro SD, Endicott SK, Province MA, Pierce JA, Campbell EJ 1991 Marked longevity of human lung parenchymal elastic fibers deduced from prevalence of D-aspartate and nuclear weapons-related radiocarbon. J Clin Invest 87:1828–1834

Sharrow L, Tinker D, Davidson JM, Rucker RB 1989 Accumulation and regulation of elastin in the rat uterus. Proc Soc Exp Biol Med 192:121–126

Shemie S, Rabinovitch M 1993 The effect of alpha-1 antitrypsin inhibition of early elastase release on the pathophysiology of progressive pulmonary hypertension. Am Rev Respir Dis 147:A495(abstr)

Strauss BH, Chisholm RJ, Keeley FW, Gotlieb AI, Logan RA, Armstrong PW 1994 Extracellular matrix remodelling after balloon angioplasty injury in a rabbit model of restenosis. Circ Res 75:650–658

Takasaki I, Chobanian AV, Sarzani R, Brecher P 1990 Effect of hypertension on fibronectin expression in the rat aorta. J Biol Chem 265:21935–21939

Todorovich-Hunter L, Johnson DJ, Ranger P, Keeley FW, Rabinovitch M 1988 Altered elastin and collagen synthesis associated with progressive pulmonary hypertension induced by monocrotaline. A biochemical and ultrastructural study. Lab Invest 58:184–195

Todorovich-Hunter L, Dodo H, Ye C, McCready L, Keeley FW, Rabinovitch M 1992 Increased pulmonary artery elastolytic activity in adult rats with monocrotaline-induced progressive hypertensive pulmonary vascular disease compared with infant rats with nonprogressive disease. Am Rev Resp Dis 146:213–223

Tozzi CA, Poiani GJ, Harangozo AM, Boyd CD, Riley DJ 1989 Pressure-induced connective tissue synthesis in pulmonary artery segments is dependent on intact endothelium. J Clin Invest 84:1005–1012

Tozzi CA, Wilson FJ, Yu SY, Bannett RF, Peng BW, Riley DJ 1991 Vascular connective tissue is rapidly degraded during early regression of pulmonary hypertension. Chest (suppl 3) 99:41S–42S

Tozzi CA, Christiansen DL, Poiani GJ, Riley DJ 1994 Excess collagen in hypertensive pulmonary arteries decreases vascular distensibility. Am J Respir Crit Care Med 149:1317–1326

Wolinsky H 1970 Response of the rat aortic media to hypertension: morphological and chemical studies. Circ Res 26:507–522

Wolinsky H 1971 Effects of hypertension and its reversal on the thoracic aorta of male and female rats. Morphological and chemical studies. Circ Res 28:622–627

Wong LCY 1994 Developmental remodelling of the internal elastic lamina of rabbit carotid arteries: effect of blood flow. MSc thesis, University of Toronto, Toronto, ON, Canada

Yamada E, Hazama F, Amano S, Sasahara M, Kataoka H 1986 Elastase, collagenase, and cathepsin D activities in the aortas of spontaneously hypertensive and renal hypertensive rats. Exp Mol Pathol 44:147–156

Yamazaki T, Tobe K, Hoh J et al 1993 Mechanical loading activates mitogen-activated protein kinase and S6 peptide kinase in cultured rat cardiac myocytes. J Biol Chem 268:12069–12076

Yazaki Y, Komuro I, Yamazaki T et al 1993 Role of protein kinase system in the signal transduction of stretch-mediated protooncogene expression and hypertrophy of cardiac myocytes. Mol Cell Biochem 119:11–16

Zhu L, Wigle D, Hinek A et al 1994 The endogenous vascular elastase that governs development and progression of monocrotaline-induced pulmonary hypertension in rats is a novel enzyme related to the serine proteinase adipsin. J Clin Invest 94:1163–1171

## DISCUSSION

*Robert:* It's true that matrix deposition during hypertension is not reversible. However, for a long time it was not customary for doctors to treat the age-dependent increase of blood pressure. During the 1970s a French gerontologist, Dr Francoise Forette, began to treat hypertension; she noticed a significant decrease in deaths from heart disease and stroke in her patients (Forette et al 1980, Panisset et al 1993, Forette & Boller 1990). In addition, there was a recent report from a Swedish group (Dahlöf et al 1991), who showed that when they treated hypertension in a large population of old people, they decreased the frequency of strokes and heart attacks by nearly 50%. It is clearly a very important issue, and great benefit is obtained just by lowering blood pressure in ageing people.

*Keeley:* I certainly did not intend to imply that treatment of hypertension was without benefit. However, in our experiments, we have never detected regression of collagen or elastin in vascular tissues, at least after up to 5 weeks of treatment with enalapril. However, Karl Weber and his colleagues have reported regression of ventricular changes, including newly deposited collagen, after longer periods of treatment with angiotensin-converting enzyme (ACE) inhibitors (Brilla et al 1991). Experiments to determine whether longer periods of treatment with ACE inhibitors (up to 25 weeks) promote regression of hypertension-induced changes in vascular elastin and collagen content are now underway in our laboratory.

When assessing regression of hypertension-induced changes in ventricular tissues, it is important to distinguish between regression of cardiac myocyte hypertrophy and regression of the additional collagen deposited. Decrease in ventricular weight after lowering of blood pressure certainly is observed with antihypertensive treatment.

*Robert:* That's true, but what I think what they saw concerns mainly the heart muscle and not the deposited collagen. One of the major problems in age-dependent heart problems is the progressive fibrosis that occurs. This is the most frequent cause of arrhythmia and sudden heart arrest (Besse et al 1993).

*Keeley:* If you look just at the weights of the ventricles of the treated rats in our experiments, it would appear that after five weeks of treatment with ACE inhibitors the ventricles have been restored to their pre-hypertensive state. However, because reversal of cardiac myocyte hypertrophy has not been accompanied by a resorption or degradation of the additional collagen deposited during hypertension, these tissues have an increased concentration of collagen: that is, they have undergone fibrosis. Therefore, while it is certainly good to have a reduction in the ventricular mass, the ventricular tissue, at least after this length of treatment, has not returned to its pre-hypertensive state in terms of connective tissue content.

*Robert:* Yes; I think the real problems are fibrosis and the loss of $\beta$ receptors during ageing (Besse et al 1993).

*Mecham:* Do you know at what stage these various agents are acting? Is it increased elastin synthesis, or synthesis of microfibrillar components or other aspects that might affect assembly?

*Keeley:* As you say, the data I have shown are expressed as production of insoluble elastin, which can be influenced by secretion and assembly of tropoelastin, as well as by tropoelastin synthesis. Indeed, as I indicated, the effects of $Ca^{2+}$ chelation and verapamil to abolish both control and stretch-induced insoluble elastin production are due to effects on secretion and/or assembly, rather than on synthesis. However, we would suggest that agents which inhibit the stretch-induced response without affecting control levels of production (for example, the blockers of $Ca^{2+}$ reuptake into the sarcoplasmic reticulum, or staurosporine, tyrphostin and NCDC) would be less likely to be mediated through effects on secretion and assembly, since it is hard to explain how these effects could then be selective for stretch-induced production. That having been said, we will only be able to provide a complete answer to your question when we move our investigations from aortic organ culture to aortic smooth muscle cell cultures in which it will be possible to measure newly synthesized intracellular tropoelastin, extracellular tropoelastin and insoluble elastin.

*Mecham:* Have you looked ultrastructurally at the elastic lamellae in the vessels of the hypertensive animals?

*Keeley:* No.

*Bailey:* I'm a little confused about the idea that newly synthesized elastin is weakly cross-linked and is therefore involved in rapid turnover. There is no evidence for this—it is generally believed that elastin matures to the desmosines very rapidly. An alternative explanation for your observation could be that stretching of the vessels by hypertension results in a slow increase in both elastin synthesis and the activity of the degradative enzymes. In this way, there could be a build up of normally cross-linked elastin, some of which is degraded by the enzymes depending on the relative proportions of the latter.

*Keeley:* In a sense, that is what I was trying to get at. In aortic tissue, the transition from tropoelastin to insoluble elastin is rapid and efficient (Keeley &

Johnson 1983). However, there is some evidence that this may not be the case in all tissues (Johnson et al 1993). I'm saying that at least in some tissues there may be mechanisms for degradation and removal of newly accumulated insoluble elastin.

*Bailey:* You've got no evidence, and I don't see how you can get evidence for that. Unlike weakly cross-linked collagen, elastin initially cross-linked by the aldol prior to desmosine formation would still be insoluble. Turnover studies using labelled elastin would be rather complex.

*Keeley:* There is some evidence, although it is all indirect, that turnover of newly accumulated elastin may take place under some circumstances. For example, remodelling of arterial vessels with growth and development must certainly involve some mechanism for controlled and selective elastin turnover. There are also several examples of mismatch between elastin synthesis and accumulation, which point to some turnover process (Strauss et al 1994). In addition, Riley and his colleagues have shown, using a hypoxia model of pulmonary hypertension in young rats, that recently accumulated insoluble elastin in pulmonary arteries can be turned over relatively rapidly (Poiani et al 1990). Therefore, although direct evidence, such as zymograms, is still lacking, I believe that this is a process which should be considered.

*Lapis:* Did I understand correctly that the increased elastin production is not associated with the formation of new elastic lamellae in the vessel wall? In human pathology, lamellar elastosis is still regarded as a reliable marker of essential hypertension in medium-sized arteries.

*Keeley:* We may be talking about different vessels at different stages of hypertension. Initially, the deposition of new elastin takes place in the interlamellar islands. Perhaps if hypertension is prolonged the large amounts of elastin may take the form of pseudolamellae. As far as I am aware, the actual number of lamellae is set early in development and does not increase.

*Winlove:* Very early in your paper you mentioned Laplace's law relating stress and strain. Of course, this only applies to thin walled vessels—in something thick and complicated like a blood vessel there would be a very complicated distribution of stresses and strains across the thickness of the wall. Do you actually have any evidence for differences in the synthetic response across the wall?

*Keeley:* No, we don't.

*Mecham:* Looking with *in situ* hybridization at pulmonary arteries, where you can maintain high pressure by hypoxia, we find that there are different populations of cells in the wall that respond to pressure change (Prosser et al 1989). We need to be cautious about thinking of the entire vessel wall as a single population of smooth muscle cells. There are different responders at different times, and the age of the vessel determines in many respects how it is going to respond. There are going to be different stresses at different points in the vessel wall, and perhaps the different responses we see might be indirect responses to these stress factors.

*Keeley:* Certainly, the further out you go radially, the greater will be the stress in the vessel wall.

*Winlove:* This assumes that the vessel is a classical engineering solid and wouldn't be true if it were, for example, 'preformed' or 'prestressed'. I think there is some evidence that blood vessels do generate such 'residual stresses' (e.g. Fung & Liu 1992).

Another question: when you do your *in vitro* experiments, you essentially have a steady force applied. *In vivo*, in hypertension, you alter not only the mean pressure but the pressure pulse amplitude and wave form. Do you think the cells are able to sense and transduce all of these more subtle variations in the forces applied to them as well?

*Keeley:* I really can't answer that. We have used a static stretch model because it is simple and easy to manipulate, and produces a consistent response in terms of increased elastin production. In an earlier model (Keeley et al 1990), we showed that the wall stress threshold for eliciting this response was similar in these *in vitro* static models to that seen in *in vivo*, pulsatile models of hypertension. Therefore, mean arterial pressure may be the most important factor. However, I would certainly not want to argue that pulse pressure effects are of no importance.

*Tamburro:* Your circular argument is that hypertension increases production of elastin and collagen, which increases the resistance to flow, increasing pressure of blood and so on. It seems to imply that you can no longer apply Laplace's law, because it is a linear equation of pressure as a function of wall stress. The other factors must be constant but, for example, in your case $\delta$ is not constant. It is a function of pressure, which is a function of stress; this implies a non-linear equation. The response could be exponential.

*Keeley:* The so-called 'vascular amplifier' hypothesis was originally proposed by Folkow (1978) and expresses the concept that decreased lumen size due to hypertrophy of the vascular wall in resistance and pre-resistance vessels would act as a positive feedback mechanism for increasing blood pressure.

*Rosenbloom:* Most of the peripheral vascular resistance occurs much further down the vascular tree. Is there any evidence for the kind of mechanism that you are postulating at the level where most of this resistance is actually taking place? While we get these nasty thickenings and responses happening in the aorta, they really don't have anything to do with regulating the blood pressure.

*Keeley:* The data I showed are for the superior mesenteric artery, which is not a resistance vessel. However, we have also been able to detect biochemically similar responses as far down the arterial tree as second-order branches of the superior mesenteric artery, which could be considered pre-resistance vessels. Below that level the evidence is all morphological, or results from physiological studies of resistance of vascular beds.

*Mungai:* Changes occur to elastin in the uterine artery during pregnancy. There is also an increase of elastic fibres in the male in the ductus deferens (the duct that delivers sperm from the testis during ejaculation) which occurs at puberty (Paniagua et al 1983). Is there any suggestion of a hormonal relationship between these factors and an increase of elastic fibres in the reproductive system?

*Starcher:* I would assume that there is a hormonal influence on the increase of elastin in the uterus during development, and then its disappearance. If you look at most systems like this there's a dramatic increase in collagen which then gets degraded. This is hormonally controlled, if I remember correctly. If you look at the ratio of collagen to elastin during this time it increases.

*Robert:* A pathologist in Germany, Dr Leutert (1964), produced data on the abundance of elastic fibres in the vocal chords. In women there are far more than in men. Again, this is probably hormonally regulated.

*Davidson:* Did the *in vitro* experiments you performed using presumably selective inhibitors show parallel effects for other matrix proteins? Were these effects unique to elastin, or were they also seen in collagen?

*Keeley:* Although I did not show you the results, for many of these agents we measured similar effects on both collagen and elastin production.

*Davidson:* You made some intriguing comments about changes in protein production or synthetic rate in the absence of changes in mRNA level. This is provocative and exciting. You suggested S6 kinase might be the mediator. Thinking about the focal adhesion kinase story and transduction through cytoskeletal stress, it came to mind that there's a very interesting story developing with actin expression and localization of actin mRNA as being a regulatory mechanism for its expression. Of course, actin is an intracellular protein, but I wonder whether it's worth looking at the localization of elastin message as a function of stress on cells?

*Keeley:* I think that would be really interesting. Rapid effects of physical stresses on protein synthesis are apparently also seen in skeletal and cardiac muscle cells, and there have been suggestions that increased protein production in these tissues may, at least initially, be a result of effects on translation rates.

*Davidson:* The kinetics you described are consistent with this. These translational effects happen within the first few minutes of introduction of the signal; you don't have to invoke any transcription factors.

*Keeley:* We do see almost a doubling of tropoelastin synthesis within 30 min of stretch of the tissues, so the response is certainly fast.

## References

Besse S, Assayag P, Delcaye C et al 1993 Normal and hypertrophied senescent rat heart: mechanical and molecular characteristics. Am J Physiol 134:H183–H190

Brilla CG, Janicki JS, Weber KT 1991 Cardioreparative effects of lisinopril in rats with genetic hypertension and left ventricular hypertrophy. Circulation 83:1771–1779

Dahlöf B, Lindholm LH, Hansson L, Schersten B, Ekbom T, Webster PO 1991 Morbidity and mortality in the Swedish trial in old patients with hypertension. Lancet 338:1281–1285

Folkow B 1978 Cardiovascular structural adaptation: its role in the initiation and maintenance of primary hypertension. Clin Sci Molec Med 55:3s–22s

Forette F, Boller F 1990 Hypertension and risk of dementia in the elderly. Am J Med 90:14S–19S

Forette F, de la Fuente J, Golmard JL, Henry JF, Hervy MP 1980 Elderly hypertension as a risk factor. Curr Concepts Hypertension & Cardiovasc Disord 1:11–14

Fung YC, Liu SQ 1992 Strain distribution in small blood vessels with zero-stress state taken into consideration. Am J Physiol 262:H544–H552

Johnson DJ, LaBourene J, Rabinovitch M, Keeley FW 1993 Relative efficiency of incorporation of newly synthesized elastin and collagen into aorta, pulmonary artery and pulmonary vein of growing pigs. Connect Tissue Res 29:213–221

Keeley FW, Johnson DJ 1983 Measurement of the absolute synthesis of soluble and insoluble elastin by chick aortic tissue. Can J Biochem Cell Biol 61:1079–1084

Keeley FW, Alatawi A, Cho A 1990 The effect of wall stress on the production and accumulation of vascular elastin. In: Tamburro AM, Davidson JM (eds) Elastin: chemical and biological aspects. Congedo Editore, Galatina, Italy, p 341–356

Leutert G 1964 Uber die histologische biomorphose der menschilden stimmlippen. Morph J 106:11–72

Paniagua R, Regardera J, Nistal M, Santamaria L 1983 Elastic fibres of the human ductus deferens. J Anat 137:467–476

Panisset M, Forette F, Boller F 1993 Isolated systolic hypertension and the brain. Cardiology in the Elderly 1:263–267

Poiani G, Tozzi C, Yohn S et al 1990 Collagen and elastin metabolism in hypertensive pulmonary arteries of rats. Circ Res 66:968–978

Prosser IW, Stenmark KR, Suthar M, Mecham RP, Parks WC 1989 Regional heterogeneity of elastin and collagen gene expression in intralobar arteries in response to hypoxic pulmonary hypertension as demonstrated by *in situ* hybridization. Am J Pathol 135:1073–1088

Strauss BH, Chisholm RJ, Keeley FW, Gotlieb AI, Logan RA, Armstrong PW 1994 Extracellular matrix remodelling after balloon angioplasty injury in a rabbit model of restenosis. Circ Res 75:650–658

# General discussion IV

## Developmental regulation of elastin production

*Parks:* As was mentioned in the discussions following Charles Boyd's and Jeff Davidson's papers (Sechler et al 1995, Davidson 1995, this volume), we have been studying the molecular mechanisms controlling the developmental expression of tropoelastin. As you well know, elastin production occurs mostly during late fetal and neonatal periods, and by maturity, assembly of elastic fibres is complete and synthesis of the precursor protein, tropoelastin, turns off. Our findings, which I shall present here, demonstrate that distinct mechanisms control elastogenesis at different stages of production.

A few years ago, we reported that the marked down-regulation of tropoelastin expression mediated by vitamin $D_3$ or by phorbol ester is controlled by a post-transcriptional mechanism leading to accelerated degradation of the mRNA (Parks et al 1992, Pierce et al 1992). These findings suggested that the normal cessation of tropoelastin expression is controlled post-transcriptionally but, because these studies were carried out in cell culture, we did not know if this mechanism represented an actual *in vivo* regulatory process or a cell culture artefact.

To assess regulatory mechanisms in intact tissue, Mei Swee, a postdoctoral fellow in my lab, developed a rapid and sensitive reverse-transcription polymerase chain reaction (RT-PCR) assay to quantify tropoelastin pre-mRNA levels as an indicator of on-going transcription. Since pre-mRNA (or hnRNA) is rapidly processed to mRNA and transported from the nucleus (Krug 1993), the level of pre-mRNA should provide an estimation of the rate of active transcription.

We isolated total lung RNA from 19 d fetal, 3 d and 11 d neonatal, and 6 month adult rats, knowing that these ages would represent distinct stages of tropoelastin expression, namely the onset, peak production and cessation. Indeed, as demonstrated by Northern hybridization, steady-state levels of tropoelastin mRNA were low in the 19 d fetal lung (shortly after tropoelastin expression begins in the rat), were about sevenfold higher in the 3 d neonatal lung, about 50-fold higher in 11 d neonate, and were markedly repressed in the adult, when expression is low to non-existent (Fig. 1).

For the RT-PCR assay, these same RNA samples were treated with RNase-free DNase to remove any contaminating DNA and were reverse-transcribed with random hexamers. A 478 bp stretch of the resultant single-stranded

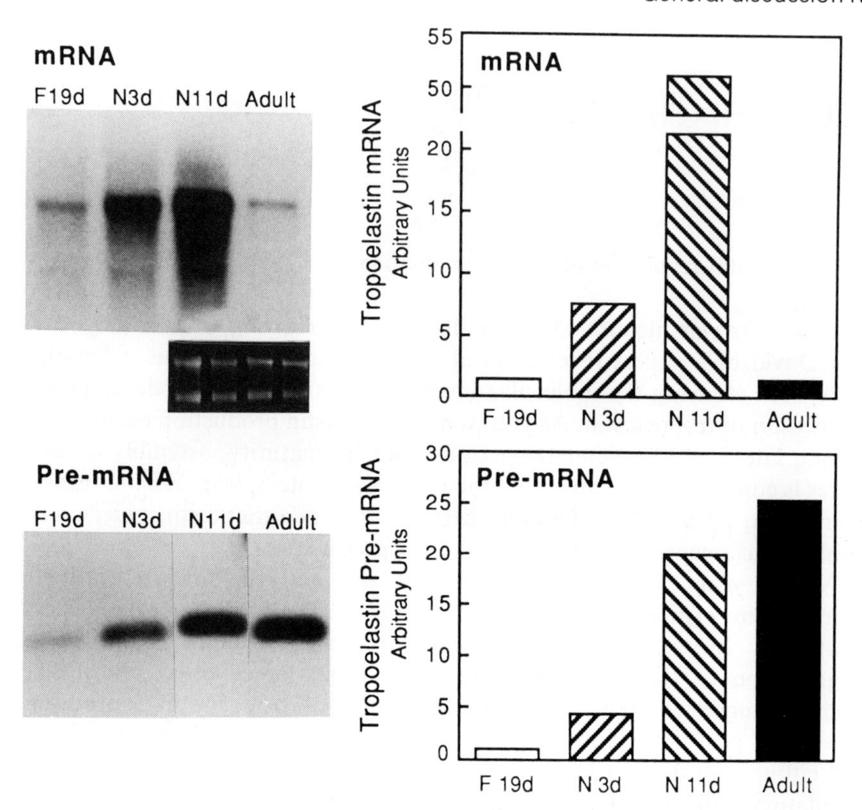


FIG. 1. Tropoelastin pre-mRNA expression persists in adult tissue. Total RNA (0.1 μg) from 19 d fetal, 3 d neonatal, 11 d neonatal and 6 month adult rat lung was analysed by Northern hybridization (*mRNA*) for tropoelastin and GAPDH mRNA or was amplified by RT-PCR using intron 35 primers (*Pre-mRNA*). Products were detected by hybridization with either $^{32}$P-labelled cDNAs or an intron 35-specific oligomer. Autoradiographic signals were quantified by densitometry and normalized to the signal of the 19 d fetal samples, which were arbitrarily set at 1. For the Northern data, the signal for tropoelastin mRNA was first normalized to that for GAPDH mRNA.

cDNA was amplified by PCR using primers specific to sequences within intron 35 of the tropoelastin gene, and the PCR product was detected by Southern hybridization with an intron-specific probe.

Initially, we optimized the RT-PCR assay by determining the relationship of signal strength to cycle number and to the amount of input RNA (Swee et al 1995). Between 21 and 29 cycles, only the 478 bp tropoelastin pre-mRNA cDNA was produced, and the signal for this product increased exponentially. With 31 or more cycles the specific tropoelastin signal increased only slightly and non-specific products appeared. At 25 cycles of amplification, the signal for intron 35 cDNA increased linearly between 0.0125 μg and 0.8 μg of

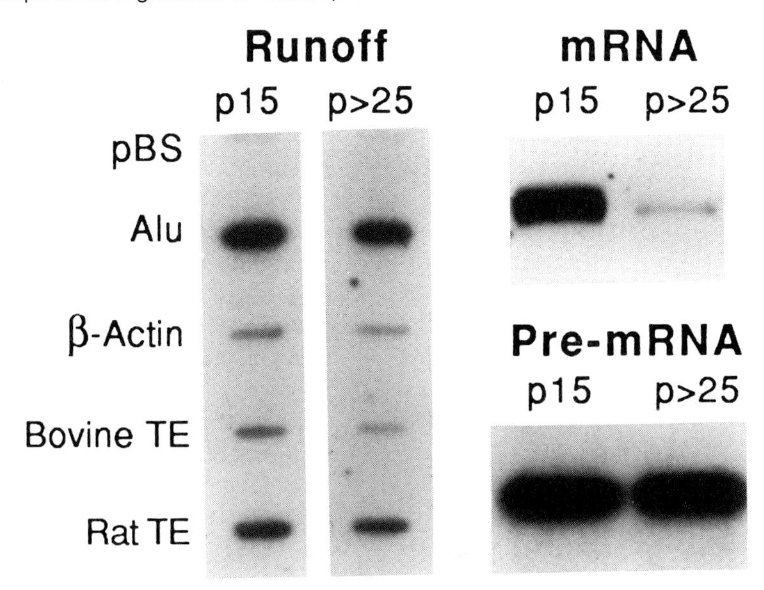


FIG. 2. RT-PCR and nuclear run-off assays both demonstrate that tropoelastin pre-mRNA expression remains elevated after down-regulation of steady-state mRNA levels. Total RNA was isolated from RFL-6 cells at passage 15 (*p15*) and from cells passed at least 25 times (*p>25*). As shown by Northern hybridization (*mRNA*), tropoelastin steady-state mRNA levels were markedly decreased in the older cells, whereas pre-mRNA levels, as determined by RT-PCR with intron 35 primers (*Pre-mRNA*), were equivalent in the two samples. Tropoelastin transcription was also assessed by nuclear run-off assay. *In vitro* transcribed $^{32}P$-labelled RNA was hybridized to parental plasmid DNA (*pBS*), a genomic fragment containing Alu I repeat sequences (*Alu*), used as an indicator of total transcription, and cDNAs for $\beta$-actin and bovine and rat tropoelastin (*TE*). The difference in signal strength with the bovine and rat tropoelastin cDNAs reflects the greater affinity of the rat pre-mRNA for its homologous cDNA. Autoradiographic exposure was 2 d for the Northern, 1 d for the RT-PCR blot, and 7 d for the run-off. (Reproduced with permission from Swee et al 1995.)

neonatal lung RNA added to the RT-PCR reaction. These findings show that the RT-PCR assay accurately reflects relative differences in the levels of pre-mRNA over at least a 64-fold range in template concentration. No product was detected in samples without reverse transcriptase, indicating the specificity of the assay and that there was no significant contamination by cellular DNA. On the basis of these results, we used 0.1 $\mu$g of DNase-treated RNA and 25 cycles for all subsequent reactions. Restriction enzyme digestion and direct sequence analyses of the RT-PCR product confirmed that it corresponded to rat tropoelastin intron 35. As an additional control, we detected no products in RT-PCR reactions with RNA isolated from PC12 cells, a rat phaeochromocytoma cell line that does not express tropoelastin mRNA, demonstrating that non-specific products were not produced in this assay.

Using this RT-PCR assay and RNA from the same animals used for Northern hybridization, we found low levels of tropoelastin pre-mRNA in the 19 d fetal samples, but about sevenfold higher levels in the 3 d neonatal RNA samples and about 20-fold higher levels in the 11 d neonate (Fig. 1). The magnitude of the increase of signal for tropoelastin pre-mRNA in the neonatal samples was essentially identical to that determined for steady-state mRNA levels, and this tight correlation indicates that gene transcription controls tropoelastin production during early lung development. The approximately twofold discrepancy between the mRNA and pre-mRNA levels in the 11 d neonatal sample suggests that multiple mechanisms, such as mRNA stability and translation efficiency, contribute to the maintenance of high levels of steady-state mRNA levels during peak periods of elastogenesis.

Surprisingly, we found that the levels of tropoelastin pre-mRNA remained elevated in adult lung even though steady-state mRNA levels were reduced about 50-fold relative to those in the 11 d neonatal sample. These same patterns of tropoelastin pre-mRNA expression were seen in five other sets of animals of all ages and in experiments using intron 21 primers. These findings demonstrate that tropoelastin transcription does not turn off at the end of elastin production and that a post-transcriptional mechanism regulates the low levels of tropoelastin mRNA in the mature tissue. The cDNA products we detected are likely not due to spliced intron fragments since these sequences are rapidly degraded once removed from the pre-mRNA (Baurén & Wieslander 1994, Jakubowski & Roberts 1994).

We used various controls to verify our findings. As a positive control, we assessed the expression of $\alpha_1$(I) procollagen pre-mRNA. Type I collagen is produced in a developmentally regulated pattern similar to that for elastin, and the expression of $\alpha_1$(I) procollagen is controlled at the level of transcription (Slack et al 1993). Using RT-PCR, we found that the expression of $\alpha_1$(I) procollagen pre-mRNA parallelled its mRNA levels in fetal, neonatal and adult lung (Swee et al 1995). Thus, amplification of pre-mRNA reflects the known transcriptional regulation of $\alpha_1$(I) procollagen.

Reflective of the *in vivo* developmental pattern of elastin production, elastogenic cells lose their capacity to express tropoelastin as they age in culture or are serially passed. To assess the regulation of this down-regulation, we isolated total RNA from RFL-6 cells, an elastogenic rat lung fibroblast cell line, at passage 15 and from cells which had been passed extensively (>25 passages). As shown by Northern hybridization, the older cells expressed much less tropoelastin mRNA than did the passage-15 cells, but the levels of tropoelastin pre-mRNA, as assessed by RT-PCR, were equivalent in the two cell populations (Fig. 2). To verify these findings, we isolated nuclei from the early and extended passage cells for nuclear run-off assay. In agreement with the RT-PCR results, the transcription rate was the same in both cell populations (Fig. 2). Although it could be argued that tropoelastin pre-mRNA is processed slowly in adult tissue, thereby leading to an accumulation of pre-spliced transcripts, these nuclear run-off data demonstrated that the persistence of pre-mRNA expression is due to continued transcription of the tropoelastin gene and is not due to age-specific processing of the primary transcript.

In summary, our findings indicate that distinct mechanisms control elastin production at different stages of development. I have shown data which demonstrate that tropoelastin pre-mRNA and mRNA levels rise in parallel during fetal and neonatal development (Fig. 1), and in Richard Pierce's presentation (Pierce et al 1995, this volume), he provided evidence that maternal administration of glucocorticoid, which accelerates fetal lung development and stimulates elastin production, mediated a proportional increase in tropoelastin pre-mRNA levels. Thus, the close correlation between mRNA and pre-mRNA levels indicates that the induction and up-regulation of tropoelastin expression during the early phases of elastogenesis is primarily controlled at the level of transcription.

Interestingly, we found that tropoelastin pre-mRNA levels in mature lung remained elevated, although the steady-state mRNA level decreased sharply from peak levels attained in neonatal lung (Fig. 1). Furthermore, we detected the same persistence of tropoelastin pre-mRNA levels in fibroblasts isolated from adult lung (data not presented) and in lung fibroblasts that were aged in culture (Fig. 2). Thus, the discrepancy between pre-mRNA and steady-state mRNA levels in fully grown tissues indicates that a post-transcriptional mechanism controls the down-regulation of tropoelastin expression.

Although multiple mechanisms can participate in the control of gene expression, production of structural proteins is primarily regulated at the level of transcription. There are, however, numerous examples of proteins whose production is primarily regulated by a post-transcriptional mechanism. Many of these products, such as cytokines, iron-metabolism proteins, oncogenes and cytoskeletal proteins, are expressed during physiological transitions or for brief periods during developmental processes, and changes in the stability of the mRNA provides a mechanism to rapidly govern protein synthesis and activity. In contrast, once the growth of elastic tissue is complete, new elastin production is not needed and thus the post-transcriptional control we describe is a novel mechanism to control the expression of a stable structural protein. It is likely that *cis* elements in tropoelastin mRNA are involved in this regulatory mechanism, and we are currently attempting to identify such sequences.

*Keeley:* Do you have any idea why elastin should be regulated this unusual way, only at the message stability level?

*Parks:* As Leslie Robert talked about in his introduction, the elastin gene is, in evolutionary terms, a young gene. Its promoter has some features that are unusual for a tissue-specific structural gene, such as multiple start sites, the lack of a consensus TATA box, and GC-rich areas. It may not have evolved 'proper' control mechanisms. Early on in development, tropoelastin expression is controlled by what we would consider to be a conventional regulatory mechanism, namely transcription. Later in life this control is no longer active, and so other mechanisms are needed.

*Keeley:* This would be particularly unexpected for aortic tissue, where the rates of production of elastin are very high during development and growth, but then fall to almost undetectable levels in the adult. It is hard to imagine that the cells would carry on transcribing the elastin gene at that rate if the message is not going to be used to make protein.

*Parks:* That is, at first glance, an understandable bias; why would a cell continue to transcribe so much RNA that it is not going to use? But transcription itself is not an inherently efficient process. For example, Darnell and his associates showed some time ago that at least 25% of the RNA produced is not destined for translation or for rRNA, and this figure accounts for intron splicing (Harpold et al 1981). It is simply transcribed and degraded. In addition, pre-mRNA includes apparently unneeded intron sequences. If you look at tropoelastin itself, it's a 3.5 kb mRNA, but its pre-mRNA is about 40 kb. That is seemingly a lot of wasted RNA. Thus, we should be a little cautious in thinking that transcription is a truly efficient process.

*Starcher:* We have been looking at the turnover of elastin by measuring desmosines in the urine, and others have looked at elastin turnover by measuring peptides using antibodies made against peptides of elastin or tropoelastin. We have measured desmosine cross-links, which will only be a measure of elastin that has been deposited and turned over. There is a discrepancy in that elastin peptides appear to turnover under conditions in which we find no desmosine. I have often thought that there was considerable tropelastin being made these conditions, which degrades and is excreted without any desmosine secretion. I think this agrees very well with your data.

*Davidson:* Our findings support this concept of post-transcriptional control. If we compare either elastin production or message levels among cutis laxa, normal and Hutchinson–Gilford progeria fibroblasts, there is a difference two orders of magnitude between both message levels and production of the protein, yet the transcription rates are similar. How universal is this uncoupling in other tissues? Is this a lung-specific phenomenon?

*Parks:* We've looked at rat skin, and the same mechanism occurs there. We've just developed the assay for bovine pulmonary artery, and it seems to be active there as well. Thus the post-transcription regulation of tropoelastin seems to be common to all elastic tissues and to all species.

## References

Bauren G, Wieslander L 1994 Splicing of Balbiani ring 1 gene pre-mRNA occurs simultaneously with transcription. Cell 76:183–192

Davidson JM, Zhang M-C, Zoia O, Giro MG 1995 Regulation of elastin synthesis in pathological states. In: The molecular biology and pathology of elastic tissues. Wiley, Chichester (Ciba Found Symp 192) p 81–99

Harpold MM, Wilson MC, Darnell JE 1981 Chinese hamster polyadenylated messenger ribonucleic acid: relationship to non-polyadenylated sequences and relative conservation during messenger ribonucleic acid processing. Mol Cell Biol 1:188–198

Jakubowski M, Roberts JL 1994 Processing of gonadotropin-releasing hormone gene transcripts in the rat brain. J Biol Chem 169:4078–4083

Krug RM 1993 The regulation of export of mRNA from nucleus to cytoplasm. Curr Opin Cell Biol 5:944–949

Parks WC, Kolodziej ME, Pierce RA 1992 Phorbol ester-mediated downregulation of tropoelastin expression is controlled by a posttranscriptional mechanism. Biochemistry 31:6639–6645

Pierce RA, Kolodziej ME, Parks WC 1992 1,25 dihydroxyvitamin $D_3$ represses tropoelastin expression by a posttranscriptional mechanism. J Biol Chem 267:11593–11599

Pierce RA, Mariani TJ, Senior RM 1995 Elastin in lung development and disease. In: The molecular biology and pathology of elastic tissues. Wiley, Chichester (Ciba Found Symp 192) p 199–214

Sechler JL, Sandberg LB, Roos PJ, Snyder I, Amenta PS, Riley DJ, Boyd CD 1995 Elastin gene mutations in transgenic mice. In: The molecular biology and pathology of elastic tissues. Wiley, Chichester (Ciba Found Symp 192) p 148–171

Slack JL, Liska DJ, Bornstein P 1993 Regulation of expression of the type I collagen genes. Am J Med Genet 45:140–151

Swee MH, Parks WC, Pierce RA 1995 Developmental regulation of elastin production. Expression of tropoelastin pre-mRNA persists after down-regulation of steady-state mRNA levels. J Biol Chem 270, in press

# Elastin in blood vessels

L. Robert, M. P. Jacob* and T. Fülöp†

*Laboratoire de Biologie Cellulaire, Equipe de Biochimie du Tissu Conjonctif, Université Paris VII-Denis Diderot, Tour 23/33, 3 Place Jussieu, 75261 Paris Cedex 05, France, *INSERM U 367, Physiologie et Pathologie Expérimentales Vasculaires, 17 rue du Four à Moulin, 75005 Paris, France and †Centre de Recherches en Gérontologie et Gériatrie, Hôpital d'Youville, Sherbrooke, Quebec J1H 4C4, Canada*

*Abstract.* Elastic fibres give blood vessels important rheological properties, such as the postsystolic elastic recoil. The age-dependent increase of $Ca^{2+}$ and lipid content, and elastolytic degradation of the fibres progressively impairs their function and produces circulating elastin peptides. Their interaction with the elastin receptor on smooth muscle cells induces not only increased cell–elastin fibre adhesion and endothelium-dependent vasorelaxation but also the release of lytic enzymes and oxygen free radicals from monocytes penetrating the vascular wall during atherogenesis. The age-dependent 'uncoupling' of the receptor has been shown to be involved in the loss of $Ca^{2+}$ homeostatic mechanisms and the progressive calcification of the vessel wall.

*1995 The molecular biology and pathology of elastic tissues. Wiley, Chichester (Ciba Foundation Symposium 192) p 286–303*

Blood vessels form a continuous tubular network for the circulation of blood as it travels from the heart, through the lungs (where gas exchange takes place), towards the periphery, through the capillaries and then back to the heart through the venous circulation. The diameter and wall construction of this tubular network varies continuously from 4–5 mm and a single cell layer surrounded by a basement membrane in capillaries, to about 2 cm and a relatively thick multicellular layer with a large amount of extracellular matrix in the large arteries. These elastic arteries are rich in elastic fibres, enabling them to perform a physiologically important function, known as the 'secondary heart' or 'windkessel' effect. The systolic blood volume dilates the proximal part of the thoracic aorta, distending its elastic meshwork. During the diastolic period, the energy-free elastic recoil of this portion of the aorta continues to pulse the blood towards the periphery. This elastic dilation/recoil continues according to a gradient along the main elastic arteries, from the thoracic aorta to the abdominal aorta, the carotids, and the iliac and femoral arteries. This pressure gradient is also reflected in the gradual attenuation of the elastic fibre content of these arteries. In the smaller, peripheral arterioles, elastin is restricted to the internal and sometimes the external limiting elastic lamina. The elastin content of blood vessels varies from nil in capillaries to about 40% of dry weight in the proximal thoracic aorta.

It has been shown over the last few decades that the elastic fibres undergo characteristic age changes which play an important role in the progressive loss of elasticity of blood vessels that occurs in the development of arteriosclerosis. This type of age-dependent modification of blood vessels is characterized by an increase of wall thickness, calcification, fragmentation of elastic fibres, diffuse lipid deposition, underlying loss of elasticity, increasing rigidity and loss of the 'secondary heart' effect. The consequence of these changes is an increasing load on the heart muscle which also loses its efficiency with ageing because of increasing fibrosis and irregularities in the rhythm of contractions. This is, in a nutshell, the basis of heart failure in old age; it is different in mechanism from local atheromatous plaque formation, which often starts early on in childhood and is susceptible to attenuation by nutritional measures and blood-lipid-correcting medication. Atheromatous plaque formation also involves modifications of the elastic fibres. It was observed by the 19th century pathologists that elastic fibre degradation precedes plaque formation. Balo (1963) first drew attention to the importance of elastic tissues in atherogenesis. He and his wife, I. Banga, discovered pancreatic elastase (Balo & Banga 1950)—a discovery which precipitated increasing interest in elastic tissues and their role in vascular pathology, and later in lung and skin pathology. The purpose of this review is to illustrate some of the mechanisms at the cellular and tissue levels which have been shown to be involved in the physiology and pathology of elastic blood vessels.

## The origin and deposition of elastic fibres in blood vessels

All three major cell types of the vascular wall—endothelial cells, smooth muscle cells and adventitial fibroblasts—can and do synthesize elastin (for a review see Robert & Hornebeck 1989). By far the largest amounts are synthesized by the smooth muscle cells (SMCs) of the media. Several isotypes of elastin mRNA and tropoelastin (the soluble precursor of elastin fibres) are produced by SMCs in culture. This heterogeneity is supposed to be due to alternative splicing of the primary transcript of the elastin gene, which can involve eight or more exons out of 36 exons of human elastin, according to results from the groups of Joel Rosenbloom and Charles Boyd (see Rosenbloom et al 1995, Sechler et al 1995, this volume). The functional significance of this process is not yet clear, but it would be very important if alternative splicing patterns were to change with age and also with elastin genotype. This process may thus turn out to be one of the most important factors behind genetic susceptibility to arterial disease.

The regulation of elastin gene expression is influenced by hormones, cytokines, growth factors and even vitamin C, which has been shown by C. Franzblau and his team to be able to decrease elastin expression. This important topic is discussed in this book by Davidson et al (1995, this volume). Such hormonal effects on elastin gene expression might well be involved in the well-documented sex differences in susceptibility to atherogenesis.

Although elastin neosynthesis is most intensive during embryonal development and early postnatal growth (Holzenberger et al 1993a), this mainly concerns the deposition of the concentric elastin lamellae of the larger blood vessels, such as the aorta. Interlamellar fibrils are continuously produced during maturation and our recent results on human skin fibroblasts show that during ageing there is an increase of elastin mRNA detectable by reverse transcriptase-PCR (Holzenberger et al 1993b). These results suggest a progressive 'de-repression' of elastin gene transcription with ageing, at least in the dermis. Histologically, it was shown, years ago, that aged skin can be rich in elastic fibres ('elastosis'), and patients with Werner's syndrome, undergoing premature ageing have a dense elastic fibre network in the skin (Robert 1994). Computerized image analysis techniques have also shown that there is an age-dependent increase of the elastic fibre network in the dermis, accompanied, however, by a steady decrease of skin elasticity (C. Robert et al 1988a,b). Similar studies on vascular elastin yielded similar results (Bouissou et al 1985). It appears, however, that the rheological properties of tissues and especially their elasticity is not a simple function of their elastin content.

## Interaction between cells and elastic fibres

Blood vessels undergo frequent volume changes according to the pulsation of the blood and also as a result of the change of tension of the SMCs produced by the mediators of the autonomous nervous system. In order to be efficient, the contractions and relaxations of SMCs have to be transmitted to the extracellular matrix of the vessel wall. This requires a dynamic coupling of cells to the surrounding matrix. As elastic lamellae are major components of the vessel wall, the question arose as to the nature of the coupling between SMCs and elastin fibres. We were able to show that SMCs, as well as fibroblasts, exhibit an inducible adhesion to elastin fibres (Hornebeck et al 1986). The addition of elastin peptides to the culture medium accelerated and intensified this process. This inductive effect of elastin degradation products on cell–elastic fibre adhesion could be ascribed to a cell membrane receptor reacting with elastin peptides and inducing the neosynthesis of a cell membrane-localized adhesive complex mediating the adhesion of cells to fibres (Robert et al 1989). The mechanism of transduction of the receptor towards the cell interior was shown to depend on the activation of a G protein, phospholipase C and phosphokinase C by the rapid mobilization of the inositol phosphate pathway

(Varga et al 1989). The demonstration of the presence of the elastin receptor on white blood cells, monocytes and polymorphonuclear leukocytes largely facilitated both its study and the exploration of its clinical significance (Fülöp et al 1986, Jacob et al 1987a). Figure 1 depicts schematically the elastin receptor and its intracellular transmission pathway. The variety of effects that can be ascribed to the activation of this receptor are shown in Table 1.

## Functional aspects of the elastin receptor

The reactions triggered by the activation of the elastin receptor vary according to the cell type concerned. Table 1 gives an overview of these effects observed in a variety of cell types. Those which are most relevant in the present context concern the cells of the normal vessel wall and the cells participating in the atherogenic processes. Endothelial cells have recently been shown to express the receptor and to produce a vasorelaxing effect on rat aorta rings in presence of elastin peptides (Faury et al 1994, 1995). This effect is dose dependent and is mediated by the release of nitric oxide (Fig. 2). The involvement of the 67 kDa subunit of the elastin receptor (Mecham et al 1989) is suggested by the

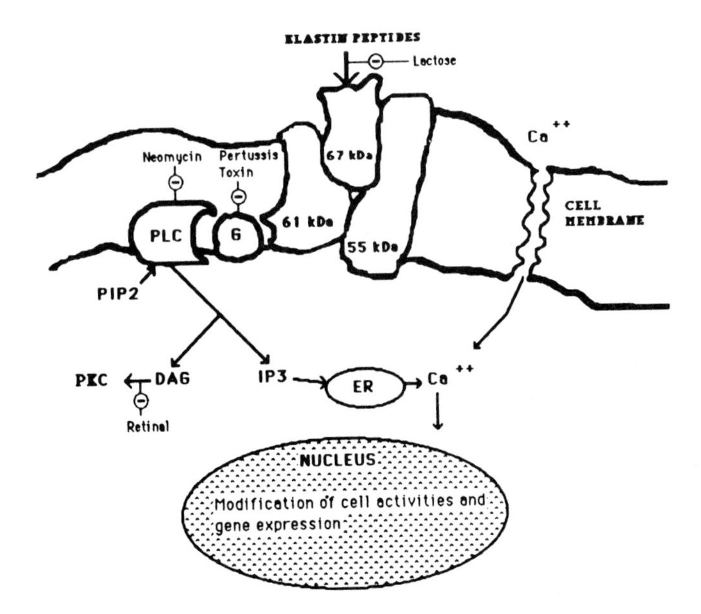


FIG. 1. Schematic representation of the elastin receptor with its three subunits and the transmission pathway of signalling. (Reproduced with permission from Robert et al 1989). DAG, diacylglycerol; PKC, protein kinase C; PLC phospholipase C; G, G protein; IP3, inositol trisphosphate; PIP2, phosphatidylinositol 4,5-bisphosphate.

**TABLE 1 Biological activities mediated by the elastin receptor**

| *Activity* | *Cell type* | *Reference* |
|---|---|---|
| Chemotactic migration | Monocytes<br>Fibroblasts<br>SMCs | Senior et al 1980, Ooyama et al 1987 |
| Modifications of ion fluxes: increase of $Ca^{2+}$, $Na^+$ influx; decrease of $K^+$ influx | Fibroblasts<br>SMCs<br>Mononuclear cells | Fülöp et al 1986, Jacob et al 1987b |
| Release of lysosomal enzymes | Mononuclear cells | Fülöp et al 1986 |
| Increased synthesis of membrane-bound matrix metalloproteases and elastases | Fibroblasts<br>SMCs | Archilla-Marcos & Robert 1993 |
| Increased oxygen consumption and free radical release | Mononuclear cells | Fülöp et al 1986 |
| Induction of adhesion of cells to elastin fibres | SMCs<br>Fibroblasts<br>Malignant cells | Hornebeck et al 1986, Tímár et al 1991 |
| Release of NO vasorelaxation | Endothelial cells | Faury et al 1994, 1995 |
| Growth factor-like activity | Fibroblasts | Ghuysen-Itard et al 1992 |

inhibition of the vasorelaxing effect of elastin peptides by lactose and laminin. A similar or identical 67 kDa subunit to the one described for the elastin receptor was claimed to react with laminin also (Mecham 1991, Mecham et al 1991). However, the inhibition of the vasorelaxing effect of elastin peptides by low concentrations of laminin ($10^{-10}$ M) suggest either a different site for receptor–laminin interaction or the reaction of the carbohydrate side chains of laminin with the lectin site of the receptor (Faury et al 1994, 1995). Smooth muscle cells were also shown to be chemotactically attracted to an elastin peptide gradient (Ooyama et al 1987). Activation of the elastin receptor has been shown to modify the relative rates of biosynthesis of matrix macromolecules (Fodil et al 1995). When added to monocytes, elastin peptides triggered the release of lytic enzymes and oxygen free radicals (Jacob et al 1987b, Fülöp et al 1986). Monocytes are also chemotactically attracted to an elastin peptide gradient (Senior et al 1980). Their presence in atherosclerotic plaques suggests an important role for the aforementioned reaction during the development of the atherosclerotic plaque. The lytic enzymes and free radicals released by elastin peptides from monocytes present

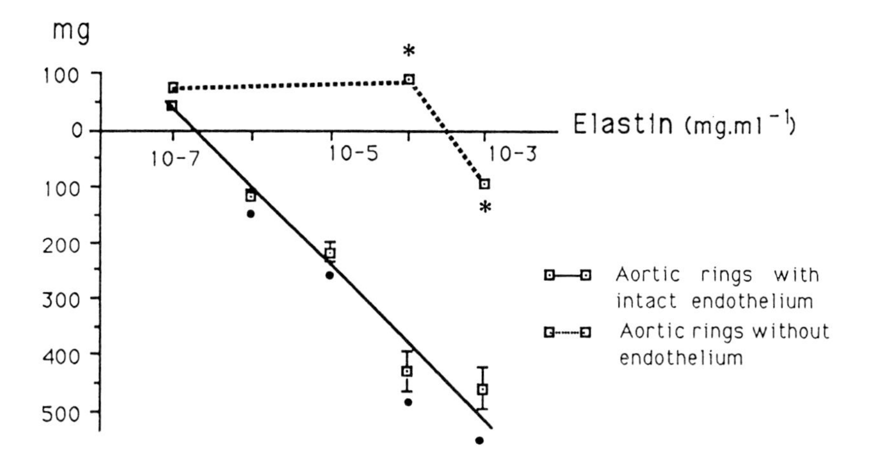


FIG. 2. Dose- and endothelium-dependent vasorelaxation of aorta rings (pre-contracted with $10^{-6}$ M noradrenaline) in presence of elastin peptides. Vertical axis represents vascular tone. (Reproduced with permission from Faury et al 1994, 1995.)

in the plaque area may well contribute to the degradation of the vascular wall and the production of more elastin peptides, therefore triggering a vicious circle entertaining the degradative process.

## Age-dependent modifications of the elastin receptor

When we tested the elastin receptor on monocytes or polymorphonuclear leukocytes from 'young' ($\leqslant$45 years) and 'old' ($\geqslant$65 years) individuals by recording the $Ca^{2+}$ transients, we found that the downhill slope of the curve following the rapid rise of intracellular free $Ca^{2+}$ decreased with ageing (Fig. 3). Similar results were observed with another receptor tested on the cells from the same individuals with the chemotactic peptide f-Met-Leu-Phe (FMLP; Fülöp et al 1990a, Varga et al 1990). This progressive age-dependent loss of the homeostatic regulation of intracellular $Ca^{2+}$ became even more evident when similar experiments were performed on polymorphonuclear leukocytes of older individuals in a geriatric hospital (Ghuysen-Itard et al 1993). About one out of three patients exhibited $Ca^{2+}$ transients suggesting a complete loss of control of intracellular $Ca^{2+}$ (Fig. 4). These patients' cells could be considered to be unable to fulfil their physiological role in host defence. If we assume a comparable age-related loss of $Ca^{2+}$ homeostatic regulation in other cell types, we have a logical explanation for the progressive increase of intracellular- and matrix-bound $Ca^{2+}$ that occurs in the ageing vascular wall. This increase also affects the elastic fibres.

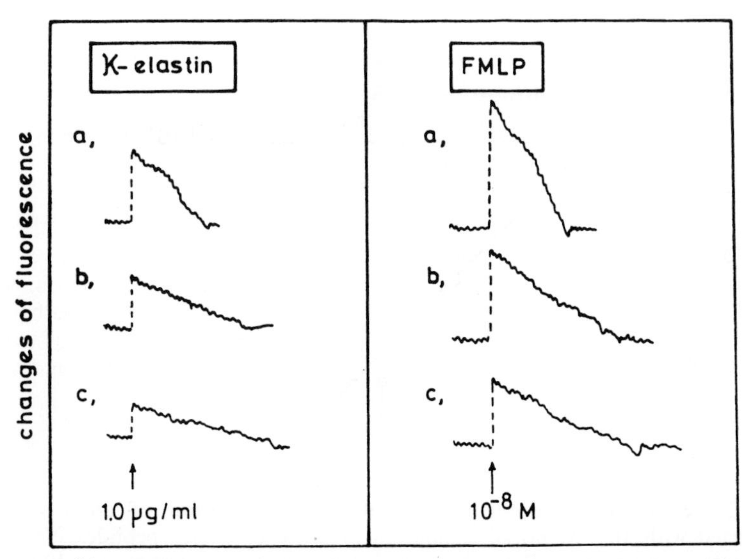


FIG. 3. $Ca^{2+}$ transients induced by κ-elastin peptides (left) and FMLP in polymorphonuclear leukocytes from (a) 'young' individuals (<45 years), (b) 'aged' individuals (65 years) and (c) from aged atherosclerotic individuals. Notice the decrease of the slope of the return to pre-excitation $Ca^{2+}$ levels with age and pathology. (Reproduced by permission from Varga et al 1988.)

As shown many years ago by Lansing, the $Ca^{2+}$ content of purified elastin fibres increases with age (see Robert & Robert 1980). It was also demonstrated that the fixation of lipids, and cholesterol and its esters is strongly potentiated by the presence of $Ca^{2+}$ (as opposed to $Na^{+}$) in the medium (Jacob et al 1983). These findings provided a physicochemical explanation for our previous histochemical observations of the strong intra- and extracellular $Ca^{2+}$ deposition in rabbit aorta after injection of elastin peptides in Freund's adjuvant (Jacob et al 1987b). These rabbits developed arteriosclerotic lesions with pronounced destruction of their elastic fibres. Strips of their aorta, when placed in a radioactive $Ca^{2+}$-containing bath, exhibited an increased $Ca^{2+}$ uptake which could be prevented by simultaneous *in vivo* treatment with calcitonin injections.

## Role of circulating elastin peptides

There have been several immunochemical demonstrations of elastin peptides in the blood circulation (see Fülöp et al 1990b). The concentrations of circulating elastin peptides varied according to the details of the procedure used (Wei et al 1993). The method we consider to be the most reliable yielded concentrations in the mg/ml range (Fülöp et al 1990b) and also showed significantly increased

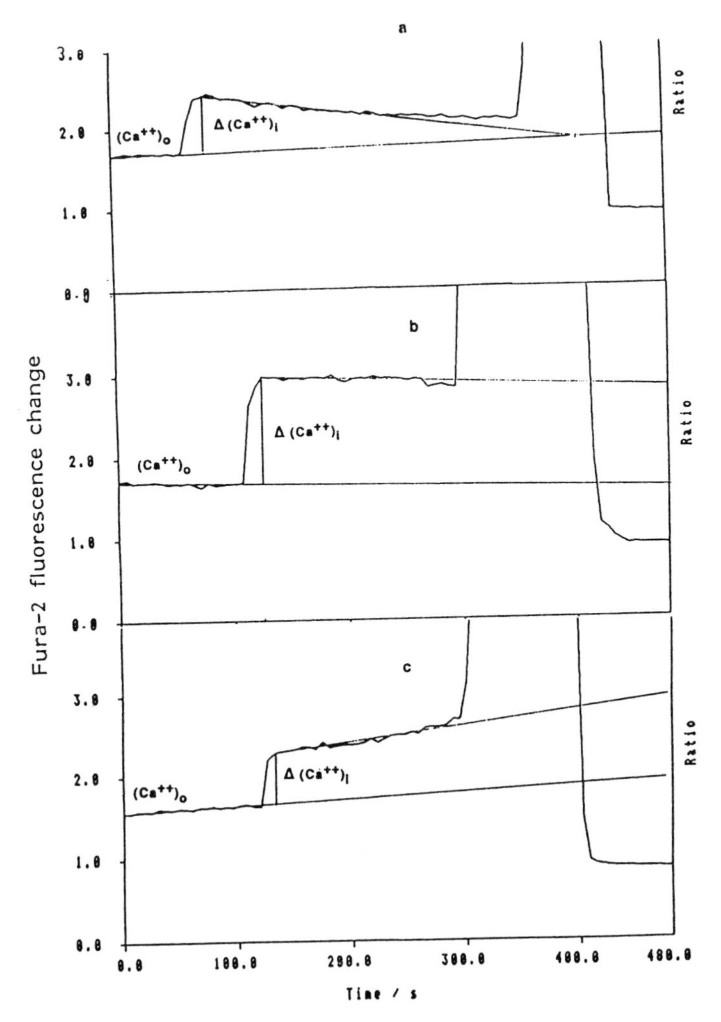


FIG. 4. Loss of $Ca^{2+}$ homeostatic mechanisms in polynuclear leukocytes in old (≥70 years) individuals. The $Ca^{2+}$ transients were recorded after addition of FMLP. The return to prestimulation levels was slow in most individuals (b) or absent (a + c) in about one-third of investigated patients. (Reproduced with permission from Ghuysen-Itard et al 1993.)

levels in obstructive arteriopathies and some hyperlipidaemias. These concentrations are in the range of saturation of the elastin receptor, as suggested by Scatchard analysis (Robert et al 1989). The origin and mechanism of production of these elastin peptides is not yet clear. They might result from the degradation of elastin fibres but they also might

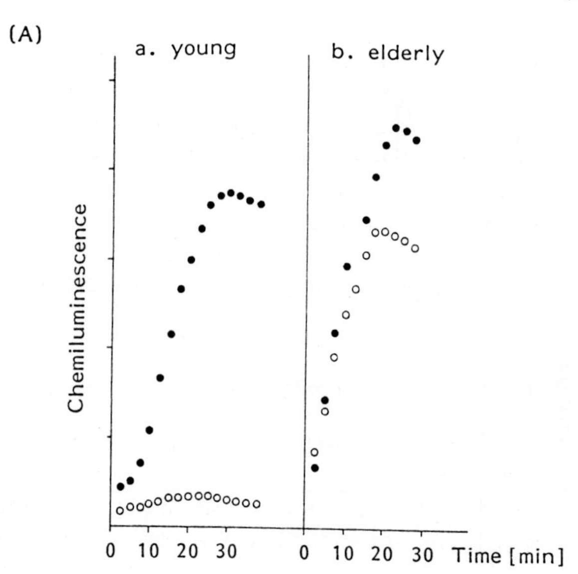
(A)
a. young
b. elderly
Chemiluminescence
0 10 20 30
0 10 20 30 Time [min]

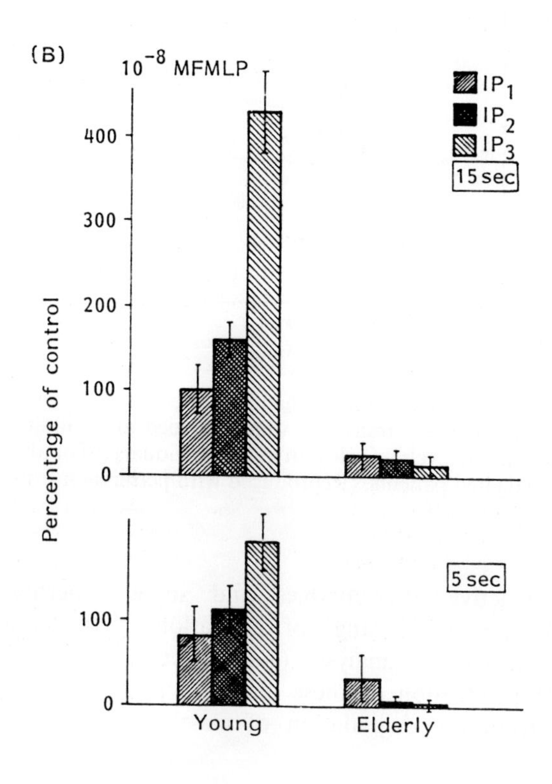
(B)
10⁻⁸ MFMLP
IP1
IP2
IP3
15 sec
5 sec
Percentage of control
400
300
200
100
0
100
0
Young
Elderly

represent products of post-synthetic degradation of the soluble precursors of elastin. Tropoelastins, rich in lysine residues, are easily attacked by a variety of proteolytic enzymes. Studies performed on more than a thousand human sera showed a wide dispersion of elastin peptide concentrations. Several hundred sera also exhibited the presence of anti-elastin antibodies (autoantibodies) of significant titres (Fülöp et al 1989). It can be assumed therefore that circulating immune complexes of elastin–anti-elastin antibodies are present in the blood of most individuals. These complexes may well contribute to the initiation of vascular lesions as observed in rabbits immunized with elastin peptides (Jacob et al 1987b).

## Pathogenic significance of the elastin receptor in the vascular wall

The aforementioned results suggest an important role for the elastin receptor in the physiopathology of the vascular wall. As mentioned in the introduction, the elastic fibre network of the vascular wall plays a significant role in the mechanical performance of the circulatory system. The emergence of the elastic fibre system in the vertebrates must have played an important role in their evolutionary success. However, from recent epidemiological studies on the ageing human populations, and especially from the Baltimore Longitudinal Study of Aging (BLSA, Shock 1977), it has become clear that the decline of the vascular and respiratory functions with age is among the fastest decrements registered by these *in vivo* clinical/physiological studies. The tissues concerned, such as the lungs and vessel walls, are rich in elastic fibres. The age-dependent loss of the physicochemical and rheological properties of elastic fibres appears to be the main factor in these processes. From detailed knowledge of the properties of elastin, its strong affinity for $Ca^{2+}$ and lipids (Robert & Robert 1980) has been suggested to underlie these processes. The progressive accumulation of lipids has also been demonstrated in human aortas (Claire et al 1976). In addition, we have demonstrated the age-dependent increase of elastase-type protease production by vascular smooth muscle cells and skin fibroblasts (Robert et al 1986). The presence of low-density lipoproteins or very

FIG. 5. Age-dependent 'uncoupling' of the elastin receptor. (A) Superoxide anion release from human monocytes in presence of elastin peptides (filled circles). In 'young' individuals this can be inhibited by the addition of pertussis toxin (open circles), a G protein inhibitor. In 'old' individuals this is without effect, but the superoxide anion release is maintained. Abscissa gives time after addition of elastin peptides. (Reproduced with permission from Varga et al 1990.) (B) The inositol phosphate pathway is rapidly mobilized after the addition of the agonist (FMLP) to 'young' mononuclear leukocytes. This mobilization is strongly reduced in cells from 'old' individuals. Inositol mono-, di- and trisphosphate concentrations were recorded at 5 and 15 seconds after addition of agonist. (Reproduced by permission from Fülöp et al 1990a.)

low-density lipoproteins (but not high-density lipoproteins) also increases protease production. Together, these findings form the elements of a vicious circle with continuous production of elastin peptides increasing with age and their sustained action on the elastin receptor resulting in progressive increase of the $Ca^{2+}$ content and 'uncoupling' of the receptor from its transmission pathway. The loss of inhibition of superoxide production in the presence of the agonist by pertussis toxin (Varga et al 1990) as well as the strong decrease of the mobilization of inositol phosphate metabolism in cells from 'old' donors (Varga et al 1988) (Fig. 5) can be interpreted as a functional uncoupling of the receptor from its membrane-localized components involved in the induction of intracellular signalling. It can be hypothesized that some processes at the level of vascular endothelium might result in an increased free-radical damage by oxygen free radicals and excessive production of nitric oxide (NO), yielding the highly toxic peroxynitrite radical (Faury et al 1995). These processes may well represent the lesion-initiating phenomena claimed by most theories of atherogenesis. These processes are part of the post-synthetic epigenetic mechanisms we proposed to be involved in the age-dependent loss of tissue function (Robert 1994).

It is proposed that the above described reactions are involved in the age-dependent modifications of the elastic arteries, *arteriosclerosis*, as well as in the local development of fibrous-lipidic plaques, *atherosclerosis*. These two processes seem to be at least partially independent. Atheromatous plaque formation can and does start early and is not strictly age-dependent; in contrast, arteriosclerosis is the progressive hardening of the elastic arteries characterized by loss of elasticity, increase in collagen: elastin ratio, and diffuse $Ca^{2+}$ and lipid depositions. The elementary physicochemical processes and the interaction between the elastin receptor and elastin peptides appear to be involved mainly in the age-dependent arteriosclerotic process. However, the presence of elastin receptor on monocytes, known to be involved in atherosclerotic plaque formation, suggests that the above process also has a role in the development of local atherofibrotic plaques.

It appears from the above considerations that the development of the elastic network in vertebrate tissues has been a vital element in their evolutionary success. The physicochemical properties of elastin and the nature of its interactions with the elastin receptor are an important factor in this respect. However, the age-dependent modifications of elastic fibres and their interaction with lipids and $Ca^{2+}$ impairs their function and accelerates their degradation. The interaction of the resulting elastin peptides with their receptor, and its progressive uncoupling during ageing with preserved free radical and lytic enzyme production, are crucial factors in the age-dependent modifications of elastic tissues, such as blood vessels, lungs and skin.

## References

Archilla-Marcos M, Robert L 1993 Control of the biosynthesis and excretion of the elastase-type protease of human skin fibroblasts by the elastin receptor. Clin Physiol Biochem 10:86–91

Balo J 1963 Connective tissue changes in atherosclerosis. Int Rev Connect Tissue Res 1:241–306

Balo J, Banga I 1950 The elastolytic activity of pancreatic extracts. Biochem J 46:384–387

Bouissou H, Pieraggi MT, Julian M 1985 Aging of elastic tissue in skin and arteries. In: Robert L, Murata K, Nagai Y (eds) Degenerative diseases of connective tissue and aging. Kodansha, Tokyo, p 203–216

Claire M, Jacotot B, Robert L 1976 Characterization of lipids associated with macromolecules of intercellular matrix of human aorta. Connect Tissue Res 4:61–71

Davidson JM, Zhang M-C, Zoia O, Giro MG 1995 Regulation of elastin synthesis in pathological states. In: The molecular biology and pathology of elastic tissues. Wiley, Chichester (Ciba Found Symp 192) p 81–99

Faury G, Ristori MT, Verdetti J, Jacob MP, Robert L 1994 Rôle du récepteur de l'élastine–laminine dans la vasorégulation. CR Acad Sci Ser III Sci Vie 317:807–811

Faury G, Ristori MT, Verdetti J, Jacob MP, Robert L 1995 Effect of elastin peptides on vascular tone. J Vasc Res 32:112–119

Fodil I, Ghuysen-Itard A, Jacob MP, Robert L 1995 Modulation of the biosynthesis of extracellular matrix by the elastin receptor. Acceleration of age-dependent modification. Gerontology, in press

Fülöp T, Jacob MP, Varga Z, Foris G, Loevey A, Robert L 1986 Effect of elastin peptides on human monocytes: $Ca^{2+}$, mobilization, stimulation of respiratory burst and enzyme secretion. Biochem Biophys Res Commun 141:92–98

Fülöp T Jr, Jacob MP, Robert L 1989 Determination of anti-elastin peptide antibodies in normal and arteriosclerotic human sera by ELISA. J Clin Lab Immunol 30:69–74

Fülöp T, Varga Z, Csongor J et al 1990a Altered phosphatidylinositol breakdown in polymorphonuclear leukocytes with aging. In: Goldstein AL (ed) Biomedical advances in aging. Plenum Press, NY, p 187–194

Fülöp T, Wei SM, Robert L, Jacob MP 1990b Determination of elastin peptides in normal and arteriosclerotic human sera by ELISA. Clin Physiol Biochem 8:273–282

Ghuysen-Itard AF, Robert L, Jacob MP 1992 Effet des peptides d'élastine sur la prolifération cellulaire. CR Acad Sci Ser III Sci Vie 315:473–478

Ghuysen-Itard AF, Robert L, Gourlet V, Berr C, Jacob MP 1993 Loss of calcium-homeostatic mechanisms in polymorphonuclear leukocytes of demented and nondemented elderly subjects. Gerontology 39:163–169

Holzenberger M, Ayer-Le Lièvre C, Robert L 1993a Tropoelastin gene expression in the developing vascular system of the chicken: an in situ hybridization study. Anat Embryol 188:481–492

Holzenberger M, Levi-Minzi SA, Herzog CP, Deak SB, Robert L, Boyd CD 1993b Quantitation of tropoelastin mRNA and assessment of alternative splicing in human skin fibroblasts by reverse transcriptase-polymerase chain reaction. PCR Methods Applications 3:107–114

Hornebeck W, Tixier JM, Robert L 1986 Inducible adhesion of mesenchymal cells to elastic fibers: elastonectin. Proc Natl Acad Sci USA 83:5517–5520

Jacob MP, Hornebeck W, Robert L 1983 Studies on the interactions of cholesterol with soluble and insoluble elastins. Int J Biol Macromol 5:275–278

Jacob MP, Moura AM, Tixier JM et al 1987a Prevention by calcitonin of the pathological modifications of the rabbit arterial wall induced by immunization with elastin peptides: effect on vascular smooth muscle permeability. Exp Mol Pathol 46:345–356

Jacob MP, Fülöp T, Foris G, Robert L 1987b Effect of elastin peptides on ion fluxes in mononuclear cells, fibroblasts, and smooth muscle cells. Proc Natl Acad Sci USA 84:995–999

Mecham RP 1991 Receptors for laminin on mammalian cells. FASEB J 5:2538–2546

Mecham RP, Hinek A, Entwistle R, Wrenn DS, Griffin GL, Senior RM 1989 Elastin binds to a multifunctional 67-kilodalton peripheral membrane protein. Biochemistry 28:3716–3722

Mecham RP, Whitehouse L, Hay M, Hinek A, Sheetz MP 1991 Ligand Affinity of the 67-kD elastin/laminin binding protein is modulated by the protein's lectin domain—visualization of elastin/laminin/receptor complexes with gold-tagged ligands. J Cell Biol 113:187–194

Ooyama T, Fukuda K, Oda H, Nakamura H, Hikita Y 1987 Substratum-bound elastin peptide inhibits aortic smooth muscle cell migration *in vitro*. Arteriosclerosis 7:593–598

Robert C, Lesty C, Robert AM 1988a Ageing of the skin: study of elastic fiber network. Modifications by computerized image analysis. Gerontology 34:91–96

Robert C, Blanc M, Lesty C, Dikstein S, Robert L 1988b Study of skin ageing as a function of social and professional conditions: modification of the rheological parameters measured with a non-invasive method: indentometry. Gerontology 34:84–90

Robert L 1994 Le vieillissement de l'homme à la cellule. Belin-CNRS, Paris

Robert L, Hornebeck W 1989 Elastin and elastases (2 vols). CRC Press, Boca Raton, FL

Robert L, Robert AM 1980 Elastin, elastase and arteriosclerosis. In: Robert AM, Robert L (eds) Frontiers of matrix biology, vol 8: biology and pathology of elastic tissues. Karger, Basle, p 130–173

Robert L, Labat-Robert J, Hornebeck W 1986 Aging and atherosclerosis. In: Gotto AM, Paoletti R (eds) Atherosclerosis reviews, vol 14. Raven Press, NY, p 143–170

Robert L, Jacob MP, Fülöp T, Tímár J, Hornebeck W 1989 Elastonectin and the elastin receptor. Pathol Biol 37:736–741

Rosenbloom J, Abrams WR, Indik Z, Yeh H, Ornstein-Goldstein N, Bashir MM 1995 Structure of the elastin gene. In: The molecular biology and pathology of elastic tissues. Wiley, Chichester (Ciba Found Symp 192) p 59–80

Sechler JL, Sandberg LB, Roos PJ et al 1995 Elastin gene mutations in transgenic mice. In: The molecular biology and pathology of elastic tissues. Wiley, Chichester (Ciba Found Symp 192) p 148–171

Senior RM, Griffin GL, Mecham RP 1980 Chemotactic activity of elastin-derived peptides. J Clin Invest 66:859–862

Shock NW 1977 Systems integration. In: Finch CE, Hayflick L (eds) Handbook of the biology of aging. Van Nostrand Reinhold, New York, p 639–665

Tímár J, Lapis K, Fülöp T et al 1991 Interaction between elastin and tumor cell lines with different metastatic potential: *in vitro* and *in vivo* studies. J Cancer Res Clin Oncol 117:232–238

Varga Z, Kovacs EM, Paragh G, Jacob MP, Robert L, Fülöp T 1988 Effect of elastin peptides and N-formyl-methionyl-leucyl-phenylalanine on cytosolic free calcium in polymorphonuclear leukocytes of healthy middle-aged and elderly subjects. Clin Biochem 21:127–130

Varga Z, Jacob MP, Robert L, Fülöp T Jr 1989 Identification and signal transduction mechanism of elastin peptide receptor in human leukocytes. FEBS Lett 358:5–8

Varga Z, Jacob MP, Csongor J, Robert L Leovey A, Fülöp T Jr 1990 Altered phosphatidylinositol breakdown after κ-elastin stimulation in PMBIs of elderly. Mech Ageing Dev 52:61–70

Wei SM, Erdei J, Fülöp T, Robert L, Jacob MP 1993 Elastin peptide concentration in human serum: variation with antibodies and elastin peptides used for the enzyme-linked immunosorbent assay. J Immunol Methods 164:175–187

## DISCUSSION

*Urry:* Which peptides in κ-elastin are involved in the receptor interactions?

*Robert:* We don't know. In our binding experiments we use elastin peptides of a homogeneous molecular size, but this is still a mixture. Some of the reactions can be obtained with the VGVPG peptide, but we are sure that this is not the only active sequence. The only extensive work we have done is screening for epitopes with monoclonal and polyclonal antibodies.

From work which was done on the insulin receptor, it can be deduced that epitope sequences may be involved in the receptor activation.

*Urry:* When we were actively working on elastin structure and function, we synthesized every one of the hydrophobic sequences between cross-links. Small amounts of each of these sequences, usually as polymers of the sequences between cross-links, are available should you wish to assay for particular biological activity. I need to warn you though that chemical synthesis has its problems with racemization. Some of the synthetic polymers would be available, too, for $Ca^{2+}$-induced cholesterol binding studies of interest to Larry Sandberg.

*Uitto:* Leslie, you described measurements of the serum metalloproteinase or metalloelastase-like enzyme activities. Were those measurements done with synthetic peptides, such as SAPNA, or are they really elastolytic enzymes capable of degrading elastic fibres?

*Robert:* The serum enzyme can also act, albeit very slowly, on elastic fibres, but we used synthetic peptides for the epidemiological studies. Our primary goal was to see whether serum elastase activity was correlated with pathology. When we found a significant correlation between elastase activity and what clinicians consider to be some of the risk factors for atherosclerosis, such as serum triglycerides and diabetes, we looked to see if there was any correlation with possible sources of metallo-elastases such as monocytes or macrophages. There was no significant correlation. Therefore the most probable source of the serum enzyme is endothelial cells (where we showed the presence of metalloendopeptidases), the liver or other organs. It was also shown that HDL carries a metalloendopeptidase, tightly linked to ApoA1 (Jacob et al 1981).

*Keeley:* One of the things that happens when you activate the focal adhesion complex is that you activate the $Na^+$ antiporter. I noticed that when you bound your elastin peptides you had an increase in $Na^+$ flux into the cell. Is this amelioride sensitive?

*Robert:* I don't know. I didn't show the results, but we have shown that the ouabaine-sensitive $K^+$ pump is inhibited by elastin peptides in a dose-dependent fashion (Jacob et al 1987).

*Mecham:* One of the issues that we've struggled with in our lab is whether or not there are multiple receptors mediating all of these effects, or if all of these effects occur through the 67 kDa receptor. We've seen that the 67 kDa receptor might act as a molecular chaperone inside the cell and you have shown many effects that are stimulated in the cell by elastin peptides. Do you have any idea which signalling protein is responsible for these effects? Is it the 67 kDa receptor, or elastonectin, or some other protein?

*Robert:* All our data favour the 67 kDa receptor. We only see the 120 kDa molecule (elastonectin) involved in the adhesion of cell membranes to the elastin fibres (Perdomo et al 1994).

*Hinek:* You painted a beautiful picture of what happens in atherosclerosis. But we also have to mention that atherosclerosis involves loss of the 67 kDa receptors. Endothelial cells and myocytes produce a lot of galacto-glycosaminoglycans, such as dermatan and chondroitin sulfate, which cause shedding of the 67 kDa elastin binding protein from the surfaces of some smooth muscle cells. Immunostaining of atherosclerotic plaques revealed a real deficiency in the cell-surface of expression of this receptor. In culture, stripping the elastin receptor by the use of lactose or chondroitin sulfate up-regulates the migration of smooth muscle cells and also up-regulates production of fibronectin (Hinek et al 1992), which many people believe provides a sliding path for migration of smooth muscle cells. The whole picture must be completed with the idea that smooth muscle cells are also losing receptor so they can't respond in the same way as cells in normal vascular wall.

*Robert:* If this turns out to be the case we'll have to modify our picture. This is not what we see in white blood cells; these don't lose their receptor, but become unresponsive to any agent which acts beyond the G protein (Varga et al 1990).

*Mecham:* You might want to look at HL-60 cells, which in their undifferentiated state don't interact with elastin—at least in the kinds of assays we use. If you differentiate them with various agents, they express an elastin-binding protein. You might be able to use that kind of a cell and modulate its ability to interact with elastin to look at some of these finer points.

*Hinek:* HL-60 cells have elastin receptor intracellularly, but they require stimulation by phorbol esters to express receptor on the cell surface.

*Robert:* We tried using another line of pre-monocytes, but this didn't work. We tested by $Ca^{2+}$ flux by immunofluorescence and adhesion experiments. By incubating these cells with elastin, we couldn't induce $Ca^{2+}$ flux to elastin peptides or adhesion to elastin fibres. Normal monocytes (according to experiments Dr Fülöp did in Hungary) do adhere to elastin fibres.

*Davidson:* Would you like to speculate on the effect of laminin as an antagonist in your relaxation study?

*Robert:* When we looked through the data we thought that perhaps we had been too careful not to add higher concentrations of laminin—we used very low concentrations ($10^{-4}$ to $10^{-6}$ mg/ml). Now we are repeating these experiments with higher concentrations. We expected to see agonist-type activity, not antagonist. Elastin peptides induced vasorelaxation and this was inhibited by laminin as well as by lactose (Faury et al 1995). This effect is mediated by nitric oxide release triggered by the activation of the receptor.

*Mungai:* Studies of migrants from rural to urban environments have found significantly higher blood pressures in the migrants compared with the rural controls (Poulter et al 1985, 1990). Rural–urban migration has also been found to be associated with a marked increase in dietary $Na^+$ and a decrease in $K^+$ intake. This has further been found to be associated with abandonment of $K^+$-rich traditional foods used by rural communities and adoption of $Na^+$-rich foods by urban migrants.

Since the $Na^+/K^+$ ratio has also been found to be as significant as the low-$Na^+$ diet in the protection against high blood pressure (Obel et al 1985), I wonder if $Na^+$ has featured at all in your studies?

*Robert:* This is exactly what we see, because the receptor pumps in $Na^+$ and excludes $K^+$. These interesting data you mention confirm nicely what we think about ageing and the elastin receptor (Varga et al 1990). We wanted to see if other cells also express the elastin receptor. We are interested in what happens to the CNS during ageing, because if a chronic treatment with elastin peptides really does cause them to release NO and superoxide and produce peroxynitrite (a very toxic free radical), we should see its toxic effect in the CNS also.

In collaboration with Dr Annie Reber in Rouen, we injected elastin peptides intraperitoneally to rats for 5 weeks and tested the visual vestibular reflex before and during treatment, and 3 weeks after treatment. In the treated animals the reflex showed accelerated ageing. The constant circulation of elastin peptides and the chronic action on the elastin receptor might result in the desensitization of the receptor with a toxic effect by sustained peroxynitrite production. This may play an important role not only in vascular ageing but in other organs such as the brain.

Hypertension, as you know, develops progressively with age, together with increased wall thickness. This is a very important mechanism and only a few people have tested $Ca^{2+}$ antagonists (Fleckenstein 1990). The Fleckensteins have advocated the use of $Ca^{2+}$ antagonists for the treatment of arteriosclerosis.

*Urry:* You raised this question about the interaction with cholesterol that induces $Ca^{2+}$ binding. We studied $Ca^{2+}$ binding to a number of the repeats of elastin, to (APGVGV), (VPGVG) and (VPGG). It does bind, especially to APGVGV, and when it binds it turns the polar carbonyls in and the hydrophobic groups out. This leaves a hydrophobic microdomain with positive central charge. So, it should induce the best binding for the bile salts for other negatively charged lipids.

*Robert:* This fits very nicely with results we obtained by analysing the lipids in purified human elastin from patients with severe atherosclerosis (Claire et al 1976). The most conspicuous change is the increase in free fatty acids, but all lipid classes increase with age and atherosclerosis.

*Urry:* Any hydrophobic molecule that can add ion pairing with the $Ca^{2+}$ should have a good affinity.

*Kagan:* I wonder if your metalloproteinase might be activated by $Ca^{2+}$, in part at least, by interaction of the metal ion with the substrate rather than with the enzyme.

*Robert:* This should be tested. In the experiments I described, we wanted to see whether the enzyme is clinically significant. Now we would like to see if the correlations we found with risk factors such as diabetes and serum triglycerides will affect the clinical outcome in four or six years (Bizbiz et al 1995).

## References


Bizbiz L, Bonithon-Kopp C, Ducimetiere P, Alperovitch A, Robert L and the EVA Group 1995 Relation of serum elastase activity to ultrasonographically assessed carotid artery wall lesions and cardiovascular risk factors. The EVA study. Atherosclerosis, submitted

Claire M, Jacotot B, Robert L 1976 Characterization of lipids associated with macromolecules of the intercellular matrix of human aorta. Connect Tissue Res 4:61–71

Faury G, Ristori MT, Verdetti J, Jacob MP, Robert L 1995 Effect of elastin peptides on vascular tone. J Vasc Res 32:112–119

Fleckenstein A 1990 History and prospects in calcium antagonist research. J Mol Cell Cardiol 22:241–251

Hinek A, Boyle J, Rabinovitch M 1992 Vascular smooth muscle cell detachment from elastin and migration through elastic laminae is promoted by chondroitin sulfate-induced shedding of the 67-kD cell surface elastin binding protein. Exp Cell Res 203:344–353

Jacob MP, Bellon G, Robert L, Hornebeck W et al 1981 Elastase-type activity associated with high density lipoproteins in human serum. Biochem Biophys Res Comm 103:311–318

Jacob MP, Fülöp T, Foris G, Robert L 1987 Effect of elastin peptides on ion fluxes in mononuclear cells, fibroblasts, and smooth muscle cells. Proc Natl Acad Sci USA 84:995–999

Obel AOK, Kofi-Tsekpo WM, Ellison RH, Mugambi M 1985 Dietary sodium/potassium ratio in salt substitute and its putative significance in essential hypertension. East Afr Med J 62:507–514

Perdomo JJ, Gounon P, Schaeverbeke M et al 1994 Interaction between cells and elastin fibers: an ultrastructural and immunocytochemical study. J Cell Physiol 158:451–458

Poulter NR, Khaw KT, Mugambi M, Peart WS, Sever PS 1985 Migration-induced changes in blood pressure: a controlled longitudinal study. Clin Exp Pharmacol Physiol 12:211–216

Poulter NR, Khaw KT, Hopwood BE et al 1990 The Kenyan Luo migration study: observation on the initiation of a rise in blood pressure. Brit Med J 300:967–972

Varga Z, Jacob MP, Csongor J, Robert L, Leovey A, Fülöp T 1990 Altered phosphatidylinositol breakdown after κ-elastin stimulation in PMBIs of elderly. Mech Ageing Dev 52:61–70

# General discussion V

## The non-enzymic glycation of elastin

*Bailey:* The non-enzymic glycation of proteins, particularly those with a long biological half-life, has generated considerable interest in view of its relevance to long-term diabetic complications (for review see Baynes & Monnier 1989). Of the structural connective tissue proteins, collagen has been intensively studied, while elastin has been totally neglected, despite the possibility of changes in the latter in the early onset of disease of the vascular system. This is particularly important since arteriosclerosis is a major cause of premature mortality in diabetic patients.

The initial reaction in non-enzymic glycation is the condensation of glucose with the ε-amino group of the peptide-bound lysine residues. Such reactions have been known for some time and have been reported to occur in various tissues during ageing and in certain pathological conditions. More recently it has been shown that subsequent oxidation of the hexosyl-lysines results in additional products some of which form intermolecular cross-links, such as the fluorescent compound pentosidine (Sell & Monnier 1989) and our recently isolated but as yet structurally uncharacterized non-fluorescent compound NFC-1 (Bailey et al 1995). However, the lysine content of elastin is very low, only five residues per molecule, compared with 35 for collagen, owing to the large number involved in the formation of the extensive network of desmosine cross-links. It is unlikely therefore that the changes would be extensive. However, any deleterious effect of glycation on the elasticity of the elastin fibres is likely to have a more dramatic effect on the properties of the vascular system than on the already virtually inextensible collagen fibres. It is therefore of some interest to demonstrate the possibility of additional cross-linking of the elastin through non-enzymic glycation.

*In vitro* incubation with glucose or ribose (250 mM for 3 days for ribose and 30 days for glucose) revealed the presence of glycated lysines. Determination of the hexosyl-lysines as furosine indicated one glycated lysine per 10 elastin molecules. This initial blocking of the ε-amino groups would alter the charge on the elastin fibre and possibly its interaction with other matrix molecules. Additional analysis for fluorescent cross-linking compounds by HPLC revealed the presence of pentosidine. The yield was approximately one pentosidine per 2000 molecules of elastin, assuming a molecular weight for elastin monomer of 68.5 kDa. This low yield can be compared to that of other tissues rich in lysine, where the yield is usually about one pentosidine per several hundred substrate molecules. Sell & Monnier (1989) proposed that pentosidine is formed through the reaction of ribose, lysine and arginine, although the same product is formed from glucose. Examination of the amino acid sequence data (Ayad et al 1994) for the potential cross-linking sites of elastin reveals only two domains with lysines (K) close to arginine (R), both of which are in the desmosine cross-linking regions, i.e. AAA**K**AAA**K**AA**K**YGA**R**PG and AAA**K**SAA**K**VAA**K**AQL**R**A. These α-helical cross-link regions would be aligned in the elastin fibre to allow the formation of intermolecular desmosine cross-links. This alignment would also permit the formation of intermolecular pentosidine cross-links. However, in mature elastic tissue most of these lysines would already be involved in desmosine cross-linking. An additional, non-cross-linking sequence involving lysine and arginine, **K**ACG**RKRK**, occurs at the C-terminus of the molecule and would also be capable of forming pentosidine. In this case the pentosidine could be formed within a single peptide chain or between peptide chains.

Analysis for the non-fluorescent putative cross-link, NFC-1, previously identified in collagen, revealed its presence at a low yield (approximately one NFC-1 cross-link per 250 elastin molecules, compared with one per five molecules in glycated collagen). The structure of NFC-1 is currently being determined and it is not therefore possible to consider its location from the amino acid sequence.

In an extension of these studies we have been investigating the increased degradation of low-density lipoproteins (LDL) by glycated collagen, which results in a higher yield of malondialdehyde (MDA), one of its major breakdown products. The MDA is a potent cross-linking agent and reacts directly with collagen, rendering it insoluble and stiffer. Elastin, along with collagen, is a major component of the arterial wall, and even though the extent of glycation is small, it could over a long period produce a similar reaction with LDL to form MDA, which in turn would cross-link the elastin fibre. We have therefore studied the possible reaction of MDA with purified elastin and demonstrated a number of fluorescent peaks on the HPLC columns that increase with reaction time. Studies are continuing to isolate and characterize these components as cross-links.

The cycle of glycation and cross-linking of the arterial collagen and elastin, oxidation of the LDL and further cross-linking by the MDA would lead not only to a slow stiffening of the arterial wall but a reduction in the already low turnover of these proteins and as a consequence allow them to undergo further glycation. This cycle of glycation and MDA cross-linking would continue. Thus both glycation and cross-linking would affect the elastic properties of the elastin and possibly contribute to early arteriosclerosis in diabetes.

The amounts of pentosidine and NFC-1 formed by glycation of elastin are certainly not very extensive compared with collagen, and it is not clear whether it would affect the physical properties of the elastin, although it certainly has the possibility of increasing several-fold over many years of ageing *in vivo*. Although I failed to find any published data on the effect of glycation on the properties of elastin, discussion with Peter Winlove at this meeting revealed that he has some preliminary data indicating glycation does indeed affect the physical properties of elastin. Clearly further studies are warranted.

*Robert:* It's interesting that in collagen the cross-linking confers resistance to collagenase, whereas cross-linking appears to make elastin more labile to elastases. Elastin is degraded faster in diabetics than in non-diabetics. There must be subtle differences in this story to explain why cross-linking by glycation stabilizes collagen and not elastin.

*Uitto:* One of the clinical features of poorly-controlled diabetes mellitus is the scleroderma-like cutaneous changes which involve thickening of the skin. We looked into this a few years ago (Perejda et al 1984). The collagen, which was non-enzymically glycosylated about two- to threefold over the background level, was more resistant to collagenase. This is one of the mechanisms behind the collagen build-up in diabetes.

*Keeley:* At the normal physiological range of pressures, most of the load and the expansion of the vessel is taken up by the elastin and not by the collagen. Therefore, increased stiffness or cross-linking of elastin would be much more crucial for changing the properties of the vessel wall than collagen. If you make collagen stiffer it's not going to matter a lot, but if you make elastin stiffer it's going to make a big difference.

*Bailey:* Yes; that is why we are looking at it.

## References

Ayad S, Boot-Handford RP, Humphries MJ, Kadler KJ, Shuttleworth CA 1994 The extracellular matrix facts book. Academic Press, London

Bailey AJ, Sims TJ, Avery NC, Halligan EP 1995 Non-enzymic glycation of fibrous collagen: reactions of glucose and ribose. Biochem J 305:385–390

Baynes JW, Monnier VM 1989 The Maillard reaction in aging, diabetes and nutrition. Alan Liss, NY

Perejda AJ, Zaragoza EJ, Eriksen E, Uitto J 1984 Non-enzymatic glycosylation of lysyl and hydroxylysyl residues in type I and type II collagens. Collagen Rel Res 4:427–439

Sell DR, Monnier VM 1989 Structural elucidation of a lysine-arginine cross-link from aging human extracellular matrix. Implication of ribose in the aging process. J Biol Chem 264:21957–21602

# Elastase in the prevention of arterial ageing and the treatment of atherosclerosis

Toshiro Ooyama* and Hiroshi Sakamato

*Department of Internal Medicine, Tokyo Metropolitan Geriatric Hospital, 35-2 Sakae-machi, Itabashi-ku, Tokyo, Japan*

*Abstract.* Arterial ageing is defined as the age-related structural and functional changes in arteries from the precapillary to the aortic level. These include atheromatous changes. Such changes can be estimated in medium-sized or larger arteries by clinical diagnostic studies, including B-mode echography of the carotid artery (thickening of the intima, plaque formation and increase in luminal diameter) and abdominal aorta (plaque formation and increase in luminal diameter), Doppler echography (pulse wave velocity; PWV) and autopsy studies. These changes include distension of the lumen, increased arterial wall thickness (which may be associated with atherosclerotic plaques) and decreased extensibility of the arterial wall. Since 1981, an anti-atherosclerotic drug containing porcine pancreatic elastase 1 (PPE1) has been used for the prevention of arterial ageing and the treatment of atherosclerosis in elderly patients in Japan. So far, the age-related increase in PWV has been found to be lower in those who take PPE1 than in controls. The atherosclerotic index of the carotid artery has also been found to be lower in subjects receiving PPE1 treatment than in control subjects. The pharmacological basis of PPE1 therapy, although paradoxical to the consensus opinion of the pathogenic role of elastase in western countries, is discussed with reference to data gathered in Japan. The pathomechanism of arterial ageing and atherosclerosis, with special reference to elastin, is reviewed along with the presentation of some of our data.

*1995 The molecular biology and pathology of elastic tissues. Wiley, Chichester (Ciba Foundation Symposium 192) p 307–320*

## A definition of arterial ageing

Ageing of the arterial wall is defined as the age-related structural and functional changes in arteries from precapillary to aortic level (Ooyama 1991, Robert & Jacotot 1994). Such changes are often visible in medium-sized or

*Present address: Department of Clinical Functional Physiology, Faculty of Medicine, Toho University, 6-11-1 Omori-nishi, Ota-ku, Tokyo, Japan.

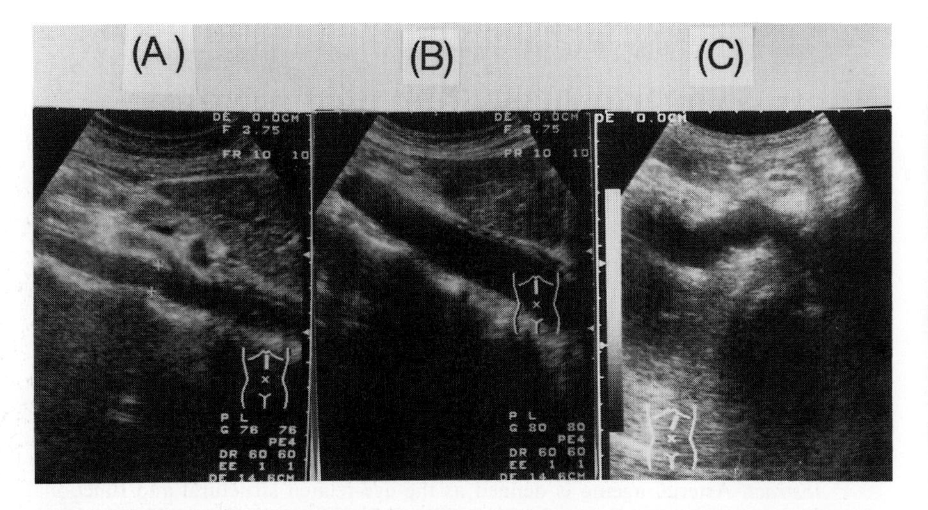


FIG. 1. B-mode echography of the human abdominal aorta in: (A) a 23-year-old woman (12 mm luminal diameter); (B) a 69-year-old woman (20 mm luminal diameter); and (C) a 89-year-old woman (22 mm luminal diameter and kinking).

larger arteries in routine clinical and pathological studies. These changes include distension of the lumen (Fig. 1) and stiffening of the arterial wall. With B-mode echography, pulse movement is observed in the arteries of young people but not in the arteries of the elderly. In the human abdominal aorta, the luminal diameter increases progressively with age from the third to the fifth decades, at which point it reaches a plateau (Fig. 2). In an autopsy study, Nakashima & Tanikawa (1971) showed that the thoracic aorta of young subjects could be extended to nearly 1.8 times its resting luminal diameter when the intraluminal pressure was increased. In contrast, the aorta from aged subjects extended only to 1.2 times the resting luminal diameter. In B-mode echography, the absence of pulse movement indicates a decrease in the elasticity of the arterial wall. Measurement of the pulse wave velocity (PWV) of an elastic artery by Doppler echography is the only method that can quantitatively estimate the extent of the decrease in the elasticity of the arterial wall. PWV increases with age (Fig. 3) (Hasegawa et al 1984).

## Arterial ageing: its clinical significance

Age-dependent modifications in the artery occur alongside atheromatous changes, such as the development of focal plaques with deposition of lipids and $Ca^{2+}$ which, in contrast to age-related arterial changes, narrow the lumen of the artery. However, the severity and extent of atherosclerosis and age-related changes in a single artery usually do not correlate with those of other arteries.

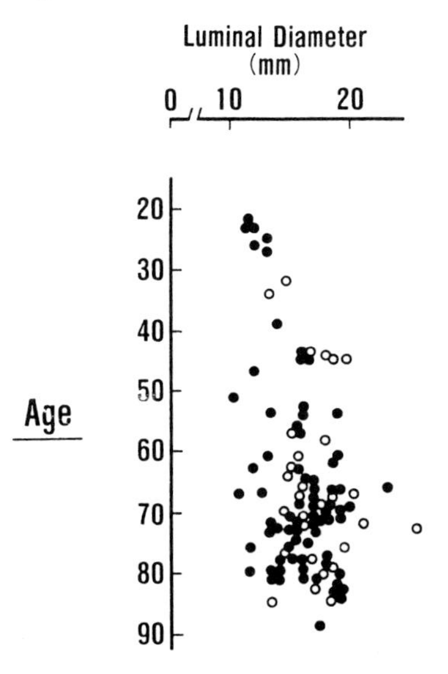


FIG. 2. Age-related changes in the luminal diameter of the abdominal aorta. Using B-mode echography, we measured the luminal diameter of the abdominal aorta at the origin of the coeliac artery at end-diastolic pressure. ○, male; ●, female.

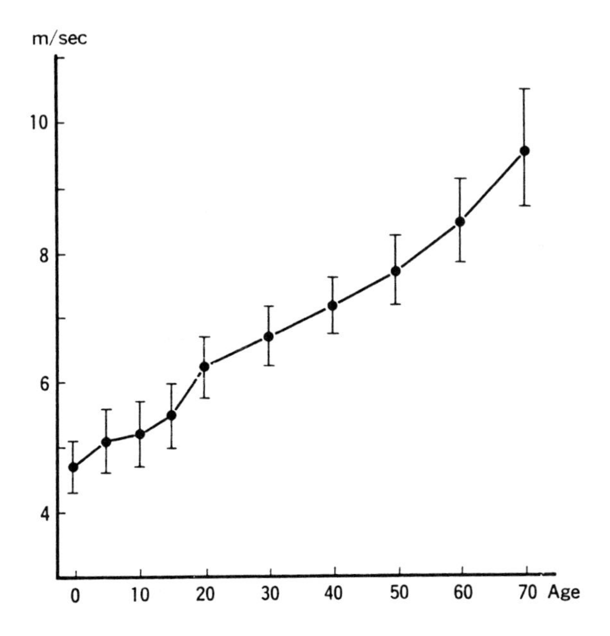


FIG. 3. Age-related increase of pulse wave velocity in the aorta. The pulse wave velocity in the aorta was measured by Doppler echography in 41 046 subjects. Results are shown as the mean ± SD.

Thus, because age-related changes in the artery are independent of atheromatous changes (which are designated as pathological phenomena), they are considered to be a physiological event. The size of organs and tissues (for instance, the kidney) decreases with age. Parameters used to assess the function of organs and tissues (for instance, creatinine clearance as a measure of renal function) also decrease with age: this loss of function is thought to be related to the decrease in size of the organs and tissues. Thus, age-related physiological and functional deterioration of organs and tissues might arise from chronic ischaemia owing to an age-related decrease in the elasticity of elastic arteries.

## Morphological characteristics of arterial ageing

Increased luminal diameter with stiffening of the arterial wall is morphologically characterized as the fragmentation of elastic fibres associated with degenerative changes in the smooth muscle cells of the media. Elastic laminae in the media, which exist in wavy arrays with reticular networks in young adults, begin to straighten and fragment from the luminal side after the fourth decade (Nakamura & Ohtsubo 1992). In addition, the elastin content decreases with age (Meyers & Lang 1946, Hosoda et al 1984). The increase in PWV with age correlates well with this age-related decrease in elastin content (Hasegawa et al 1984). Thus, quantitative and qualitative changes in elastic fibres with age might result in luminal distension of the elastic artery with stiffening.

## What causes the fragmentation of elastic fibres?

Continuous stimuli, including pulse movement and the binding of substances with a strong affinity for elastin, such as lipoproteins, $Ca^{2+}$, anti-elastin antibodies and endogenous elastases, might cause elastic fibres to fragment. The stiffening observed in the arterial wall, however, is not due solely to the degradation of the elastin network, but is also caused by variations in the supramolecular organization of the connective tissue components induced by several factors (Spina et al 1983).

## Elastin selectively modulates the biological responses of smooth muscle cells

Soluble elastin peptides induce a chemotactic response in smooth muscle cells (SMCs), but this response is inhibited when the filter is precoated with soluble elastin peptides (Ooyama et al 1987). The chemotactic response of SMCs induced by platelet-derived chemotactic factor (PDCF) is also blocked by filter-bound elastin peptides (Fig. 4). This inhibitory effect appears to be selective for SMCs because the chemotactic response of polymorphonuclear leukocytes induced by f-Met-Leu-Phe is not impeded by the filter-bound elastin peptides. The inhibitory effect is abolished by treatment of the elastin-bound

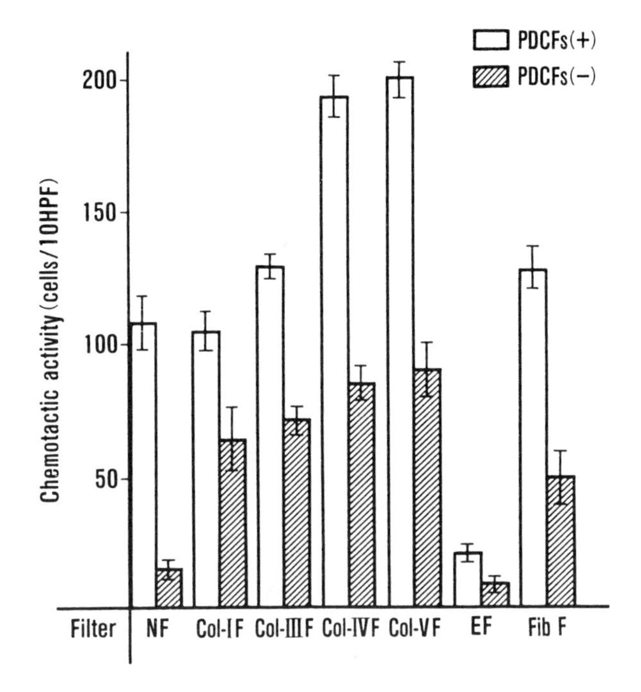


FIG. 4. Inhibition of the chemotactic response of smooth muscle cells by substratum-bound elastin. The effect of a filter-bound elastin peptide on smooth muscle cell migration was tested in a modified Boyden's chamber. Migration of cultured rat smooth muscle cells was inhibited by filter-bound elastin (EF). The mean ± SEM is given for three experiments. NF, normal filter; Col, collagen; F, fibronectin; PDCF, platelet-derived chemotactic factor; 10HPF, 10 high-power fields ($\times 400$ by light microscopy).

filter with low density lipoproteins (LDL) and anti-elastin antibody (Ooyama et al 1992).

Elastin binds preferentially to lipoproteins (Noma et al 1981). Plasma fibronectin, which binds to elastin through a hydrophobic interaction (Harumiya et al 1993), might act by the same mechanism. In addition, if SMCs are cultured with hyperlipaemic serum, elastin coated on the culture dishes modulates the SMC phenotype from modified to foam cells (Grande et al 1987). In elastin-coated dishes, the phenotypic change of SMCs from the contractile to the modified type is significantly retarded (T. Ooyama, H. Sakamoto, unpublished results 1992). There are some data to indicate that elastin selectively modulates the various biological responses of SMCs. Thus, the deteriorating effect of continuous stimulation on elastic fibres previously described might also affect the biological responses of SMCs.

## Porcine pancreatic elastase 1 as an anti-atherosclerotic drug

### *The history*

A report suggesting the causative deficiency of pancreatic elastase in atherogenesis (Balo & Banga 1953) stimulated a Japanese pharmaceutical company, Eisai, to isolate porcine pancreatic elastase 1 (PPE1) and develop it as an anti-atherosclerotic drug. PPE1 was designed as a drug to make up for the deficiency of a pancreatic exocrine enzyme, elastase, present in atherosclerosis. At that time, this was analogous to the case of the pancreatic endocrine hormone, insulin, used in the treatment of diabetes mellitus. The Ministry of Health of Japan, however, did not officially certify PPE1 as an 'anti-atherosclerotic drug' because no such drug classification was legally recognized. However, PPE1 has also been found to lower the level of cholesterol in the serum of animals fed a high-cholesterol diet. Fortunately for Eisai, there was a legal classification of 'hypocholesterolaemic drugs' which are the most common drugs used to treat atherosclerosis. Thus, Elaszym (commercial name) which contains PPE1, was certified as a 'hypocholesterolaemic drug' in 1981 and has now been used for over 13 years. Long-term administration of Elaszym to elderly patients has been found to modulate not only their age-related increase in blood cholesterol but also their age-related increases in pulse wave velocity (PWV) and intimal thickening of the carotid artery. Now about 200 000 patients per year take Elaszym in Japan.

### *The paradox*

The pharmacological basis of Elaszym treatment, however, is complicated by the lack of a rational explanation for its mechanism of action. For instance, the basal level of endogenous immunoreactive pancreatic elastase 1 (IRPE1) in the serum of elderly patients with atherosclerotic disease is not lower than that in healthy elderly people. The level of serum IRPE, most of which is in a complex of PE1 and $\alpha_1$ protease inhibitor, increases with age (Fig. 5). There is also no clinical evidence for a correlation between the severity of atherosclerosis and a decrease in pancreatic exocrine function. At the same time in a western country (USA), polymorphonuclear leukocyte elastase was considered to be a causative enzyme in the destruction of elastic fibres in pulmonary tissue (Janoff 1985). The consensus on the pathological role of polymorphonuclear leukocyte elastase in elastin research, therefore, was that elastase was harmful. Therefore, elastase therapy in Japan, where it was considered to be beneficial for the treatment of arterial ageing and atherosclerosis, appeared paradoxical.

### *The present situation*

*Chemistry of Elaszym.* Elaszym is given as a tablet for oral administration containing 5 mg of purified PPE1, equivalent to 1800 elastase units. The daily oral dose is 3–6 tablets.

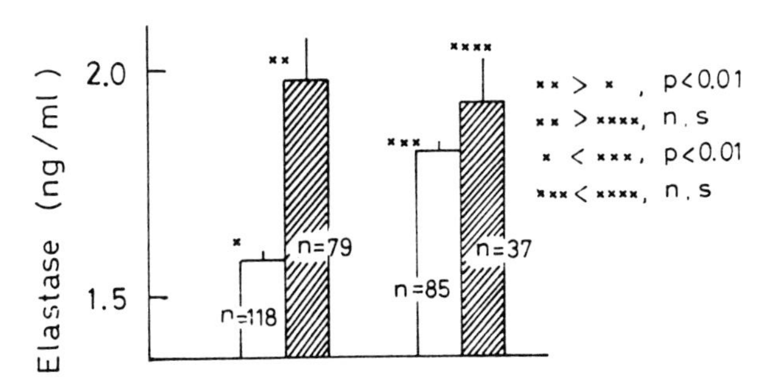


FIG. 5. Age-related increase in the serum level of immunoreactive pancreatic elastase 1. The serum concentration of immunoreactive pancreatic elastase 1 was measured by radioimmunoassay. Left, female; right, male; ▭, below 50 years; ▨, over 50 years; n.s., not significantly different.

*Absorption of Elaszym.* It has been shown that PPE1 given parenterally to rats and rabbits is absorbed from the intestinal mucosa with its size and activity unchanged, and forms a complex with $\alpha_2$ macroglobulin ($\alpha_2$ M) and $\alpha_1$ protease inhibitor ($\alpha_1$ PI) (Katayama & Fujita 1972a,b, 1974a,b,c). Once PPE1 is absorbed it enters the portal vein and the lymphatic system.

*Absolute quantity of PPE1 in blood and its metabolic half-life.* The absolute quantity of PPE1 in the blood of healthy volunteers 12 hours after taking Elaszym, as measured by ELISA, is approximately 0.8 ng/ml (Fig. 6) (Kouno et al 1991). The metabolic half-life of PPE1 in the blood is 17–18 hours.

*Elastase activity of the blood.* The serum elastolytic activity and serum succinyltrialanine paranitroanilide (Suc-(Ala)$_3$ PNA) hydrolysing activity are significantly lower in the elderly than in children and young adults (Usami et al 1985). The endogenous inhibitory capacity of the serum toward pancreatic elastase is significantly elevated in atherosclerotic patients (Bihari-Varga et al 1984). The Suc-(Ala)$_3$ PNA hydrolysing activity of plasma from rabbits receiving daily intraperitoneal administrations of 5 mg/kg of PPE1 increases to about 2.5 times the level before administration (Ooyama et al 1989). The level of immunoreactive elastin peptides in the plasma of elderly patients before and after the administration of Elaszym (3–6 tablets per day) for 3 months, as measured in the laboratories of Dr L. Robert and Dr J. Rosenbloom, do not change (T. Ooyama, H. Ito, J. Rosenbloom, L. Robert, unpublished results 1990).

*Pharmacology of Elaszym.* The hypocholesterolaemic and anti-atherosclerotic effects of Elaszym are explained by the proteolytic effects of

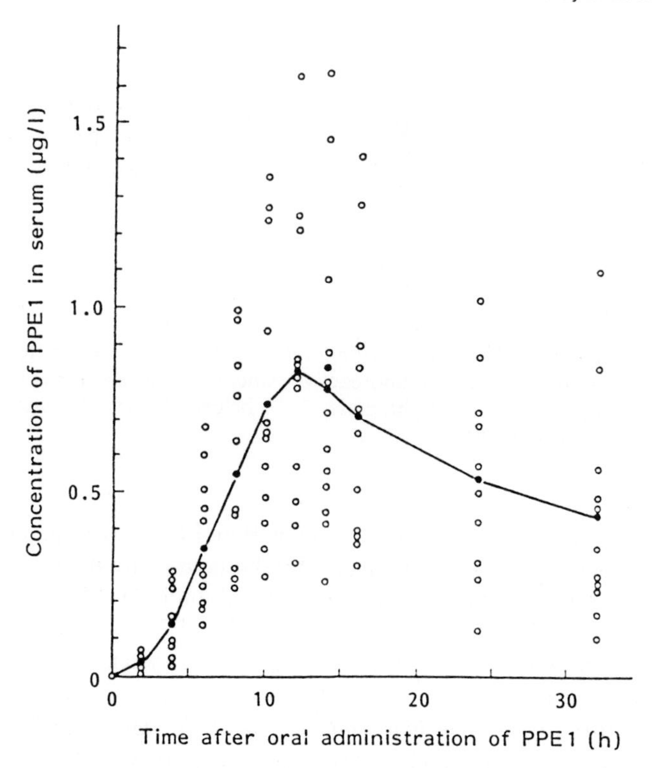


FIG. 6. The serum concentrations of immunoreactive PPE1 in 12 healthy male subjects aged 20–28 years after oral administration of six tablets of Elaszym (32 mg PPE1) was measured by ELISA. Open circles indicate the concentration of PPE1 in each subject. Closed circles indicate the mean concentration of PPE1 in serum. Reproduced with permission from Kouno et al (1991).

PPE1 on: (1) L-CAT activity; (2) LDL metabolism with an increase in ApoC2 and a decrease in ApoC3; (3) lipoprotein lipase (LPL) activity; (4) the secretion of bile acids from the liver; and (5) elastin metabolism in the arterial wall.

*Clinical effect.* Long-term administration of Elaszym to elderly patients produces a beneficial modulation in: (1) the increase in cholesterol levels (Naito et al 1972); (2) the age-related increase in PWV (Fig. 7) (Hasegawa et al 1984); and (3) the age-related increase in the atherosclerotic index as measured by B-mode echography as the intimal-medial thickness of the carotid artery (Y. Matsumoto, personal communication).

*Effects of PPE1 on arterial metabolism.* Immunizing rabbits with elastin peptides for six months induces the fragmentation of elastic fibres in the aortic

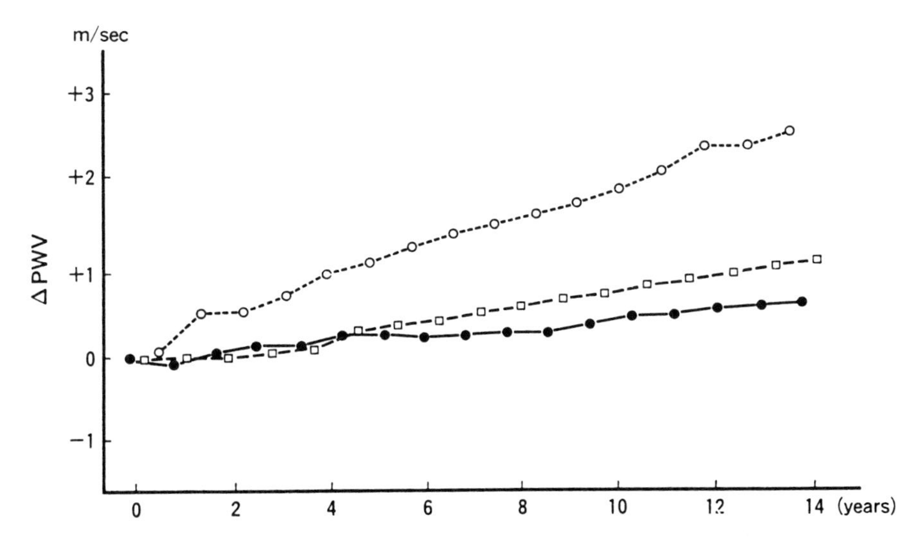


FIG. 7. Change in pulse wave velocity with time in the atherosclerosis group after the administration of Elaszym. □– – –□, healthy control ($n$=259); ○- - -○, atherosclerosis group ($n$=58); ●——●, atherosclerosis plus Elaszym group ($n$=126).

wall (Jacob et al 1984, Ooyama et al 1989). This fragmentation is prevented by the simultaneous administration of PPE1 (Table 1). Administration of PPE1 also prevents the stiffening of the aortic wall in rabbits who have been fed a high cholesterol diet (Hayashi et al 1987), the production of atherosclerotic lesions in rats fed an atherogenic diet (Katsunuma et al 1983) and the increase in the intimal thickness of the aorta of rabbits after balloon catheter injury (Saeki & Sueishi 1987).

*Mechanism of action.* Normally, when elastase is introduced into blood vessels, it is bound to $\alpha_1$PI and $\alpha_2$M. The $\alpha_1$PI–elastase complex is incorporated into the elastic lamina through the endothelium by pinocytosis and is capable of directly attacking the elastic fibres in the blood vessel wall (Tujii et al 1988). Why exogenous PPE1 can selectively modulate the amount of degenerated elastic fibres in the arterial wall is unclear at the moment.

## Conclusion

The action of PPE1 as an anti-atherosclerotic drug is paradoxical to the effect of polymorphonuclear leukocyte elastase, which has a pathological role in tissue injury. The long-term administration of PPE1 to elderly patients, however, has been found to modulate not only the age-related increase in blood cholesterol but also the age-related increases in PWV and intimal thickening of the carotid artery.

**TABLE 1 Morphometric analysis of fragmented elastic fibres of the aortic media in rabbits immunized with elastin peptides and the effect of simultaneous elastase administration**

| *Group* | *n* | *Percentage of short fibres ± SEM* | *Welch's t-test value* | |
|---|---|---|---|---|
| Control | 5 | 57 ± 1 | 3.94 | 0.76 |
| Elastin | 5 | 67 ± 2 | | |
| Elastin + elastase | 5 | 55 ± 2 | 3.90 | |

Rabbits were immunized with elastin peptides with or without simultaneous PPE1 treatment, for a period of six months. The percentage of short fibres to all fibres was compared among the three experimental groups by Welch's *t*-test (significant difference, $p < 0.0025$).


## *Acknowledgements*

The authors thank Dr M. Hasegawa of Toho University, Dr Y. Matsumoto of Osaka University and Mr M. Asou of Eisai Pharmaceutical Company for their cooperation in the preparation of this manuscript.


## References


Balo J, Banga I 1953 Change in elastase content of the human pancreas in relation to atherosclerosis. Acta Physiol Acad Sci Hung 4:187–197

Bihari-Varga M, Keller L, Landi A, Robert L 1984 Elastase-type activity, elastase inhibitory capacity, lipids and lipoproteins in the sera of patients with ischemic vascular disease. Atherosclerosis 50:273–281

Grande J, Davis HR, Bates S, Matthews MB, Glagov S 1987 Effect of an elastin growth substrate on cholesterol ester synthesis and foam cell formation by cultured aortic smooth muscle cells. Atherosclerosis 68:87–93

Hasegawa M, Arai C, Abe N et al 1984 A biophysical study of antisclerotic agents—long-term effect of elastase as measured by aortic pulse wave velocity method. J Jpn Gerontol Soc 21:115–123

Harumiya S, Jung SK, Sakano Y, Fujimoto D 1993 Interaction of human plasma fibronectin with α-elastin. J Biochem 113:710–714

Hayashi K, Takamizawa K, Nakamura T, Kato T, Tsushima N 1987 Effects of elastase on the stiffness and elastic properties of arterial walls in cholesterol-fed rabbits. Atherosclerosis 66:259–267

Hosoda Y, Kawano K, Yamasawa F, Ishii T, Shibata T, Inayama S 1984 Age-dependent changes of collagen and elastin content in human aorta and pulmonary artery. Angiology 35:615–621

Jacob MP, Hornebeck W, Lafuma C, Bernaudin JF, Robert L, Godean G 1984 Ultrastructural and biochemical modifications of rabbit arteries induced by immunization with soluble elastin peptides. Exp Mol Pathol 41:171–190

Janoff A 1985 Elastases and emphysema: current assessment of the protease–antiprotease hypothesis. Am Rev Respir Dis 132:417–433

Katayama K, Fujita T 1972a Studies on biotransformation of elastase. 1. Transport of $^{131}$I-labeled elastase across rat intestine *in vitro*. Biochim Biophys Acta 288:172–180

Katayama K, Fujita T 1972b Studies on biotransformation of elastase. 2. Intestinal absorption of $^{131}$I-labeled elastase *in vivo*. Biochim Biophys Acta 288:181–189

Katayama K, Fujita T 1974a Studies on biotransformation of elastase. 3. Effects of elastase-binding proteins in serum on disappearance of $^{131}$I-labeled elastase from blood. Biochim Biophys Acta 336:165–177

Katayama K, Fujita T 1974b Studies on biotransformation of elastase. 4. Tissue distribution of $^{131}$I-labeled elastase and intracellular distribution in liver after intravenous administration in rats. Biochim Biophys Acta 336:178–190

Katayama K, Fujita T 1974c Studies on biotransformation of elastase. 5. Degradation of injected $^{131}$I-labeled elastase subcellular particles of rat liver. Biochim Biophys Acta 336:191–200

Katsunuma H, Shimizu K, Ebihara T, Iwamoto T, Kiyokawa M 1983 Anti-atherosclerotic action of elastase with special reference to its effect on elastic fibers. Age Ageing 12:183–194

Kouno T, Yuzuriha T, Hasegawa J, Ishikawa E 1991 Concentration of hog elastase I in serum of healthy subjects after oral administration measured by sensitive two-site enzyme immunoassay. Clin Chem Commun 4:123–133

Meyers VC, Lang WW 1946 Some chemical changes in the human thoracic aorta accompanying the ageing process. J Gerontol 1:441–444

Naito T, Tohno T, Iwabuchi T et al 1972 A double blind study of effects of elastase on abnormal serum lipid levels. Ikagaku no Ayumi 82:848–859 (in Japanese)

Nakamura H, Ohtsubo K 1992 Ultrastructure appearance of atherosclerosis in human and experimentally induced animal models. Electron Microsc Rev 5:129–170

Nakashima T, Tanikawa J 1971 Study of human aortic distensibility with relation to atherosclerosis and ageing. Angiology 22:447–490

Noma A, Takahashi T, Wada T 1981 Elastin–lipid interaction in the arterial wall. 2. *In vitro* binding of lipoprotein-lipids to arterial elastin and the inhibitory effect of high-density lipoproteins on the process. Atherosclerosis 38:373–382

Ooyama T 1991 Ageing of the blood vessel. J Jpn Atheroscler Soc 19:123–127

Ooyama T, Fukuda K, Oda H, Nakamura H, Hikita Y 1987 Substratum-bound elastin peptide inhibits aortic smooth muscle cell migration *in vitro*. Arteriosclerosis 7: 593–598

Ooyama T, Fukuda K, Masuda S, Nakamura H 1989 Administration of elastase blocks the formation of fragmented elastic fibers in aorta of rabbit. Artery 16:293–311

Ooyama T, Sakamoto H, Fukuda K et al 1992 LDL eliminates the inhibitory effect of human aortic elastin on cultured rat aortic smooth muscle cell migration *in vitro*. J Jpn Atheroscler Soc 20:703–770

Robert L, Jacotot B 1994 Extracellular matrix in atherogenesis. Role of the elastin receptor. J Jpn Atheroscler Soc 21:613–618

Saeki T, Sueishi K 1987 Antiatheromatous properties of elastase in cholesterol-fed rabbits and *ex vivo* suppressive effect of elastase on smooth muscle cell proliferation in the presence of hypercholesterolemia. Acta Pathol Jpn 37:1423–1432

Spina M, Garbisa S, Hinnie J, Hunter JC, Serafini-Fracassini A 1983 Age-related changes in composition and mechanical properties of the tunica media of the upper thoracic human aorta. Arteriosclerosis 3:64–76

Tujii T, Katayama K, Naito I, Seno S 1988 The circulating antitrypsin–elastase complex attacks the elastic lamina of blood vessels. An immunohistochemical study. Histochemistry 88:443–451

Usami E, Seyema Y, Yamashita S 1985 The relation between the elastase activity and the protease inhibitor level in human serum. Clin Chem 14:13–18 (in Japanese)

## DISCUSSION

*Starcher:* I was amazed when I heard that elastase could be absorbed through the gut and then be utilized as a protease later on. You suggested that this was because it bound to an inhibitor but was still effective in getting to the arterial walls. This brings to mind some studies carried out over the years which suggest that acute pancreatitis can lead to emphysema (Geokas et al 1968, Loeven 1969, Lungarella et al 1985). Serum elastase levels can reach 50–100 times the normal levels in these patients. What happens in the lungs of patients treated with your PPE1? If anything, the lungs should be more susceptible to elastase than the arterial wall.

*Ooyama:* We haven't looked at this specifically, but to my knowledge treatment with Elaszym has no side effects on the lungs. This is because of the low levels of elastase involved, compared with the huge levels produced in pancreatitis, which are 100–1000 times higher.

As I mentioned, the level of PPE1 in the blood in patients after they had taken Elaszym, as measured by RIA, was in the range of 1–5 ng/ml. This was roughly the same level as that of the endogenous pancreatic elastase 1 in normal subjects, and does not have any effect on the lungs. But the levels in patients with pancreatitis were usually in excess of several micrograms per ml. This would result in pulmonary emphysema.

*Parks:* Since elastase is a protein, it must be susceptible to degradation by digestive enzymes, such as pepsin in the stomach and the various pancreatic enzymes released into the duodenum. Are you measuring intact elastase in the blood, or are you looking at fragments of it?

*Ooyama:* I didn't say much about the absorption of elastase from the duodenum, so I might have confused you. The drug is coated with a sugar to protect it against gastric juices and digestive enzymes. When the drug is given parenterally to rats and rabbits, it is absorbed from the intestinal mucosa, its size and activity unchanged. Once it is absorbed, it enters the portal vein and the lymphatic system. But degradation might also occur, in which case we will be looking at both the intact molecule and fragments.

*Parks:* Are these enzymically active?

*Ooyama:* Technically, it is very difficult to prove that the endogenous elastase in the blood of patients is still active. To test this we would need to isolate the protein or fragments of it from the blood—these are present in nanogram quantities per ml, insufficient for isolation.

Parks: You could use your antibody against PPE1 as an affinity reagent to isolate whatever fragment has retained antigenicity.

*Ooyama:* We could, but we haven't done it yet. We were unable to demonstrate an increase of elastolytic activity together with an increase of elastin peptides in the blood, although there was an increase in immunoreactive elastase in the blood of patients after they had taken Elaszym.

*Hinek:* Since you have the immunodetectable elastase protein, or a fragment of the protein, but no elastolytic activity, the beneficial effect of the treatment must stem from a property of the protein separate from its enzymic activity. Let me offer one explanation. Even if you have a fragment or inactive enzyme, the N-terminus of your pancreatic elastase may still be free. In a recent study (Hinek & Rabinovitch 1994), we noticed tremendous homology between the 67 kDa elastin receptor and the N-terminus of pancreatic elastase. Therefore, your drug might be acting as a soluble receptor which binds to and scavenges floating elastin peptides which are released by endogenous elastases during the development of atherosclerosis.

*Robert:* Quite a few years ago, Jerry Turino (Colombia University) did his sabbatical with us in Paris. We injected pancreatic elastases intravenously into puppies. The highest amount was equivalent to circulating $\alpha_1$PI (or $\alpha_1$ antitrypsin) and we still found extensive lesions in several tissues, especially in the vessel wall. It appears therefore that some activity can be retained even if the enzyme is combined with its inhibitor (Turino et al 1974).

*Kagan:* The question of whether the administered elastase is functional or not is obviously crucial with regards to its effect in the arterial wall. It would be of interest to determine the effect of treatment with elastase that has been chemically inactivated, for example. It would be surprising if there was functional activity in absorbed elastase since there is an excess of $\alpha_1$ antitrypsin, which forms an irreversible complex at the active site of elastase.

*Rosenbloom:* If I understood you correctly, there is a substantial group of patients who have been treated for over 13 years with Elaszym. This is quite a long time, but you didn't mention anything about the relative mortality between the atherosclerotic control group and those who were treated. Have you any data on the mortality or the incidence of heart attacks in these patients?

*Ooyama:* No, we don't have those data. But I think your question raises a key point as far as our study is concerned with the therapeutic effect of exogenous elastase on arterial ageing and atherosclerosis. I hope we will be able to answer this question in the near future.

*Mecham:* We've actually had some experience with PPE1. Several years ago I had the opportunity to visit Dr Ooyama and attend a meeting sponsored by Eisai, where they presented data on the therapeutic use of this drug. They described results from studies on the inhibition of vascular remodelling and pulmonary hypertension, which is one of my interests. I must admit I was a bit sceptical when I saw the data, so I offered to test PPE1 in one of our animal models: a bovine model of neonatal pulmonary hypertension. When they are exposed to hypobaric hypoxia, the animals become suprasystemic and within days the pulmonary blood vessels begin to thicken and there is right heart hypertrophy. Eventually, the animals die of right heart failure. Our experiment involved administering PPE1 by intramuscular injection; we set it up with normoxic controls and hypoxic hypertensive animals. Much to my surprise we found that the animals given the elastase were hypertensive, but did not remodel their vessels. There was no right heart hypertrophy and the pulmonary vessels retained their vasoreactivity. Our colleagues in Denver were equally surprised, because they have never seen anything that would do this. The experimented was repeated with the same result. I can't explain what happened mechanistically, but the elastase had a definite protective effect.

*Kimani:* Onset of hypertension triggers a very intense intimal reaction, characterized by intimal hyperplasia and subsequent deposition of elastin. Could it be that this is the elastin that is being digested by elastase in your preparation?

There is also evidence showing that the development of such intimal foci is inhibited by the use of anti-inflammatory and antiproliferative drugs, such as colchicine, aspirin, cortisone and pencillamine (Hollander et al 1974). How does the use of elastase in your studies compare with the effects of these drugs?

*Ooyama:* At present, we can't answer your questions, but we should certainly investigate these issues experimentally.

## References

Geokas MC, Murphy DR, McKenna RD 1968 The role of elastase in acute pancreatitis. I. Intrapancreatic elastolytic activity in bile-induced acute pancreatitis in dogs. Arch Path (Chicago) 86:117–126

Hinek A, Rabinovitch M 1994 67-kD elastin binding protein is a protective "companion" of extracellular insoluble elastin and intracellular soluble tropoelastin. J Cell Biol 126:563–574

Hollander W, Kramsch DM, Franzblau C, Paddock JR, Colombo MA 1974 Suppression of atheromatous plaque formation by anti-proliferative and anti-inflammatory drugs. Circ Res 34 & 35(Suppl 1):131–141

Loeven WA 1969 Human pancreatic elastolytic enzymes and atherosclerosis and lung emphysema in elderly people. J Atheroscler Res 10:379–390

Lungarella G, Gardi C, De Santi MM, Luzi P 1985 Pulmonary vascular injury in pancreatitis: evidence for a major role played by pancreatic elastase. Exp Mol Pathol 42:44–59

Turino GM, Hornebeck W, Robert L 1974 *In vivo* effects of pancreatic elastase I. Studies on the serum inhibitors. Proc Soc Exp Biol Med 146:712–717

# Interaction of tumour cells with elastin and the metastatic phenotype

J. Tímár, Cs. Diczházi, A. Ladányi, E. Rásó, W. Hornebeck*, L. Robert† and K. Lapis

*1st Institute of Pathology & Experimental Cancer Research, Semmelweis University of Medicine, Budapest, H-1085, Hungary, *Université de Paris XII, Créteil 94010 Cedex, France and †Laboratoire de Biologie Cellulaire, Université de Paris VII, 3 Place Jussieu, Paris 75261 Cedex 05, France*

*Abstract.* It is now well established that the interaction of tumour cells with elastin is important during invasion and metastasis. This is due to the fact that the elastin receptor complex is widely expressed by tumour cells and is overexpressed in highly metastatic variants. There is evidence that the elastin receptor complex is associated with a signal system involving G proteins, phospholipase C, the phosphoinositol cycle and protein kinase C. Therefore, activation of the elastin receptor system results in activation of protein kinase C-dependent cellular processes such as enzyme secretion and migration. Accordingly, soluble elastin can be used *in vivo* to interfere with tumour cell dissemination into elastin-rich tissues such as lung, skin or blood vessels. The importance of elastin–tumour cell interactions is emphasized by the observation that the 67 kDa receptor for laminin may well be identical to the 67 kDa elastin receptor of the elastin receptor complex. Interference with the function of this receptor system by the use of both laminin peptides and elastin ligands may provide the basis for a novel and more powerful antimetastatic intervention.

*1995 The molecular biology and pathology of elastic tissues. Wiley, Chichester (Ciba Foundation Symposium 192) p 321–337*

Tumour cell dissemination involves sequential proliferative and non-proliferative events leading to tumour progression. Proliferation takes place predominantly at the primary and secondary sites while the non-proliferative steps consist of series of events characterized by cell–cell and cell–extracellular matrix (ECM) interactions. Tumour cell–ECM interactions involve adhesion to, digestion of and migration on complex matrix materials (Liotta & Stetler-Stewenson 1991). Vascular walls provide natural barriers to migrating cells, including tumour cells, which have to (at least partially) be destroyed before migration can occur. Elastin is found in close physicochemical proximity to the basement membrane of endothelial cells and smooth muscle cells. Therefore, migrating tumour cells must cross both basement membranes and the adjacent elastin-containing matrix during intra- and extravasation. Tumour cell–ECM interactions are mediated by specific ECM receptors at the cell surface, which can be classified into three groups: the integrins (Hynes 1992), non-integrins (Castronovo 1993) and proteoglycans (Ruoslahti 1988). It is now well accepted that tumour cell–basement membrane interactions are of primary importance during invasion (Liotta & Stetler-Stewenson 1991). However, owing to the close proximity of elastin to the basement membrane it seems logical to consider tumour cell–elastin interactions as equally important in organs that contain considerable amounts of elastin, such as lungs, skin and vessel walls. Here we intend to summarize our knowledge on tumour cell–elastin interactions from the point of view of tumour dissemination.

## Tumour cells express elastin receptor(s) similarly to normal cells

Specific elastin-binding proteins were first observed on elastin-producing cells (fibroblasts and smooth muscle cells) (Hornebeck et al 1986) with apparent molecular masses of 120, 67, 60 and 45 kDa. One of these proteins, the 120 kDa species, was termed elastonectin and, unlike the others, proved to be inducible (Hornebeck et al 1986). From these elastin-producing cells the 67 kDa elastin receptor was later isolated. It proved to be a galactoside-binding lectin (Hinek et al 1988, Mecham et al 1989a) which is also associated with a 61 and a 55 kDa transmembrane protein. The 67 kDa protein was recently shown to be related or identical to the enzymically inactive alternatively spliced variant of $\beta$-galactosidase, a surface-associated protein (Hinek et al 1993).

Murine 3LL carcinoma variant M27 was the first tumour cell type in which an elastin peptide (VGVAPG)-binding receptor was demonstrated to be a 59 kDa membrane protein involved in tumour cell chemotaxis (Blood et al 1988). Subsequently, 3LL murine carcinoma cells (Robert et al 1989, Fig. 1a–c, Table 1), A-2058 human melanoma cells (Mecham et al 1989a, Table 2), and their variants with differing metastatic potential, have been shown to bind soluble and insoluble elastin peptides. In contrast to the situation with normal cells elastin binding to tumour cells does not exhibit a lag period (Robert et al 1989), indicating that the elastin-binding receptor system may be structurally or functionally different in tumour cells. Biochemical analysis of elastin binding proteins in 3LL-HH tumour cells indicates that both elastonectin as well as the 67 kDa elastin receptor are present at the cell surface (Fig. 2) and can be eluted by lactose. This suggests that, as with normal cells, the receptor complex is a galactoside lectin (Hinek et al 1988). The 67 kDa laminin receptor is chemically and immunologically identical to the 67 kDa elastin receptor (Mecham et al 1989a). Unfortunately, the molecular biological approach did not provide conclusive proof of the identity of the two receptors (Grosso et al 1991). However, tumour cell

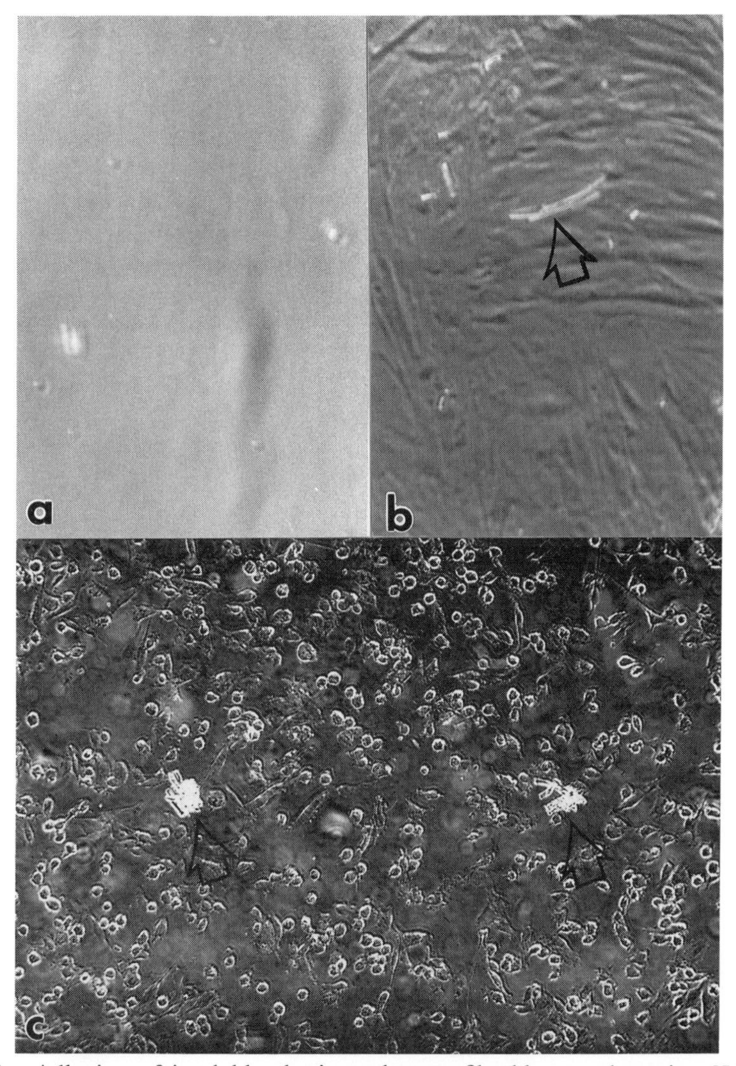

FIG. 1. Adhesion of insoluble elastin to human fibroblasts and murine 3LL-HH carcinoma cells *in vitro*. Phase contrast microscopy. Confluent monolayers of cells were incubated with 1 mg/ml insoluble elastin at 37 °C for 1 h (human fibroblasts) or 30 min (3LL-HH cells). Elastin fibres are arrowed. (a) Plastic surface; (b) human fibroblasts; (c) 3LL-HH tumour cells. (Magnification × 200.)

laminin receptors can be detected by anti-elastin receptor antibodies and tumour cell elastin receptors recognize elastin (Mecham et al 1989b). Therefore, it seems that both normal cells and tumour cells use the same 67 kDa receptor to recognize laminin and elastin.

**TABLE 1 Adhesion of insoluble elastin to 3LL tumour cell variants of differing metastatic potential**

| *Incubation time (min)* | *3LL-LM* | *3LL-HM* |
|---|---|---|
| 15 | 1990 ± 185 | 2750 ± 194* |
| 30 | 2100 ± 203 | 4250 ± 412* |
| 60 | 2200 ± 211 | 3750 ± 322* |
| 120 | 1250 ± 106 | 1350 ± 142 |

Tumour cells were incubated with $^{3}$H-labelled insoluble elastin (1 mg/ml; $10^5$ cpm/$10^6$ cells) at 37 °C as described in Tímár et al (1991). Data are expressed in cpm/$10^5$ cells ± SEM; $n = 3$.
3LL-LM, variant with low metastatic potential; 3LL-HM, highly metastatic variant.
*$P < 0.05$ by ANOVA test.

## Differences in elastin binding among tumour cells with varying metastatic potential

Ample data are available regarding the functional significance of laminin receptors in tumour cells and their particular importance in tumour invasion (Castronovo 1993). Tumour cells selected for increased adherence to laminin are characterized by increased metastatic potential. Metastatic breast and colon carcinoma cells exhibit increased 67 kDa elastin/laminin receptor expression both at the mRNA and protein levels (for references see Castronovo 1993). Data obtained from murine 3LL and human A-2058 melanoma variants (Tables 1, 2 and Tímár et al 1991) all indicate that as with laminin binding, elastin binding potential is also increased in tumour cell

**TABLE 2 Adhesion of insoluble elastin to A-2058 human melanoma cell variants of differing metastatic potential**

| *Incubation time (min)* | *A-2058S* | *A-2058M* |
|---|---|---|
| 15 | 3223 ± 201 | 8235 ± 421* |
| 30 | 0 | 7811 ± 571* |
| 60 | 0 | 1804 ± 201* |
| 120 | 0 | 201 ± 159 |

Tumour cells were incubated with $^{3}$H-labelled insoluble elastin (1 mg/ml; $10^5$ cpm/$10^6$ cells) at 37 °C as described in Tímár et al (1991). Data are expressed in cpm/$10^5$ cells ± SEM; $n = 3$.
A-2058S, variant with low metastatic potential; A-2058M, highly metastatic variant.
*$P < 0.05$ by ANOVA test.

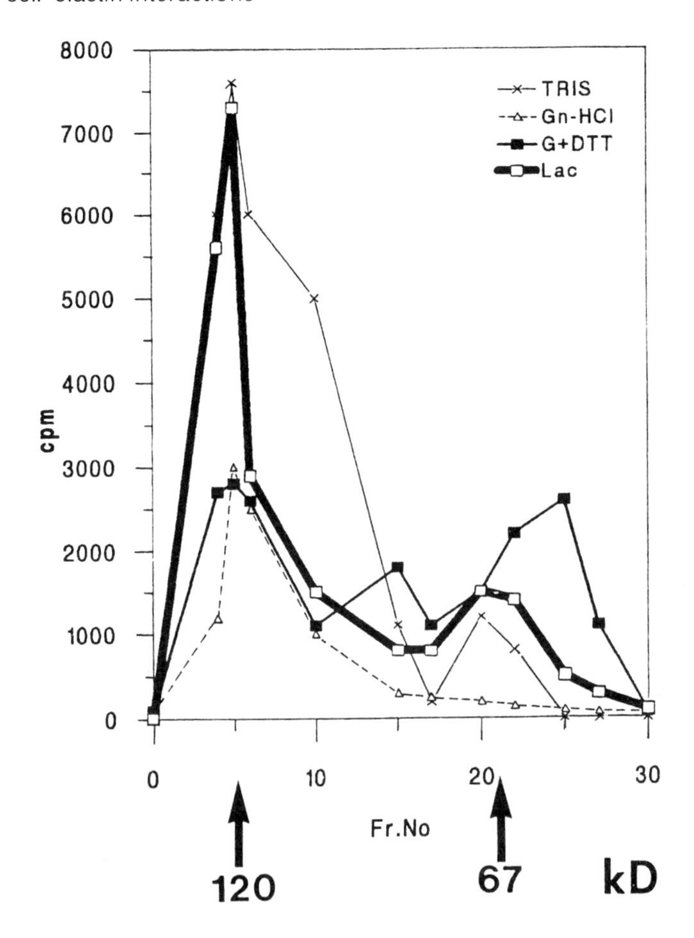


FIG. 2. Isolation of membrane proteins from 3LL-HH tumour cells incubated with 1 mg/ml insoluble elastin for 30 min. A confluent layer of tumour cells was labelled with [$^{35}$S]methionine for 6 h, then elastin for 1 h and the plasma membrane fraction was isolated as described (Hornebeck et al 1986). The extract was separated by a Sepharose column and the radioactivity was recorded using the following eluents: TRIS-HCl buffer (TRIS); 4 M guanidine hydrochloride (Gn-HCl); 4 M guanidine hydrochloride + dithiothreitol (G + DTT); and 1 M lactose. Elastin receptor complex is eluted by guanidium chloride/ dithiothreitol (Hornebeck et al 1986) and lactose (Lac) (Hinek et al 1988). Note that both eluents result in peaks of two predominant protein species; 120 kDa (elastonectin) and 67 kDa (elastin receptor). Fr. No, fraction number; cpm, counts per minute.

variants with increased metastatic potential while it is almost undetectable in non-metastatic variants (Tímár et al 1991). This is also reflected by the increased $K_d$ values of the binding of soluble and insoluble elastin to tumour cells (Blood et al 1988, Robert et al 1989, Tímár et al 1991). Importantly, unlike normal cells (Hornebeck et al 1986), tumour cells constitutively express insoluble elastin binding potential. Furthermore, peak levels of insoluble elastin binding by tumour cells occur between 15–30 min at 37 °C; these can be further increased by treatment with soluble elastin (Tímár et al 1991). The initial peak of elastin binding is followed by a similarly dynamic release of insoluble elastin from the surface of metastatic tumour cells (Tables 1, 2). This post-binding release of elastin may be due to either the release of the elastin–receptor complex from, or the activation of elastinolytic activities at, the tumour cell surface.

**Signalling through the elastin receptor complex in tumour cells**

Though the 67 kDa laminin/elastin receptor is not a transmembrane protein (Grosso et al 1991) the elastin-binding membrane receptor complex, which consists of an elastin receptor and elastonectin, has been shown to be involved in the activation of various intracellular signals. Binding of soluble elastin to monocytes induced a transient rise in the intracellular cAMP level followed by a continuous rise in the cGMP level (Fülöp et al 1986). It was also shown that the later effect of the binding of elastin peptides to monocytes, smooth muscle cells and fibroblasts induces a rise in the intracellular $Ca^{2+}$ level that peaks at 30 min. This process is calmodulin dependent (Jacob et al 1987). In metastatic 3LL cells, unlike their non-metastatic counterparts, soluble elastin also triggered a rise in intracellular $Ca^{2+}$ with an earlier peak at 15 min (Tímár et al 1991). We have analysed the effect of various signal transduction inhibitors on the binding of insoluble elastin to highly metastatic 3LL cells. Surprisingly, we found that inhibition of pertussis toxin (PT)-sensitive G proteins (by PT), phospholipase C (PLC) (by neomycin) and diacylglycerol (DAG) production (by retinal) all stimulated the initial binding of insoluble elastin to tumour cells (Fig. 3). Interestingly, at a later point (60 min) G protein and DAG inhibition decreased the elastin binding, indicating a dual role for these signal transduction elements in insoluble elastin-binding. Protein kinase C (PKC) involvement in soluble elastin binding was originally described by Blood & Zetter (1989). It was shown that elastin binding to non-metastatic 3LL cells can be induced by activation of PKC. These data indicate that the ligation of the elastin receptor complex activates G protein(s), the PLC-dependent inositol cycle (causing release of $Ca^{2+}$ from the intracellular store) and finally PKC. The mechanism of the binding of insoluble elastin versus soluble elastin to tumour cells seems to be different in the initial phase of the process, where inhibitory G proteins may be involved. It seems that in tumour cells with low metastatic potential and low elastin binding, this activation process does not work efficiently.

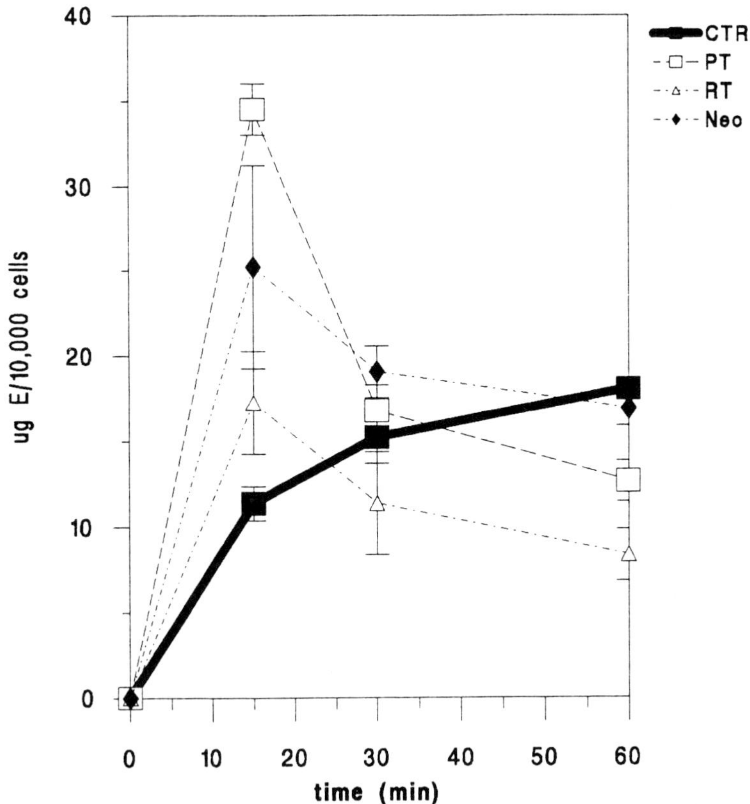


FIG. 3. Adhesion of insoluble $^{3}$H-labelled dispersed elastin fibres to 3LL-HH cells *in vitro*; effect of signal transduction modulators. Tumour cell monolayers were incubated with 1 mg/ml insoluble [$^{3}$H]elastin (E) and were treated with pertussis toxin (PT; G protein inhibitor), Neomycin (Neo; PLC inhibitor) and retinal (RT; PKC inhibitor) for 60 min and the bound [$^{3}$H]elastin to the cell layer was determined as described (Tímár et al 1991). Data are expressed in bound $\mu$g elastin/$10^4$ cells and are means ± SD ($n = 3$). Note that all inhibitors stimulated the adhesion of elastin at the earlier phase of incubation, and then RT and PT proved to be inhibitory at the stage of maximal elastin adhesion. CTR, control.

## Function of elastin interactions in tumour cells

### *Release of lytic enzymes*

Tumour cells contain various lytic enzymes which are involved in the degradation of the surrounding matrix. The major classes of these are the metalloproteinases (including type IV collagenase; Liotta et al 1982), cathepsins (Sloane 1992) and enzymes of the plasminogen activator system (Blasi & Verde 1990). The question is, however, how the release of such enzymes is regulated in tumour cells. Recent papers have indicated that PKC is

**TABLE 3 Saturable adsorption of heparin and oleoyl-heparin on insoluble fibrous elastin**

| | *Glycosaminoglycan concentration (μg/mg elastin)* | | | | |
|---|---|---|---|---|---|
| | *5* | *15* | *50* | *100* | *250* |
| Heparin | 1.9 | 3.1 | 4.8 | 9.7 | 14.8 |
| Oleoyl-heparin | 4.9 | 13.1 | 40.6 | 71.8 | 124.2 |

Data are expressed in μg bound glycosaminoglycan/mg elastin and represent means ($n=3$, SEM < 10%). Experiments were performed for 3 h at 37 °C in PBS (pH = 7.4) with continuous agitation. The supernatant was removed and the residual radioactivity of the labelled glycosaminoglycans was counted after solubilization of the elastin by porcine pancreatic elastase.

involved in the release of cathepsin B (Honn et al 1994) and elastase (Zeydel et al 1986) from tumour cells, whereas lipoxygenase metabolites were shown to induce the release of type IV collagenase (Reich et al 1988). Various enzymes are able to degrade elastin fibres, including serine proteases (cathepsins) and metalloproteinases (Robert et al 1980). The binding of κ-elastin to monocytes (Fülöp et al 1986), smooth muscle cells and fibroblasts (Jacob et al 1987) results in the release of elastase (in monocytes), serine protease (in smooth muscle cells), metalloproteinases (in fibroblasts) and β-glucuronidase (in monocytes), indicating that a signalling system regulating the release of lysosomal enzymes is activated by the interaction of the elastin–elastin receptor complex.

Assorted data indicate that various tumour cell types (rat mammary adenocarcinoma [Zeydel et al 1986], human bladder cancer [Grant et al 1989], mouse lung carcinoma [Tímár et al 1991] and melanoma [Yusa et al 1989]) express elastinolytic activities. However, mammary adenocarcinomas and bladder carcinomas contain serine proteinase while the lung carcinoma expresses metalloproteinase. Experimental data from both the rat mammary carcinoma system (Zeydel et al 1986) and the murine 3LL cells (Tímár et al 1991) indicated that the secretion of elastinolytic activity correlated with the invasive phenotype of tumour cells.

### *Regulation of serine protease-type elastases by heparin derivatives*

Various type I serine proteases are able to degrade elastin, such as leukocyte and pancreatic elastase, and cathepsin G, the latter of which has recently been shown to be overexpressed in various malignancies (Robert et al 1980). Since tumour cells express serine protease-type elastinolytic activity during invasion, inhibition of the function of this enzyme(s) may have an anti-invasive effect. Therefore, the design of elastase inhibitors is of particular significance. It has been shown that various glycosaminoglycans (GAGs) (Redini et al 1988) and fatty acids (Stanislawsky & Hornebeck 1988) are inhibitors of leukocyte

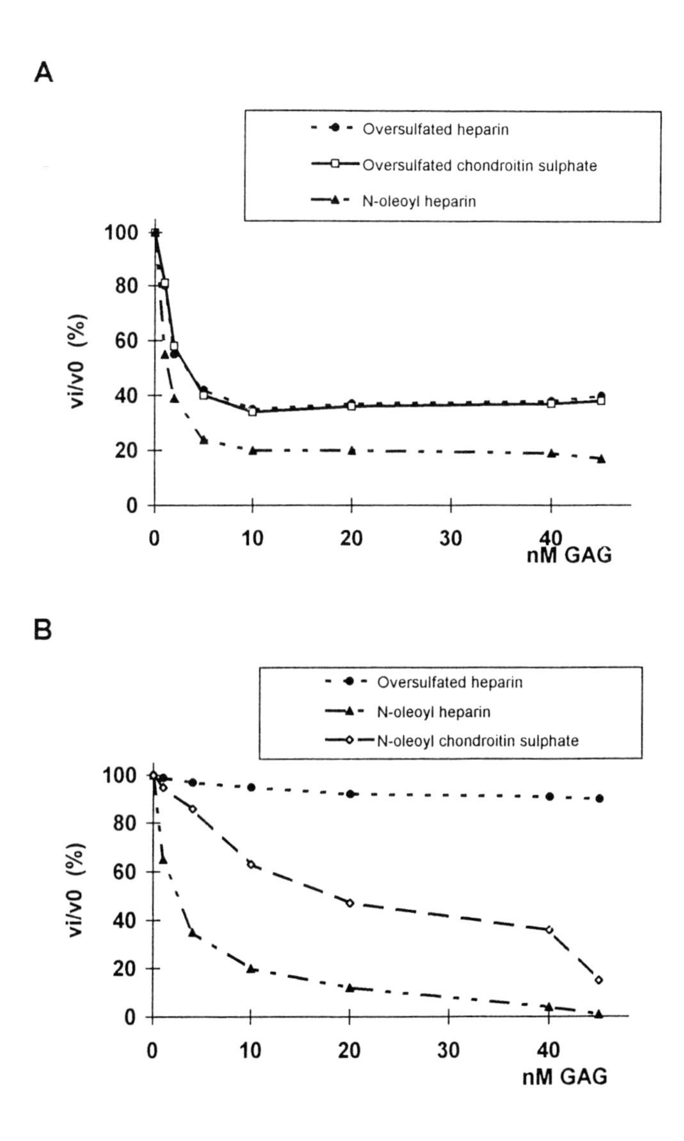


FIG. 4. Effect of lipo-glycosaminoglycans (GAGs) on the activity of various elastases *in vitro*. (A) Human leukocyte elastase (substrate, 0.05 mM MeO-Suc-Ala-Pro-Val-pNA). (B) Cathepsin G (substrate, 0.2 mM Suc-Ala-Ala-Pro-Phe-pNA). Buffer, 0.1 M TRIS-HCl; pH 8.0; ionic strength 0.05 M; temp. 37 °C (Baici et al 1993). Data points are expressed as residual activity (% of velocity without inhibitor). Note that *N*-oleoyl-heparin proved to be the most potent inhibitor of elastase among the GAG derivatives.

**TABLE 4 Effect of soluble elastin on the proliferation of 3LL-HM cells *in vitro***

| *Soluble elastin (μg/ml)* | *Cell number × $10^6$* |
|---|---|
| 0 | 4.7 ± 0.34 |
| 10 | 4.3 ± 0.13 |
| 100 | 3.2 ± 0.2* |
| 1000 | 2.4 ± 0.6* |

Adherent tumour cells were incubated with soluble elastin (κ-elastin; $M_r = 75000$) at 37 °C for 48 h *in vitro*. Data are expressed as number of cells/well ± SEM; $n = 3$.
*$P < 0.05$ by ANOVA test.

elastase. It is claimed that inhibitors with large hydrophobic residues selectively protect the elastic fibres against elastinolysis. On the basis of these findings, we synthesized fatty acid–GAG conjugates and tested them on various serine proteases. We found that *N*-oleoyl-heparin has a higher inhibitory potential against leukocyte elastase and cathepsin G than heparin or over-sulphated heparin derivatives (Fig. 4). Oleoyl-heparin also proved to be effective against pancreatic elastase, trypsin and chymotrypsin, but proved to be ineffective against subtilisin (Baici et al 1993). Oleoyl-heparin, unlike heparin itself, exhibited saturable binding potential for elastin fibres (Table 4) and protects these fibres from elastinolysis. Therefore, it seems that specific inhibition of serine proteases with elastase activity can be achieved by oleoyl-heparin derivatives. These conjugates may well be the first members of a new anti-protease family which could be tested for the inhibition of metastasis formation in elastin-rich organs.

### *Chemotaxis*

Tumour cell migration is the key event during the invasion of the degraded extracellular matrix (Nabi et al 1992). It is induced by soluble matrix peptides acting on matrix receptors, therefore some integrin and non-integrin type matrix receptors function as chemoattractant receptors responsible for the induction of tumour cell migration (Singer & Kupfer 1986). Soluble elastin is a powerful chemoattractant for leukocytes and monocytes in physiological conditions. Soluble forms of elastin, such as α-elastin (Yusa et al 1989), κ-elastin (Tímár et al 1991), tropoelastin (Yusa et al 1989), as well as the elastin peptide VGVAPG (Yusa et al 1989), stimulate tumour cell chemotaxis. The chemotactic response to elastin, shown to be dependent on PKC activity (Blood et al 1988), was similar to the random migration of tumour cells induced by autocrine motility factor (AMF) (Tímár et al 1993). Furthermore, the chemotactic response of tumour cells to soluble elastin positively correlates with the lung colonization potential, both in experimental melanoma (Blood & Zetter 1989) and in lung carcinoma (Tímár et al 1991) models. Therefore it can be stated that the binding of elastin, secretion of elastase and chemotaxis towards a soluble elastin gradient are all properties of tumour cells metastasizing to the lung.

### *Cell proliferation*

Extracellular matrix molecules have recently been shown to be able to modulate cell proliferation (Hynes 1992, Schwartz 1993). They exert their effect on the cell through the appropriate receptors. In the case of integrins, the signalling system coupled to them is believed to play an inhibitory role in cell proliferation (Schwartz 1993). This is due to the activation of FAK kinase as well as to the activation of the $Na^+$ antiporter. In the case of transformed cells, the decreased expression of fibronectin receptor and the consequent lack of activation of FAK kinase is supposed to be responsible for an increase in proliferative activity (for references see Schwartz 1993). Meanwhile, increased expression of various other matrix receptors including integrins (e.g. the $\beta_3$ class; Honn & Tang 1992), the 67 kDa laminin receptors (Castronovo 1993) and elastin receptor complex (Blood et al 1988, Tímár et al 1991), has been demonstrated primarily on highly invasive tumour cells. Unlike integrin ligands, laminin fragments were shown to stimulate tumour cell proliferation (Castronovo 1993). On the contrary, soluble elastin (75 kDa) at high but non-toxic concentrations efficiently inhibits the proliferation of highly metastatic 3LL tumour cells *in vitro* (Table 4), indicating that the signal transmitted by the ligated elastin receptor complex serves as antimitogenic signal. It is of special importance that in the ECM fibrillar elastin is associated with decorin and TGF-$\beta$1 (Heine et al 1990). Tumour cells solubilizing insoluble elastin may release TGF-$\beta$ from this extracellular reservoir, a process which may also modulate tumour cell proliferation.

## Modulation of tumour cell lung colonization by elastin

Since the interaction of tumour cells with extracellular matrix molecules is a key step in the metastatic cascade, abrogation of this interaction with soluble ligand(s) *in vivo* may represent a new and specific antimetastatic therapy. A synthetic peptide representing the active site of the integrin ligand, fibronectin, (Humphries et al 1986), the laminin receptor ligands, YIGS (Iwamato et al 1987) and the heparin binding domains (McCarthy et al 1988) have all been shown to be effective in inhibiting lung colonization by murine melanomas. Since lung is one of the elastin-rich organs and the primary target for tumour dissemination, we have examined the *in vivo* effect of the elastin peptide VAPG and soluble $\kappa$-elastin on the lung colonization of highly metastatic murine lung

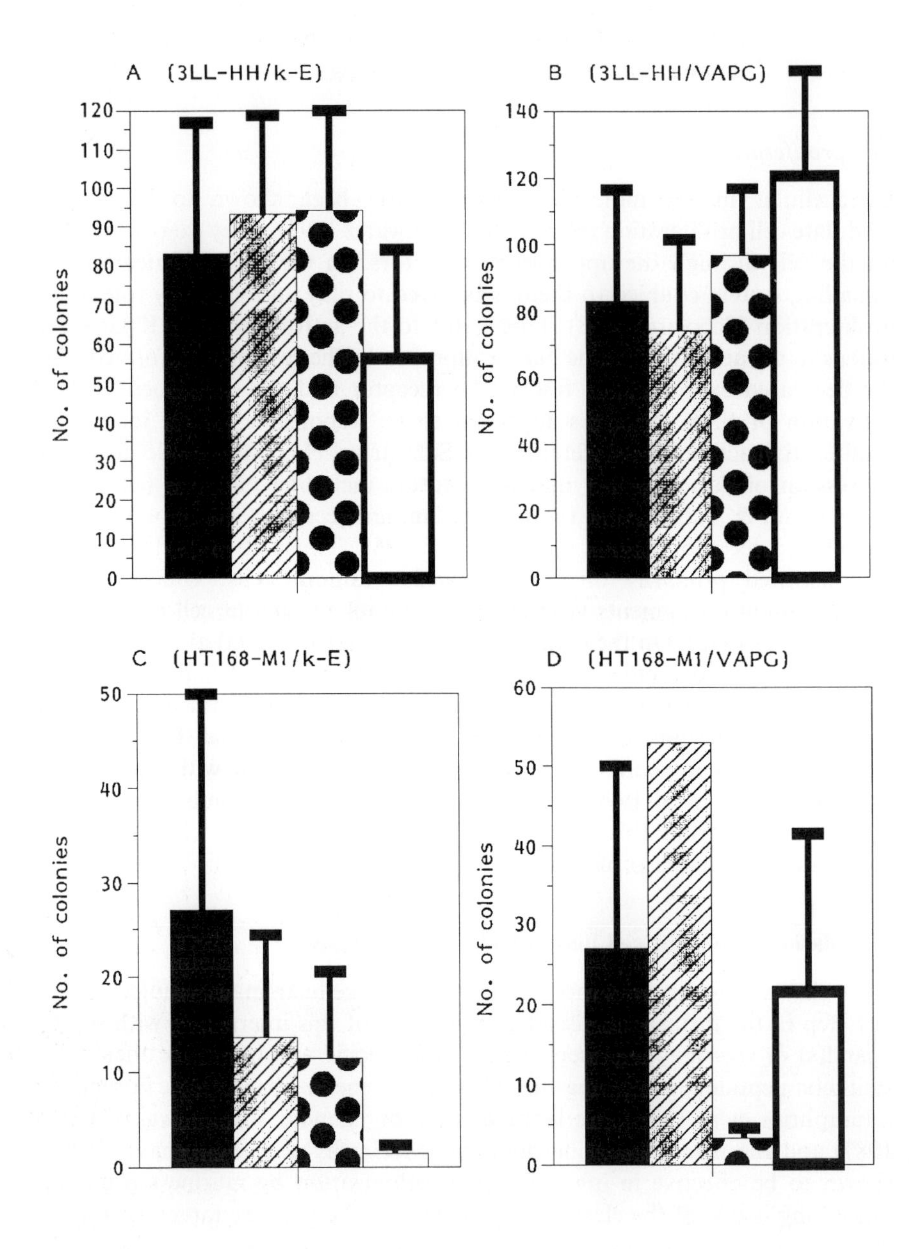
A (3LL-HH/k-E)
B (3LL-HH/VAPG)
C (HT168-M1/k-E)
D (HT168-M1/VAPG)
No. of colonies
0
10
20
30
40
50
60
70
80
90
100
110
120
140

carcinoma and human melanoma. We have used various approaches to interfere with the function of the elastin receptor complex at the tumour cell surface. We have preincubated tumour cells with the elastin ligand *in vitro* before injecting it into the circulation (pre-treatment). In another treatment, we injected tumour cells into the circulation with the elastin ligand itself (*in vivo* treatment). We also studied a combination of the two protocols. In the case of highly metastatic murine 3LL cells, soluble κ-elastin used in the combination treatment protocol significantly inhibited the formation of lung colonies (Fig. 5a) in accordance with previous results (Tímár et al 1991). Interestingly, neither type of treatment with VAPG peptide inhibited the formation of lung metastases and combination treatment with this peptide was even stimulatory (Fig. 5b). In the case of human melanoma HT168-M1 cells, κ-elastin, applied in the combination protocol, significantly inhibited the formation of lung metastases (Fig. 5c) and in this model the VAPG elastin peptide administration *in vivo* was also comparably inhibitory (Fig. 5d). These data all indicate that the interaction of tumour cells with elastin during metastasis to an elastin-rich tissue is significant and can be used as a new target for specific intervention.

## Conclusion

Experimental data show that tumour cells express specific elastin-binding proteins in the form of the elastin receptor complex. This complex contains the 67 kDa elastin/laminin receptor, which is a peripheral membrane protein. The elastin receptor complex is overexpressed in more-metastatic murine and human tumour cells, indicating that it may play a role in tumour dissemination. The elastin receptor complex has been shown to be coupled to G proteins, PLC and PKC, which provide signals for proteolytic enzyme release as well as for chemotaxis. *In vivo* administration of elastin peptides inhibited lung colonization by both murine and human tumour cells indicating that tumour cell–elastin interactions are necessary steps for tumour dissemination in an elastin-rich organ.

---

FIG. 5. Effect of elastin peptides on the lung colonization of murine 3LL-HH carcinoma and human melanoma HT168-M1 cells. Tumour cells were injected into the tail veins of mice (3LL-HH, $5 \times 10^4$/animal; HT168-M1, $5 \times 10^5$/animal) and the lung colonies were determined on the 21st (3LL-HH) or 30th (HT168-M1) day. Treatments: *in vitro* preincubation (*in vitro* pretreatment, cross-hatched bar) of tumour cells with VAPG peptide or soluble κ-elastin (κ-E) (1 mg/ml) for 90 min at 37 °C; injection (*in vivo* treatment; spotted bar) of VAPG or elastin with tumour cells into tail vein (1 mg/animal); combination (*in vitro* + *in vivo* treatment; empty bar) of the two treatment protocols (preincubation and inoculation of elastin peptides). Control cells (solid black bar) were preincubated with medium alone and were injected into tail vein in medium. Data are means ± SD ($n = 8$).

*Acknowledgements*

This work was supported by OTKA T6336 (J. T.), NATO Linkage Grant (LG921311 to J. T.), FIRCA (T-285-01 to J. T.) and OTKA 1131 (to K. L.).

## References

Baici A, Diczházi C, Neszmélyi A, Moczar E, Hornebeck W 1993 Inhibition of the human leukocyte endopeptidases elastase and cathepsin G and of porcine pancreatic elastase by *N*-oleoyl derivatives of heparin. Biochem Pharmacol 46:1545–1549

Blasi F, Verde P 1990 Urokinase-dependent cell surface proteolysis and cancer. Semin Cancer Biol 1:117–126

Blood CH, Sasse J, Brodt P, Zetter BR 1988 Identification of a tumor cell receptor for VGVAPG, an elastin-derived chemotactic peptide. J Cell Biol 107:1987–1993

Blood CH, Zetter BR 1989 Membrane bound PKC modulates receptor affinity and chemotactic responsiveness of Lewis lung carcinoma sublines to an elastin-derived peptide. J Biol Chem 264:10614–10620

Castronovo V 1993 Laminin receptors and laminin binding proteins during tumor invasion and metastasis. Invasion & Metastasis 13:1–30

Fülöp T, Jacob MP, Varga Z, Foris G, Loevey A, Robert L 1986 Effect of elastin peptides on human monocytes: $Ca^{2+}$ mobilization, stimulation of respiratory burst and enzyme secretion. Biochem Biophys Res Commun 141:92–98

Grant AJ, Russell PJ, Raghavan D 1989 Elastase activities of human bladder cancer cell lines derived from high grade invasive tumors. Biochem Biophys Res Commun 162:308–315

Grosso LE, Park PW, Mecham RP 1991 Characterization of a putative clone for the 67-kilodalton elastin/laminin receptor suggests that it encodes a cytoplasmic protein rather than a cell surface receptor. Biochemistry 30:3346–3350

Heine UI, Wahl SM, Munoz EF, Allen JB 1990 Transforming growth factor $\beta 1$ specifically localizes in elastin during synovial inflammation: an immunoelectron microscopic study. Arch Geschwulstforsch 60:289–294

Hinek A, Wrenn DS, Mecham RP, Barondes SH 1988 The elastin receptor: a galactoside-binding protein. Science 239:1539–1541

Hinek A, Rabinovich M, Keeley F, Okamuroaho Y, Callahan J 1993 The 67-kD elastin/laminin-binding protein is related to an enzymatically inactive, alternatively spliced $\beta$-galactosidase. J Clin Invest 91:1198–1205

Honn KV, Tang DG 1992 Adhesion molecules and tumor cell interaction with endothelium and subendothelial matrix. Cancer Metastasis Rev 11:353–375

Honn KV, Tímár J, Rozhin J et al 1994 A lipoxygenase metabolite 12-(S)-HETE stimulates protein kinase C-mediated release of cathepsin B from malignant cells. Exp Cell Res 212:120–130

Hornebeck W, Tixier JM, Robert L 1986 Inducible adhesion of mesenchymal cells to elastic fibres: elastonectin. Proc Natl Acad Sci USA 83:5517–5520

Humphries MJ, Olden K, Yamada KM 1986 A synthetic peptide from fibronectin inhibits experimental metastasis of murine melanoma cells. Science 233:467–470

Hynes RO 1992 Integrins: versatility, modulation and signaling in cell adhesion. Cell 69:11–25

Iwamoto Y, Robey FA, Graf J et al 1987 YIGSR, a synthetic laminin pentapeptide, inhibits experimental metastasis formation. Science 238:1132–1134

Jacob MP, Fülöp T, Foris G, Robert L 1987 Effect of elastin peptides on ion fluxes in mononuclear cells, fibroblasts, and smooth muscle cells. Proc Natl Acad Sci USA 84:995–999

Liotta LA, Thorgeirsson UP, Garbisa S 1982 Role of collagenases in tumor cell invasion. Cancer Metastasis Rev 1:277–288
Liotta LA, Stetler-Stewenson WG 1991 Tumor invasion and metastasis—an imbalance of positive and negative regulation. Cancer Res 51:5054–5059
McCarthy JB, Skubitz APN, Palm SL, Furcht LT 1988 Metastasis inhibition of different tumor types by purified laminin fragments and a heparin binding fragments of fibronectin. J Natl Cancer Inst 80:108–116
Mecham RP, Hinek A, Entwistle R, Wrenn DS, Griffin GL, Senior RM 1989a Elastin binds to a multifunctional 67-kilodalton peripheral membrane protein. Biochemistry 28:3716–3722
Mecham RP, Hinek A, Griffin GL, Senior RM, Liotta LA 1989b The elastin receptor shows structural and functional similarities to the 67-kDa tumor cell laminin receptor. J Biol Chem 264:16652–16657
Nabi IR, Watanabe H, Raz A 1992 Autocrine motility factor and its receptor: role in cell locomotion and metastasis. Cancer Metastasis Rev 11:5–20
Redini F, Tixier JM, Petitou M, Choay J, Robert L, Hornebeck W 1988 Inhibition of leukocyte elastase by heparin and its derivatives. Biochem J 252:515–519
Reich R, Stratford B, Klein K, Martin GR, Mueller RA, Fuller GC 1988 Inhibitors of collagenase IV and cell adhesion reduce the invasive activity of malignant tumour cells. In: Metastasis. Wiley, Chichester (Ciba Found Symp 141) p 193–210
Robert L, Bellon G, Hornebeck W 1980 Characterisation of different elastases. Their possible role in the genesis of emphysema. Bull Eur Physiopathol Respir 16:199–206
Robert L, Jacob MP, Fülöp T, Tímár J, Hornebeck W 1989 Elastonectin and the elastin receptor. Pathol Biol 37:736–741
Ruoslahti E 1988 Structure and biology of proteoglycans. Annu Rev Cell Biol 4:229–255
Schwartz MA 1993 Signaling by integrins: implication for tumorigenesis. Cancer Res 53:1503–1506
Singer SJ, Kupfer A 1986 The directed migration of eucaryotic cells. Annu Rev Cell Biol 2:337–365
Sloane BF 1992 Cathepsin B and its endogenous inhibitors: the role in tumor malignancy. Cancer Metastasis Rev 9:333–352
Stanislawski L, Hornebeck W 1988 Effect of sodium oleate on the hydrolysis of human plasma fibronectin by proteinases. Biochem Int 16:661–670
Tímár J, Lapis K, Fülöp T et al 1991 Interaction between elastin and tumor cell lines with different metastatic potential: *in vitro* and *in vivo* studies. J Cancer Res Clin Oncol 117:232–238
Tímár J, Siletti S, Bazaz R, Raz A, Honn KV 1993 Regulation of melanoma cell motility by the lipoxygenase metabolite 12-(S)-HETE. Int J Cancer 55:1003–1010
Yusa T, Blood CH, Zetter BR 1989 Tumor cell interaction with elastin. Implications for pulmonary metastasis. Annu Rev Respir Dis 140:1458–1462
Zeydel M, Nakagawa S, Biempica T, Takahashi S 1986 Collagenase and elastase production by mouse mammary adenocarcinoma primary cultures and cloned cells. Cancer Res 46:6438–6445

## DISCUSSION

*Hinek:* How did you prepare the elastin membranes that you used in the adhesion assay? Did you just use elastin fibres?

*Lapis:* We have used bovine ligamentum nuchae elastin fibres. The elastin was labelled with tritiated sodium borohydride ($NaB_3H_4$) (Hornebeck et al 1986).

*Hinek:* Is it possible, therefore, that your elastic fibres contain components other than elastin?

*Robert:* They are highly purified bovine ligamentum nuchae or human aorta elastin fibres micronized by sonication.

*Hinek:* Did you try using lactose to prevent attachment? This would tell you for sure whether or not you are dealing with the 67 kDa elastin receptor. This is also another way to do your *in vivo* experiment. You tried using $\kappa$-elastin to saturate elastin receptors or injected tumour cells to prevent metastasis formation. If indeed everything goes through the 67 kDa receptor, you should get the same effect with lactose which causes shedding of elastin receptors from the cell surfaces but is probably less harmful than a high concentration of elastin-derived peptides (Hinek 1994).

*Lapis:* That is an interesting idea; we will certainly try it.

*Hinek:* What kind of elastase did the tumour cells secrete? Have you characterized the enzyme?

*Lapis:* The tumour cells' elastolytic activity was produced by a stromelysin-type metalloproteinase (Tímár et al 1991).

*Robert:* It dissolves fibrous elastin; it is quite a potent elastase of the metalloendopeptidase type (Hornebeck et al 1989, Homsy et al 1988).

*Hinek:* The reason I ask these questions is because we did a similar study using freshly isolated cells from human tumours. We have now collected about 50 cases of solid lung tumours and non-Hodgkin lymphomas (Kossakowska et al 1995, Urbanski et al 1994). We have found that both kinds of tumour cells attach very nicely to pure elastin membrane prepared in a similar way to yours, and we could prevent this attachment with lactose. But when we left them without lactose, both types of tumour cell degraded elastin and penetrated into the elastin membranes. Despite the fact that non-Hodgkin lymphomas and pulmonary carcinomas secrete different elastolytic enzymes, both kinds of tumour cell used their 67 kDa elastin binding receptor for their initial attachment to the elastin substrate.

*Tamburro:* How exactly did you prepare your elastin membranes?

*Hinek:* I took sheep aorta, boiled it in 0.1 N NaOH and washed this repeatedly with 1 M NaCl and water. I ended up with a three-dimensional network on which I plated my cancer cells. I noticed the cells started to migrate, but this wasn't simple migration: they started to 'eat' the elastin. When we tried to characterize the enzymes they secreted, we found that the lung carcinoma cells secreted a serine elastase which was inhibitable with $\alpha_1$ antitrypsin, but the non-Hodgkin lymphoma cells secreted 92 kDa gelatinase which was inhibitable with *O*-phenantroline.

*Urry:* Why did you choose to use the elastin peptide VAPG for your studies, rather than VGVAPG or some other permutation or truncated sequence?

*Lapis:* We chose the shortest repetitive elastin sequence.

*Urry:* We supplied Bob Senior and his group in St. Louis with peptides that included all the permutations of the hexamer APGVGV. They found that the VGVAPG permutation was slightly more effective than the others. Then we supplied them with the truncated sequences with pentamers, tetramers and trimers, to see which sequence might have the greater activity, but I don't recall anything special about VAPG.

*Mecham:* Dr Leonard Grosso used an epitope library to try and understand what the 67 kDa protein binds to. He found that the motif XGXXPG (where X can be anything hydrophobic) is the preferred recognition motif. It has a wide range of binding specificities, as long as the peptides fit that definition. The ability to recognize multiple hydrophobic sequences is a common property for chaperone proteins, but the PG suggests that there's more to it than just the hydrophobic sequence.

*Urry:* The thing that struck me with respect to having PG at the C-terminus is that if you add an additional residue, the PG inserts a $\beta$-turn. So now you confer a conformational feature on it which may not be what the receptor site requires. If there is no residue following the G, then I think that might be helpful. Other than that it would be the specific geometry of the receptor site that requires the two hydrophobic residues for adequate affinity.

## References


Hinek A 1994 Nature and multiple functions of the 67 kD elastin/laminin binding protein. Cell Adhes Commun 2:185–193

Homsy R, Pelletier-Lebon P, Tixier JM, Godeau G, Robert L, Hornebeck W 1988 Characterization of human skin fibroblast elastase activity. J Invest Dermatol 91:472–477

Hornebeck W, Tixier JM, Robert L 1986 Inducible adhesion of mesenchymal cells to elastic fibres: elastonectin. Proc Natl Acad Sci USA 83:5517–5520

Hornebeck W, Kadar A, Adnet JJ, Robert L, Mohacsy J 1989 Elastin in human breast tumours. In: Robert L, Hornebeck W (eds) Elastin and elastases. CRC Press, Boca Raton, FL, p 203–216

Kossakowska EA, Edwards DR, Hinek A, Lim MS, Urbanski SJ 1995 Extracellular matrix degradation by human large cell non-Hodgkin's lymphomas. Int J Surg Pathol (suppl) 2:300

Tímár J, Lapis K, Fülop T et al 1991 Interaction between elastin and tumor cell lines with different metastatic potential: *in vitro* and *in vivo* studies. J Cancer Res Clin Oncol 117:232–238

Urbanski SJ, Luider J, Weaver M, Gelfand GA, Maitland A, Hinek A 1994 Elastolytic activity in primary pulmonary carcinomas. Lung Cancer (suppl) 11:22

# A role for neutrophil elastase in solar elastosis

Barry Starcher and Matt Conrad

*Department of Biochemistry, The University of Texas Health Center at Tyler, PO Box 2003, Tyler, TX 75710, USA*

*Abstract.* Hairless (SKH-1) mice were mated with beige (C57BL/bb) mice to produce a hairless mouse deficient in neutrophil elastase (hhbb). These mice were exposed to 0.09 J UVB radiation for 5 months to see if neutrophil elastase was an important factor in the development of solar elastosis. Analysis of peritoneal neutrophils confirmed that the hhbb mouse was deficient in elastase, retaining only 10% of the activity of the normal littermates (hhHb). Skin myeloperoxidase activity was equally elevated in all the mice receiving UVB indicating a similar influx of inflammatory cells. The absolute breaking strength of the skin in both the hhBb and hhbb mice was not altered by UVB treatment over the 5 month exposure period. Elastin quantitated biochemically as desmosine, or visualized histologically, was increased following UVB exposure in the normal mice. In the elastase-deficient mice, however, the elastin fibres appeared to be unaffected by exposure to UVB radiation at this level. The results suggest that neutrophil elastase is an important mediator in the development of solar elastosis resulting from continued exposure to UVB.

*1995 The molecular biology and pathology of elastic tissues. Wiley, Chichester (Ciba Foundation Symposium 192) p 338–347*

For several years we have been interested in the mechanism of the remodelling of elastin fibres under normal and pathological conditions. Elastin is an extremely inert protein that exhibits limited turnover in most tissues under normal conditions. In the studies that are presented here we have used solar elastosis as a model to study the potential role of neutrophil elastase in elastin fibre remodelling amid conditions of chronic inflammation.

One of the most prominent manifestations of photodamage is a significant accumulation of thickened, tangled elastin fibres in the upper dermis. Fibroblasts are numerous and hyperplastic and there is a noticeable accumulation of partially degranulated mast cells and neutrophils. The principal source of skin elastase is probably the neutrophils (1) and mast cells (2) which infiltrate throughout the dermal layer following UVB exposure. We propose that solar elastosis may result from a cycle of elastase-mediated elastin fibre injury, followed by elastin synthesis and repair. The net result over time could be an accumulation of irregular, thickened elastin fibres.

To explore the role of elastase in this process we crossed the beige mouse, which carries a mutation on the 13th chromosome that results in a functional deficiency of elastase and cathepsin G in the azurophilic granules of neutrophils (Vassalli et al 1978, Takeuchi et al 1986), with the SKH-1 hairless mouse, to produce a hairless mouse that is deficient in neutrophil elastase. We postulated that if elastase is an important factor in solar elastosis, removal of the primary source of the protease from the sites of inflammation may protect the elastin component of the skin matrix when these animals are exposed to UVB.

## Elastase-deficient hairless mice

Utilizing the phenotypic expressions of beige coat colour and hairlessness, augmented with characterization of the mast cell granules, we were effective in selecting hairless mice that were homozygous for the beige mutation of elastase deficiency (hhbb). Confirmation of the neutrophil elastase deficiency in the hhbb mice was obtained from peritoneal neutrophils taken 18 h after an intraperitoneal injection of 10% Freund's adjuvant in a solution of 0.1% oyster glycogen. The number of neutrophils recovered ranged from $14–21 \times 10^6$ per mouse and represented 65–92% of the total leukocytes in the peritoneal lavage. There was an average of 89% reduction in neutrophil elastase activity in the hhbb mice compared with their normal littermates (Fig. 1). There were no differences in the total number of neutrophils recovered from the hhBb and the hhbb mice.

The mice were exposed to UV radiation using a dermalight 2001 equipped with a UVB filter. This source approaches the sun spectrum with 90% UVA and 10% UVB. Groups of five mice were exposed for 30 sec with the lamp positioned 20 cm above the mice to provide 0.09 J/cm$^2$ of radiation three times per week. There was minimal erthyema and no burning or scarring with this dosage. Age-matched non-irradiated mice functioned as controls.

## Mechanical strength

The increase in skin thickness of the hhbb mice following exposure to UVB was analogous to the increase observed in the hhBb littermates or the original SKH-1 mice. There was a significant loss of tensile strength (gm/cm$^2$) in the skin of hairless mice that were exposed to UV radiation when compared with non-irradiated controls. However, when the mechanical strength of the skin was expressed in absolute values per 3 mm wide strip of tissue, irrespective of the thickness, there was no actual loss of skin strength during the first 6 months

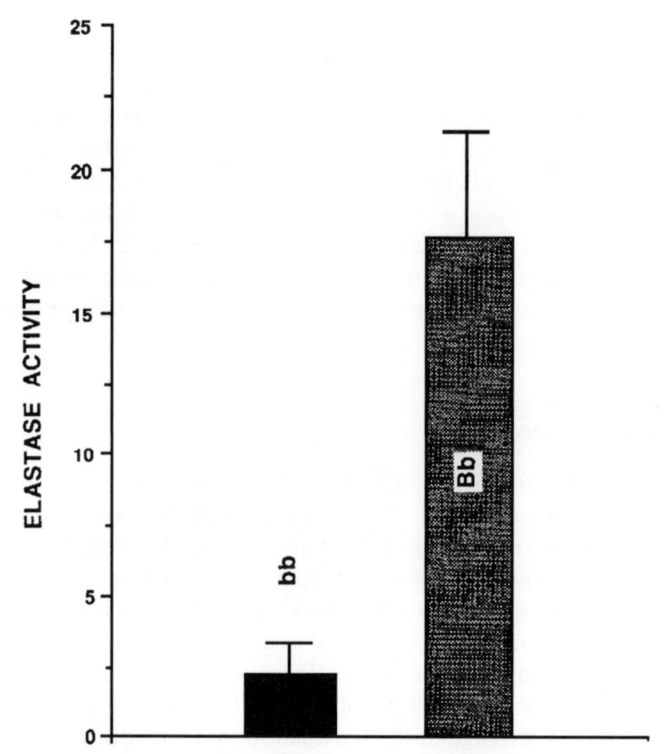


FIG. 1. Neutrophil elastase activities of the hairless beige cross (hhbb) and their normal littermates are shown with standard deviation ($n=5$). Activity is expressed as change in optical density at 405 nm per $10^6$ neutrophils.

of exposure and a cumulative UVB dose of 6 J/cm$^2$. We had originally believed that the UVB damage to mouse skin elastin would affect both the function and strength of the skin. Our results suggest that this is not the case. In mouse skin the relative amounts of elastin are very low when compared with human skin and the contribution of elastin to the mechanical strength of the skin appears to be minimal.

**Biochemical measurements**

In our studies, UVB exposure did not result in altered skin collagen levels as measured by hydroxyproline/cm$^2$ of skin. This was true for both the elastase-deficient mice as well as their normal littermates. As we had anticipated, elastin, as measured by desmosine content, was significantly increased following UV exposure in the normal hairless mice (Fig. 2). There were,

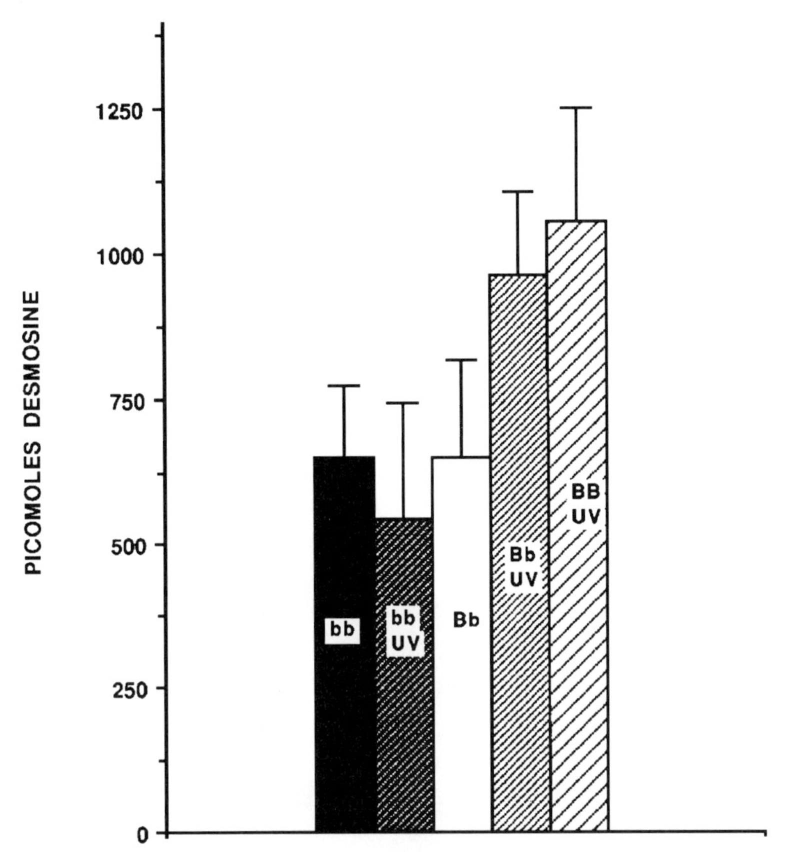


FIG. 2. Total picomoles of desmosine in 3 mm × 2 cm strips of skin from hairless-beige (bb), normal littermates (Bb) and SKH-1 (BB) with and without 5.4 J of UVB exposure over 5 months. Each value is the average of two samples from five mice with standard deviation.

however, no significant changes in the skin desmosine content of the UVB-irradiated mice carrying the beige defect. This suggests that the deficiency in neutrophil elastase prevented the elastin fibres from being exposed to continuous injury and repair.

Another possibility was that there was an impaired inflammatory reaction in the mice carrying the beige defect. As an indicator of the extent of inflammation, we measured myeloperoxidase activity in skin punches removed from the backs of the mice exposed to UVB for 5 months. There was a prominent and parallel increase in enzyme activity in both the elastase-deficient mice and their normal littermates (Fig. 3).

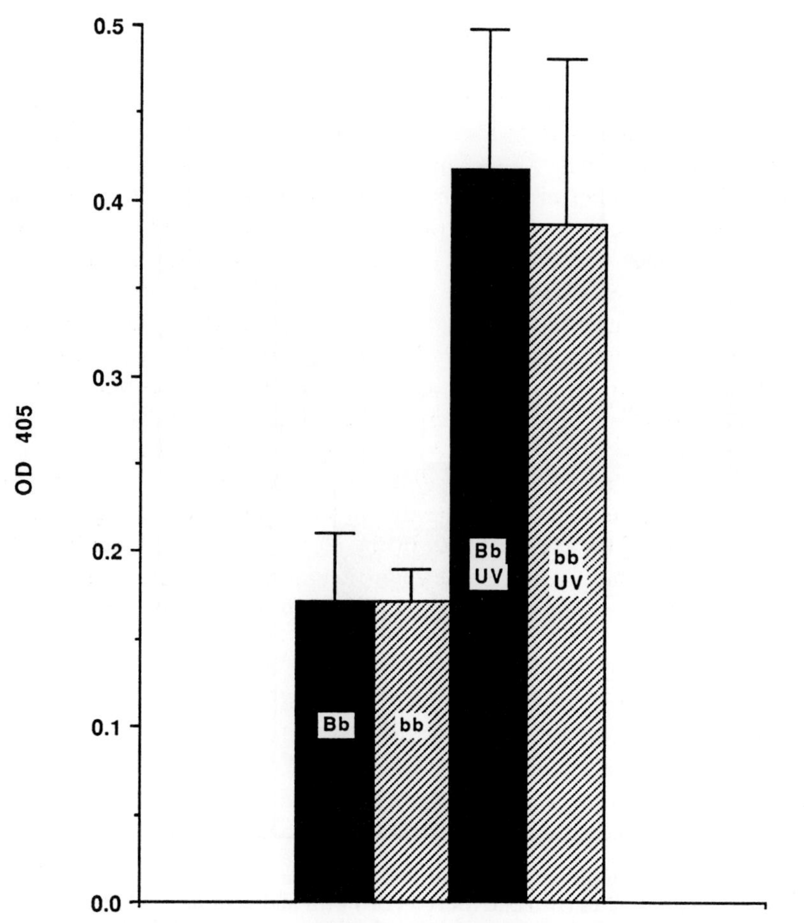


FIG. 3. Myloperoxidase activity expressed as the change in optical density (OD) at 405 nm. Each value represents the average of two 3 mm skin punches from the back of five elastase-deficient (bb) and five normal littermates (Bb) with and without 5.4 J of UVB exposure over 5 months with standard deviation.

## Histology

Histologically, there were no obvious differences in the skin from our elastase-deficient mice, their normal littermates and descriptions in the literature for SKH-1 hairless mice (Bissett et al 1987, Kligman et al 1982, 1985). Fibroblasts, lymphocytes and mast cells were scattered sparingly throughout the dermal layer. In the non-irradiated skin, the elastin was present as relatively thin and sporadic fibres in the upper dermis. The fibres were particularly radiating out from around the sebaceous glands. In some sections, it appeared that the elastase-deficient mice had an increased number of elastin fibres compared with their normal littermates, but this varied considerably between samples. The epidermis was thin and uniform in thickness. The only notable difference between the hhbb and their hhBb littermates following UVB irradiation was changes in elastin fibres. After 5 months of UVB exposure, there was a major increase in the number of elastin fibres lying under the basement membrane in the papillary dermis of the normal mice. Isolated masses of elastin fibre aggregates were scattered throughout the dermis, especially in areas immediately surrounding the sebaceous glands. In marked contrast, the elastase-deficient mice showed no apparent changes in the elastin component of the extracellular matrix following UV exposure. The elastin in most sections of the upper dermis remained as thin fibres lying parallel to the epithelium. There were some areas of apparent increased elastogenesis around the sebaceous glands but this was difficult to confirm as a result of the wide variability throughout the skin. The UVB damage to the normal mice appeared to penetrate the entire thickness of the skin, resulting in a thickening and increase in the number of the elastin fibres in the elastic sheath that lies beneath the panniculus carnosus muscle. This elastin-rich area in the elastase-deficient mice was unaffected. In both strains of mice there was a notable increase in skin thickness in response to UVB irradiation. The epidermis became irregular and increased to several times its normal thickness. Dermal cysts proliferated to form up to five rows and the width of the upper dermis was increased. There was an inflammatory infiltration of lymphocytes, macrophages and mast cells which occurred to about the same extent in both the elastase-deficient mice and their normal littermates.

One of the questions that we examined was whether the UV-induced increase in elastin was restricted to the papillary dermis or whether it was a deep injury and uniformally distributed throughout the depth of the skin. In order to achieve this we quantitated elastin (desmosine) in serial sections starting from the epidermis. Skin punches (4 mm) were taken from sun-damaged human skin as well as UVB-irradiated mouse skin. These were fixed and paraffin embedded. Sections (20 $\mu$m) were cut, starting from the epidermis and continuing through the depth of the skin. Every five serial sections were pooled so that each value represented 100 $\mu$m of skin. These were then extracted with xylene, hydrolysed and analysed for desmosine content. Figure 4 illustrates the uniform distribution of desmosine starting just below the basement membrane and continuing throughout the depth of the sun-damaged human skin sample. Hairless mice, however, exhibited a very different picture. As expected, there was a significant increase in elastin throughout the papillary dermis area in the UVB-irradiated mouse skin. However, in the lower dermis—the area that contains the empty hair follicles—there was an even larger amount of elastin which had not been evident by normal elastin staining. To illustrate this point, we pooled the desmosines into four areas representing the

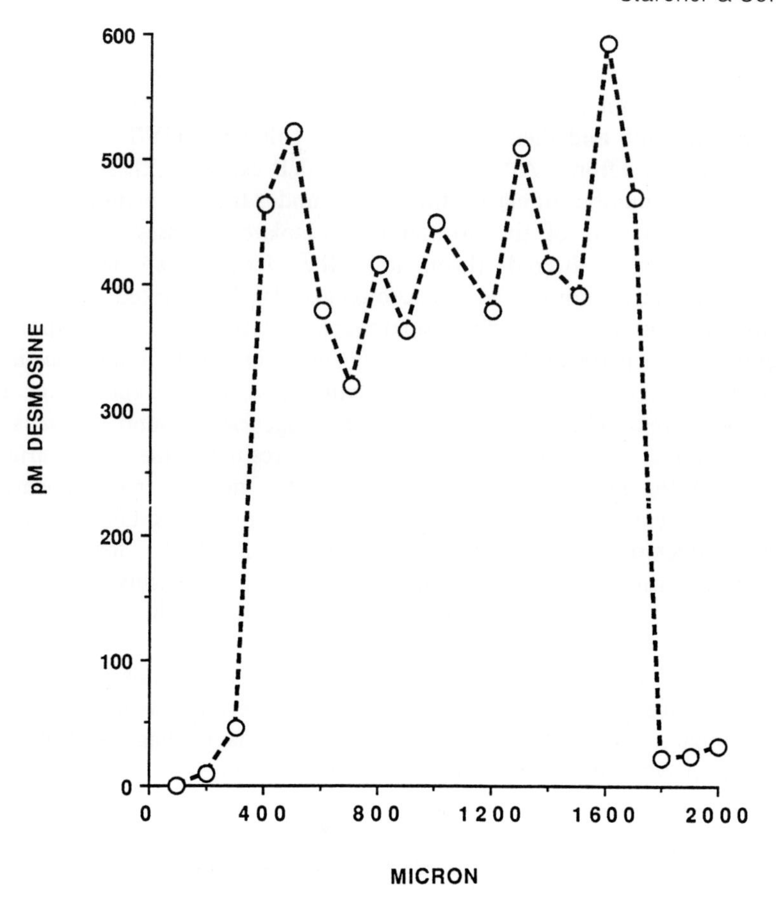


FIG. 4. Distribution of desmosine in sun-damaged human skin. Each value represents a 100 $\mu$m section of skin starting from the epithelial surface.

epidermis, upper dermis, the follicle-containing lower dermis, and the panniculus muscle layer and its underlying elastin layer (Fig. 5). The area immediately surrounding the hair follicles is rich in elastin fibres and as the numbers and size of the follicles increase in response to UVB, there appears to be a corresponding increase in total elastin. This occurs in the normal, as well as the elastase-deficient mice, and as far as we can tell there was no abnormal deposition of elastin fibres around the empty follicles during this growth process. This raises a question as to the validity of the hairless mouse as a model for human solar elastosis. By restricting our comparisons to the

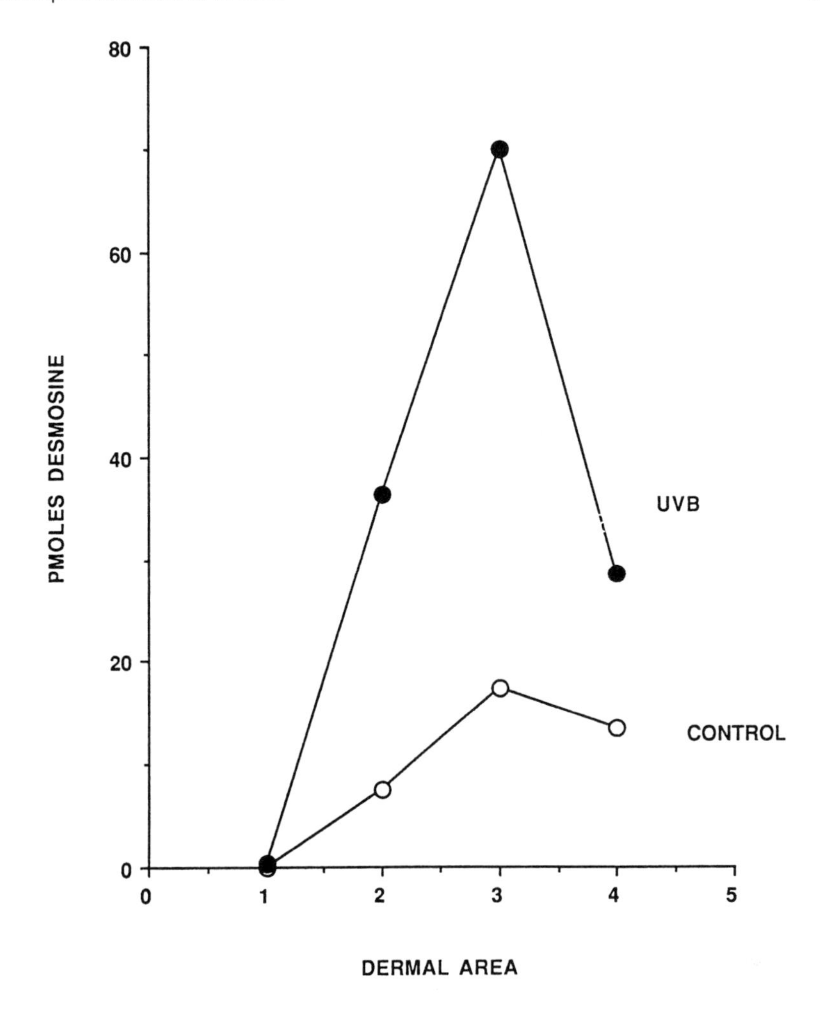


FIG. 5. A comparison of desmosine distribution in control (○) and UVB-irradiated (●) mouse skin. Each point represents the total desmosine recovered in the epithelium (1), papillary dermis (2), lower dermis (3) and panniculus muscle areas (4).

papillary dermis, we feel confident in concluding that neutrophil elastase activity is essential for the development of the abnormal accumulation of elastin fibres that occurs during UVB exposure.

The results from the animal model suggest that the reduced elastase load in the beige mice moderated the restructuring of the elastin fibres in the skin during chronic inflammation induced by UVB irradiation. Under normal circumstances, UVB probably initiates a sequence of events leading to an inflammatory cell influx to the area, degranulation and release of elastase, and in some instances, partial disaggregation of elastin fibres. This is followed by elevated elastin synthesis and attempts to repair the injured fibres. Since elastin fibres are extremely stable through three-dimensional cross-linking, they may be only partially removed. Continual elastase injury and repair eventually leads to the accumulation of the enlarged and fragmented elastin fibres observed in solar elastosis.

One noteworthy observation was that while 100% of the UVB irradiated hhBb and SKH-1 mice developed squamous cell carcinomas by the end of the 5 month experiment, none of the elastase-deficient mice developed these tumours. Preliminary evidence suggests this may be related to the role of neutrophil elastase in removing the 70 kDa TNF receptors.

## References

Bissett DL, Hannon DP, Orr TV 1987 An animal model of solar-aged skin: histological, physical, and visible changes in UV-irradiated hairless mouse skin. Photochem Photobiol 46:367–378

Kligman LH, Akin FJ, Kligman AM 1982 Prevention of ultraviolet damage to the dermis of hairless mice by sunscreens. J Invest Dermatol 78:181–189

Kligman LH, Akin FJ, Kligman AM 1985 The contributions of UVA and UVB to connective tissue damage in hairless mice. J Invest Dermatol 84:272–276

Takeuchi K, Wood H, Swank RT 1986 Lysomal elastase and cathepsin G in beige mice. J Exp Med 163:665–677

Vassalli JD, Granelli-Piperno A, Griscelli C, Reich E 1978 Specific protease deficiency in polymorphonuclear leukocytes of Chediak-Higashi syndrome and beige mice. J Exp Med 22:1285–1290

## DISCUSSION

*Uitto:* You are probably right when you say that mouse skin is not a very good model for photodamage in human skin. First, mouse skin starts off with an elastin concentration about 1/10th that of human skin and, secondly, mouse skin is so thin that the UV radiation goes all the way through, while in human skin it pretty much stops at the level of upper papillary or mid-reticular dermis.

*Mungai:* What happens to melanocytes in solar elastosis? Is there any change in production of melanin?

*Starcher:* I really don't know. In the human situation we see the melanocytes but I've never looked at what happens to them.

*Uitto:* In the human situation this treatment would simply increase the melanin production by melanocytes. I don't know whether this would happen in mice. Did you find any melanomas in the mice?

*Starcher:* No. It's very rare to find a melanoma in the mouse: all the tumours we see are squamous cell carcinomas.

*Mecham:* Is there any evidence that the desmosine content of elastin is different for new and old elastin? In other words, if new elastin has less desmosine, might you be underestimating the amount of elastin that is present by relying on desmosine? There is some suggestion that the desmosine content of elastin goes up with age: I'm just wondering if that is important in your kind of assay.

*Starcher:* I don't think so. Our desmosine analysis agreed with our histology, unless you are suggesting that there's also a difference between old and new elastin histochemically.

I'm not sure that the desmosine content goes up with age in our model. If it does, the increase has to be fairly small and the increase in elastin fibres we are looking at has occurred over 5 months—these should be completely cross-linked by then.

*Urry:* Molecules, such as elastin, that can undergo the inverse temperature transition and self assemble would seem to me to be fully capable of self-assembling and forming fibres. What they need is direction as to where to form them. They can assemble on fibres that are already in place, but otherwise the growing fibre seems to have no direction as to how to lay down with respect to the needs of a tissue undergoing repair. Is it fair to say it is not so much how elastin assembles into fibres (we believe it can self assemble), but rather it needs instructions as to where to assemble?

*Starcher:* In the skin I had thought that we would get around this directional problem, but it does seem that way. Elastin may need the correct initiation site to lay down new fibres that are functional.

*Urry:* And yet it does form fibres, it just doesn't do it in the appropriate places.

*Pasquali-Ronchetti:* Are these newly formed fibres normal ultrastructurally?

*Starcher:* I don't know. Most of the studies on them have been histochemical—I can't think of any electron micrographs of elastosis in the mouse.

*Uitto:* In humans there have been lots of studies at the ultrastructural level on the elastotic material; it has no resemblance to the structure of normal elastic fibres (Uitto et al 1989).

## Reference

Uitto J, Fazio MJ, Olsen DR 1989 Molecular mechanisms of cutaneous aging: Age-associated connective tissue alterations in the dermis. J Am Acad Dermatol (Suppl) 21:614–622

# Closing reflections

Leslie Robert

*Laboratoire de Biologie Cellulaire, Université Paris VII, 3 Place Jussieu, F-75005 Paris, France*

A final word about this meeting which, judging from comments from many of the participants, I believe was a great success. That it was so was due to the fortunate conjunction of a number of factors. First of all, the format of the meeting which encouraged intensive, frank and wide-ranging discussions among a small group of experts in a friendly and relaxed atmosphere. The highly professional support of the Ciba Foundation's staff was a crucial factor in this respect. Special mention should also be made of the beautiful setting chosen for the meeting which was greatly appreciated by all. Above all, we valued the opportunity for an exceptionally fruitful encounter with our Kenyan colleagues, Professors Mungai and Kimani and the staff of the Kenya Medical Research Institute and the Commission for Higher Education which was an additional fascinating opportunity arising out of this meeting.

Since the beginning of the 1970s, when only a few teams were engaged in elastin research, activity in the elastin field has increased dramatically and the scientists who gathered for this Ciba Foundation symposium, from all over the world, were abundant proof of the burgeoning interest in this unique protein. The meeting celebrated, in a way, the first centenary of elastin research. The progress over the last 20 years has been spectacular as will be evident to all those who read this Ciba volume with presentations ranging from molecular biophysics, through protein biochemistry, cell and molecular biology to the pathology and treatment of connective tissue disease. Among the most important aspects of elastin research are the sequencing of the elastin gene from several species, the study of its regulation during development and ageing, the mechanism of fibre formation with the contribution of the 'microfibrils' (especially fibrillin), and the modifications of these processes in a score of genetic and acquired diseases. Finally, the demonstration of cell–elastin interactions definitively eliminated the psychological barrier between cell biology and elastin chemistry, largely attributable to the belief that such a resistant protein can only be inert and uninteresting to cells.

For all these reasons this meeting and the resulting book represent a landmark in elastin research and will bear witness both to the rapid progress that has been achieved in elastin research and also to the many uncertainties and controversies that remain.

# Index of contributors

*Non-participating co-authors are indicated by asterisks. Entries in bold type indicate papers; other entries refer to discussion contributions.*

*Indexes compiled by Liza Weinkove*

# Subject Index